Body Condition and Productivity, Health and Welfare

Body Condition and Productivity, Health and Welfare

Editors

María-Luz García
María-José Argente

MDPI • Basel • Beijing • Wuhan • Barcelona • Belgrade • Manchester • Tokyo • Cluj • Tianjin

Editors
María-Luz García
Universidad Miguel
Hernández de Elche
Spain

María-José Argente
Universidad Miguel
Hernández de Elche
Spain

Editorial Office
MDPI
St. Alban-Anlage 66
4052 Basel, Switzerland

This is a reprint of articles from the Topical Collection published online in the open access journal *Animals* (ISSN 2076-2615) (available at: https://www.mdpi.com/journal/animals/topical_collections/Body_Condition_and_Productivity_Health_and_Welfare).

For citation purposes, cite each article independently as indicated on the article page online and as indicated below:

LastName, A.A.; LastName, B.B.; LastName, C.C. Article Title. *Journal Name* **Year**, *Volume Number*, Page Range.

ISBN 978-3-0365-6239-1 (Hbk)
ISBN 978-3-0365-6240-7 (PDF)

Contents

Article

Does the Degree of Fatness and Muscularity Determined by Ultrasound Method Affect Sows' Reproductive Performance?

Damian Knecht, Sebastian Środoń and Katarzyna Czyż *

Institute of Animal Breeding, Faculty of Biology and Animal Sciences, Wroclaw University of Environmental and Life Sciences, 50-375 Wroclaw, Poland; damian.knecht@upwr.edu.pl (D.K.); sebastian.srodon@o2.pl (S.Ś.)

* Correspondence: katarzyna.czyz@upwr.edu.pl

Received: 2 April 2020; Accepted: 30 April 2020; Published: 3 May 2020

Simple Summary: One of the factors strongly affecting the profitability of animals breeding is their reproductive performance. The measurement of gilts and sows' fatness and muscularity levels can be a useful tool in forecasting their future reproductive parameters, and thus possible selection with respect to these parameters, which is a significant issue from a practical breeders' point of view. The aim of the study was an evaluation of the fatness and muscularity of pure-bred and hybrid gilts measured using an Aloka SSD-500 device in three consecutive parities, and the relationship between these features and reproductive performance parameters such as litter traits at birth and weaning, pregnancy length and weaning to service interval. It was observed that fatness degree affected the reproductive parameters of females. The females with higher values of fatness traits achieved better farrowing rate and higher numbers of born piglets, and decreased mortality, higher gains of piglets and higher body weight at weaning, as well as a shortening of the weaning-to-service interval were noted. Intramuscular fat content did not affect reproductive parameters. Muscularity also affected reproductive performance parameters, except gestation period. The lowest reproductive indices were found in females with too-high muscularity levels.

Abstract: The fatness and muscularity of Polish Landrace, Polish Large White gilts and sows and their hybrids were determined on the basis of ultrasound measurements in three consecutive parities, and then the relationship between these parameters and reproductive performance was established. Ultrasound measurements demonstrated the highest fat thickness in first parity and the highest fat area over LD muscle in hybrid gilts (PL × PLW). Pure-bred gilts were characterized by poorer muscularity. Fatness level affected the reproductive parameters of females in which the thickness of backfat in UP2 point was above 22.25 mm, the thickness of backfat in UP4 point was above 17.36 mm and the fat area over LD muscle was above 25.81 cm^2. These females achieved better farrowing rates and higher numbers of born piglets. Decreased mortality, higher gains of piglets and higher body weight at weaning were observed, and the weaning to service interval was shortened in fatter females. Intramuscular fat content did not affect reproductive parameters. Muscularity negatively affected reproductive performance parameters, except gestation period. Too-high muscularity was related to the lowest levels of reproductive indices. The analysis of gilts and sows' fatness and muscularity levels can help to predict their reproductive performance in the future and thus optimize production results.

Keywords: gilts; litter; parity; sows; ultrasonography; Aloka SSD-500

---

1. Introduction

The efficiency of reproduction and the level of productivity of sows determine to a large extent both the profitability of production and the quality of piglets for fattening. Therefore, the variables influencing the reproductive potential of sows should be regularly examined. Intensive selection of pigs to improve their leanness has contributed to a significant increase in this index, not only in Poland, but also in other countries. Analyzing the changes in the slaughter traits of gilts of the maternal Polish Large White (PLW) and Polish Landrace (PL) breeds in recent years, it can be observed that the meat content in the carcass increased to the level of about 60% [1].

In order to maximize sow productivity, it is important to know about the growth potential, slaughter potential and the impact of improvement on reproductive performance. The relationships between the degree of muscularity and fatness of gilts and their further use is not clear, showing both positive and negative effects [2]. An important issue, therefore, seems to be the precise demonstration of the effect of growth parameters on the reproductive use of females.

Many researchers have searched for the current response of reproductive functions to slaughter potential raising. Some emphasize that reproductive traits should be taken into account in the development of breeding strategies aimed at improving fattening and slaughter traits. Many authors confirm the negative impact of selection focused solely on improving fattening and slaughter traits, which, in turn, leads to a deterioration in some reproductive performance parameters, and this process is most clearly evident in gilts [3,4]. The reproductive performance may be adversely affected by both an insufficient and excessive degree of fatness and muscularity [5]. According to Tummaruk et al. [3], low fat reserves in sow's body has a negative impact on future reproductive performance results, which may cause, e.g., faster culling due to poor animal condition and lactation disturbances.

A higher growth rate can accelerate the first estrus symptoms. It should be emphasized that this is an extremely important issue for the proper management and use of the economic opportunities of the basic stock. It should be noted that, as a result of long-term and unilateral selection for leanness, the level of fat cover of gilts at the time of their first mating or insemination decreased significantly. This is related, inter alia, to the negative leanness–fatness correlation, which means that with the increase in meat content, the gilt's fatness level is subject to a decrease [6,7].

According to Quinton et al. [8], the most important reproductive indicators from an economic point of view are the number of piglets born alive and the number of piglets at weaning, and the weaning weights at the same time. One of the most important objectives in both the breeding and keeping of pigs is to improve the population in terms of economically significant parameters, which undoubtedly include features related to reproductive performance. The level of these features may be related to the degree of fatness and muscularity at the time of first mating or insemination of gilts [9].

The aim of the study was determination of fatness and muscularity of pure-bred gilts and hybrids on the basis of intravital ultrasound measurements in three consecutive parities, and then to define the relationship between the degree of fatness and muscularity of gilts and sows and their reproductive performance determined on the basis of litter parameters at birth and weaning and the length of the weaning to service interval.

2. Materials and Methods

The experiment was performed in the years 2013–2014 in an industrial pig-fattening farm situated in Poland, in the Opolskie Voivodeship on 450 gilts, maintained in identical production conditions. The gilts were divided into three groups depending on the genotype: 150 Polish Landrace (PL) gilts, 150 Polish Large White (PLW) gilts, 150 hybrid gilts (PL × PLW). Gilts at the time of insemination were at similar weight (about 129 kg), age (about 252 days, and had a similar average daily gain from birth to the first insemination (about 513 g). The gilts were used for research purposes from a one to three parity. The gilts and sows were kept in individual pens with an area of 1.20 m^2 equipped with a slatted floor until 4 weeks after insemination. In the period between 4 weeks after insemination and 1 week before farrowing, the females were group-housed in a pen with 25 females on a partially slatted floor.

The area of the pen for gilts was 1.64 m^2/animal, including 0.95 m^2 of solid floor. The area of the pen for sows was 2.25 m^2/animal, and the solid floor area was 1.30 m^2.

The animals were fed in accordance with the recommendations of the Jan Kielanowski Institute of Animal Physiology and Nutrition—National Research Institute of the Polish Academy of Sciences, according to the Pig Nutrition Standards [10], with constant access to water. The feed dose for the female until 90 days of pregnancy contained 11.8 MJ of metabolic energy, 143.6 g of digestible protein, 6.4 g of lysine, 5.3 g of methionine with cystine, 8.2 g of calcium, 6.3 g of phosphorus and 2.7 g of sodium. The feed dose for gilt/sow over 90 days of pregnancy and lactating gilt/sow was 12.5 MJ of metabolic energy, 170.2 g of digestible protein, 10.1 g of lysine, 6.5 g of methionine with cystine, 9.9 g of calcium, 7.2 g of phosphorus and 3.6 g of sodium. Until the 90th day of pregnancy, each female received about 2.80 kg of feed per day, while after this period their diet was increased to 3.30 kg.

The animals were maintained in conditions meeting the requirements of the Ordinance of the Minister of Agriculture and Rural Development of 28 June 2010 on the minimum conditions for maintaining farm animal species [11]. The study was carried out in accordance with the laws and regulations of Poland, Act of 15 January 2015, on the protection of animals used for scientific and educational purposes [12].

Females in estrus were identified twice a day with the help of a search boar (visual and olfactory contact). The estrus was diagnosed by a standing reflex after a back pressure test. The detection of estrus started from the 170th day of gilts' life. The gilts were inseminated in the third estrus. Both gilts and sows were inseminated twice: the first time 12 h after the onset of estrus symptoms, and the second time 12 to 18 h after the first insemination. All females were inseminated with Duroc × Pietrain (D × P) boar semen. Litter equalization was performed within the genetic lines.

Ultrasound evaluation of fatness and muscularity was performed during insemination procedures using the Aloka SSD-500 device equipped with a 17 cm UST-5044 head with a frequency of 3.5 MHz. The use of the apparatus in combination with dedicated computer software enabled the measurements described in Table 1. Linear and surface measurements were performed according to the methodology provided by Tyra et al. [13], while intramuscular fat measurement was performed using the method provided by Schulte et al. [14].

Table 1. Linear, surface and quality measurements taken with Aloka SSD-500 ultrasound system.

Measurement	Description of the Measurement Point
UP2	Backfat thickness at point P2—backfat thickness at the last rib, at the junction of the thoracic and lumbar vertebrae—3 cm from the midline of the spine (mm).
UP4	Backfat thickness at point P4—backfat thickness at the last rib, at the junction of the thoracic and lumbar vertebrae—8 cm from the midline of the spine (mm).
UPT	Fat cross-sectional area on the surface of the Longissimus dorsi muscle (cm^2).
UTŚ	Intramuscular fat content in the Longissimus dorsi muscle (%).
UP4M	Depth of Longissimus dorsi muscle at point P4—height of eye of the loin, at the junction of the thoracic and lumbar vertebrae—8 cm from the midline of the spine (mm).
USLD	Width of the Longissimus dorsi muscle (mm).
UPLD	Cross-sectional area of the Longissimus dorsi muscle (cm^2).

The next stage of the study was an evaluation of the reproductive performance of gilts, in which the degree of fatness and muscularity were determined using the Aloka SSD-500 apparatus. The following reproductive indices were analyzed: farrowing rate based on two insemination procedures outcomes (%), number of piglets born in total (head), number of alive-born piglets (head), number of dead-born piglets (including stillborn ones, dead during parturition and mummified fetuses) (head), number of weaned piglets (head), mortality of piglets until weaning day (head), total litter weight at 1st day (kg), average piglet weight at 1st day (kg), total litter weight at weaning (kg), average piglet weight at weaning (kg), average body weight gains of piglets from birth to weaning (g), gestation period (days), weaning to service interval (days).

For each ultrasound measurement, after taking into account the distribution of its variables in the population, two levels of the examined factor were determined on the basis of the mean and standard deviation, in such a way that there is an equal number of individuals within each subgroup, within each breed. Next, it was analyzed whether the level of fatness and muscularity at the moment of insemination can influence the reproductive performance results in terms of the examined indices. It was analyzed whether there are statistical differences between the female genotypes with respect to different degrees of fatness and muscularity.

In order to check the repeatability of the results, each ultrasound measurement was performed three times, then the arithmetic mean of the three measurements was calculated. The data were checked for normality of distribution using Kolmogorov–Smirnov test (K-S) with Lilliefors correction. In addition, Brown–Forsythe test (B-F) was used to determine whether the distributions of variables have the same variance. The numerical material was compiled statistically using STATISTICA ver. 10 software (StatSoft Poland Ltd., Krakow, Poland). Verification of the data concerning the degree of fatness and muscularity determined during insemination was performed on the basis of two-factor analysis of variance. The research model was as follows:

$x_{ijk} = \mu + \alpha_i + \beta_j + (\alpha\beta)_{ij} + \varepsilon_{ijkl};$
x_{ijk}—value of the dependent variable (fatness and muscularity degree);
μ—overall average;
α_i—main effect of the i-th breed;
β_j—main effect of j-th insemination order;
$(\alpha\beta)_{ij}$—effect of i-th breed interaction with j-th insemination order;
εijk—random error with a normal distribution with an average of zero and a variance σ^2.

Reproductive performance was studied using the three-factor analysis of variance (ANOVA). The first factor was the breed of females participating in the experiment. The second factor was the parity (except for the parameter of farrowing rate, where the order of insemination was taken into account). The third factor was the degree of fatness and muscularity. In the model, the degree of fatness was used interchangeably with the degree of muscularity. The following mixed model was used:

$x_{ijk} = \mu + \alpha_i + \beta_j + \gamma_k + (\alpha\beta)_{ij} + (\alpha\gamma)_{ik} + (\beta\gamma)_{jk} + (\alpha\beta\gamma)_{ijk} + \varepsilon_{ijkl};$
x_{ijk}—value of the dependent variable (reproductive parameter);
μ—overall average;
α_i—main effect of the i-th breed;
β_j—main effect of the j-th parity or insemination order;
γ_k—main effect of the k-th level of fatness or muscularity;
$(\alpha\beta)_{ij}$—effect of the i-th breed interaction with the j-th parity or insemination order;
$(\alpha\gamma)_{ik}$—effect of the i-th breed interaction with the k-th level of fatness or muscularity;
$(\beta\gamma)_{jk}$—effect of the j-th parity or insemination order interaction with the k-th level of fatness or muscularity;
$(\alpha\beta\gamma)_{ijk}$—effect of the i-th breed interaction with the j-th parity or insemination order and the kth level of fatness or muscularity;
ε_{ijk}—random error with a normal distribution with an average of zero and a variance σ^2.

3. Results

The results of fatness measurements (UP2, UP4, UPT and UTŚ) carried out by ultrasound method are presented in Table 2. In the first parity, the highest backfat thickness was noted in hybrid gilts (PL × PLW) ($p \leq 0.05$). It should be emphasized that, after the first farrowing, during the second parity, both hybrid and PLW sows had less backfat thn PL sows ($p \leq 0.05$). Moreover, statistically significant differences in the thickness of backfat were found for the first and second parity in the case of inter-breed hybrids (PL × PLW) ($p \leq 0.05$).

Table 2. Results of fatness measurements carried out using Aloka SSD-500 (breed and parity).

Parameter UP2						
Parity	Breed					
	Polish Landrace (PL)		Polish Large White (PLW)		PL × PLW	
	$\bar{x}$ (mm)	SD	$\bar{x}$ (mm)	SD	$\bar{x}$ (mm)	SD
1	22.26	±4.70	21.17 ^{y}	±3.99	23.08 ^{a,x}	±3.89
2	23.29 ^{x}	±4.02	21.19 ^{y}	±4.16	21.18 ^{b,y}	±4.79
3	22.24	±4.76	22.93	±4.52	22.94	±4.43
Parameter UP4						
Parity	Breed					
	PL		PLW		PL × PLW	
	$\bar{x}$ (mm)	SD	$\bar{x}$ (mm)	SD	$\bar{x}$ (mm)	SD
1	17.54	±3.60	16.35 ^{y}	±4.21	18.56 ^{a,x}	±3.44
2	18.98 ^{x}	±3.34	15.67 ^{y}	±4.03	16.26 ^{b,y}	±4.15
3	17.65	±3.65	17.07	±3.89	18.12	±3.38
Parameter UPT						
Parity	Breed					
	PL		PLW		PL × PLW	
	$\bar{x}$ (cm^2)	SD	$\bar{x}$ (cm^2)	SD	$\bar{x}$ (cm^2)	SD
1	25.19 ^{y}	±6.61	23.34 ^{Y}	±6.85	28.20 ^{a,X,x}	±7.43
2	27.23 ^{X}	±7.45	23.02 ^{b,Y}	±6.45	25.10 ^{b}	±7.11
3	26.46	±8.96	25.96 ^{a}	±7.66	27.83	±6.78
Parameter UTŚ						
Parity	Breed					
	PL		PLW		PL × PLW	
	$\bar{x}$ (%)	SD	$\bar{x}$ (%)	SD	$\bar{x}$ (%)	SD
1	1.67	±0.66	1.68	±0.60	1.71	±0.67
2	1.71	±0.78	1.73	±0.74	1.75	±0.72
3	1.77	±0.83	1.79	±0.80	1.82	±0.87

$\bar{x}$—mean value. SD—standard deviation. a,b—in the same column, denote statistically significant differences between parity for a given breed, with $p \leq 0.05$. x,y—in the same line, denote statistically significant differences between breeds, for a given parity, with $p \leq 0.05$. X,Y—in the same line, denote statistically significant differences between breeds, for a given parity, with $p \leq 0.01$.

Hybrid gilts had a higher backfat thickness during the first parity compared to PLW sows ($p \leq 0.05$). After the first farrowing, in the second parity, the thickest backfat was found in PL sows ($p \leq 0.05$). Statistical analysis confirmed the differences in backfat thickness in hybrid females between the first and second parity ($p \leq 0.05$).

During the first parity, the largest area of fat over Longissimus dorsi muscle was recorded in hybrid gilts (PL × PLW) compared to PL ($p \leq 0.05$) and PLW ($p \leq 0.01$). At the moment of the second insemination, the highest results in the range of fat area were noted in PL sows ($p \leq 0.01$). It should be emphasized that, in the case of the PLW breed and the parameter of fat surface area over LD muscle, statistically significant differences were found between the second and third parity ($p \leq 0.05$). Statistical differences were also confirmed for this parameter in PL × PLW females between the first and second parity ($p \leq 0.05$).

The percentage of intramuscular fat in the Longissimus dorsi muscle was calculated using Designer Genes Technologies software compatible with the Aloka SSD-500 ultrasonograph. The analysis of numerical data did not reveal any statistically significant differences, both between the breeds ($p > 0.05$) and related to the parity ($p > 0.05$). The recorded intramuscular fat content was very similar for all studied genetic groups. It should be emphasized, however, that the intramuscular fat content increased gradually with subsequent parities. The limiting level of significance in the case of the greatest differences between the mean values in the case of the UTŚ parameter was $p = 0.12$.

The results of ultrasound measurements of muscularity level (UP4M, USLD, UPLD) are presented in Table 3. In the first parity, the highest depth of LD muscle was observed in hybrid gilts (PL × PLW), however, compared to other breeds it was not statistically significant ($p > 0.05$). The highest values of UP4M parameter were also observed in hybrid sows in the case of second and third parity, but, in this case, the observed differences in relation to PLW and PL sows were statistically confirmed ($p \leq 0.05$).

Table 3. Results of muscularity measurements carried out using Aloka SSD-500 (breed and parity).

Parameter UP4M						
Parity	Breed					
	PL		PLW		PL × PLW	
	$\bar{x}$ (mm)	SD	$\bar{x}$ (mm)	SD	$\bar{x}$ (mm)	SD
1	57.85 ^{b}	±6.16	57.73 ^{B}	±6.04	59.94 ^{B}	±6.75
2	59.52	±6.52	59.42 ^{b,y}	±7.99	63.47 ^{x}	±5.83
3	62.59 ^{a,y}	±6.62	64.53 ^{A,a}	±5.62	66.80 ^{A,x}	±4.23
Parameter USLD						
Parity	Breed					
	PL		PLW		PL × PLW	
	$\bar{x}$ (mm)	SD	$\bar{x}$ (mm)	SD	$\bar{x}$ (mm)	SD
1	131.68 ^{b}	±9.01	131.83 ^{B}	±7.06	132.86 ^{B}	±8.97
2	135.10	±8.26	134.32 ^{b}	±7.19	135.38 ^{b}	±8.52
3	139.40 ^{a}	±8.32	141.35 ^{A,a}	±7.84	142.76 ^{A,a}	±9.24
Parameter UPLD						
Parity	Breed					
	PL		PLW		PL × PLW	
	$\bar{x}$ (cm^2)	SD	$\bar{x}$ (cm^2)	SD	$\bar{x}$ (cm^2)	SD
1	59.37 ^{b}	±9.16	58.90 ^{B}	±8.92	62.51 ^{b}	±8.82
2	61.51	±10.01	61.59	±9.38	65.16	±9.93
3	66.58 ^{a}	±10.49	67.70 ^{A}	±11.40	69.58 ^{a}	±10.26

$\bar{x}$—mean value. SD—standard deviation. a,b—in the same column, denote statistically significant differences between parity for a given breed, with $p \leq 0.05$; A,B—in the same column, denote statistically significant differences between parity for a given breed, with $p \leq 0.01$; x,y—in the same line, denote statistically significant differences between breeds, for a given parity, with $p \leq 0.05$.

Statistical analysis demonstrated a number of differences between subsequent parities in all examined pig breeds. In the case of the PL breed, the statistically confirmed difference in LD muscle depth between the first and third parity was shown ($p \leq 0.05$). It was observed, in the case of PLW sows, that the depth of the Longissimus dorsi muscle measured in the third parity was significantly higher compared to the results obtained in the first and second parities ($p \leq 0.01$ and $p \leq 0.05$, respectively). Moreover, statistically confirmed differences in LD muscle depth were found between the hybrid gilts inseminated for the first time and the hybrid sows inseminated for the third time ($p \leq 0.01$).

In the case of the USLD parameter, no statistically confirmed inter-breed differences were found ($p > 0.05$). A number of statistically confirmed differences related to the parity were observed. A significant increase in Longissimus dorsi muscle width was observed in each of the breeds studied in subsequent parities.

The highest value of the USLD parameter in the PL breed was recorded in the third parity ($p \leq 0.05$), and the lowest value in the first one ($p \leq 0.05$). A similar trend was observed in the analysis of LD muscle width data for the sows of the PLW breed and hybrid females (PL × PLW). In both cases, the highest results in the range of this parameter were recorded at the third farrowing ($p \leq 0.01$). At the first parity, the width of Longissimus dorsi muscle was the lowest ($p \leq 0.01$). Moreover, statistically confirmed lower values of the width of Longissimus dorsi muscle were observed for the PLW breed and hybrids (PL × PLW) in the second parity compared to the third one ($p \leq 0.05$).

It should be noted that the analysis of the UPLD parameter did not show any inter-breed differences ($p > 0.05$). However, differences in the surface area of the Longissimus dorsi muscle were observed between the first and third parity in all experimental breeds. The largest area of LD muscle was recorded at the third parity in hybrid sows (PL × PLW) ($p \leq 0.05$). The lowest values in the UPLD parameter range, below 60 cm^2, were recorded in pure-bred PL gilts and PLW gilts inseminated for the first time ($p \leq 0.05$; $p \leq 0.01$, respectively).

Table 4 shows the influence of the degree of fatness determined by ultrasound method on reproductive parameters. The relationship between backfat thickness and fat surface over the Longissimus dorsi muscle and reproductive indices was statistically confirmed. Females in which the backfat thickness at UP2 was above 22.25 mm, the backfat thickness at UP4 was above 17.36 mm and the fat area above Longissimus dorsi muscle was above 25.81 cm^2 at the time of insemination were characterized by the best reproductive parameters. Gilts and sows classified in these groups achieved the highest, at almost 92%, farrowing rate compared to the groups with lower UP2, UP4 and UPT parameters ($p \leq 0.05$). In addition, the highest number of piglets born in total (UP2 and UP4 $p \leq 0.05$; UPT $p \leq 0.01$), the number of alive-born piglets (UP2 and UP4 $p \leq 0.05$; UPT $p \leq 0.01$), the lowest number of dead-born piglets (UP2 and UP4 $p \leq 0.05$; UPT $p \leq 0.01$), the highest number of weaned piglets ($p \leq 0.01$) and the lowest mortality of piglets until the weaning day ($p \leq 0.05$) were observed in these females.

The heaviest litters were produced ($p \leq 0.05$; $p \leq 0.01$) and reared ($p \leq 0.01$) by females which backfat thickness (UP2 and UP4) and fat area over LD muscle (UPT) at the time of insemination were higher than the mean characteristic of the studied population. Piglets from these gilts and sows had a higher body weight at birth ($p \leq 0.01$). A similar trend was observed in the mean piglets' weight at weaning (UP2 $p \leq 0.05$; UP4 and UPT $p \leq 0.01$) and mean daily gains (UP2 and UPT $p \leq 0.05$; UP4 $p \leq 0.01$). It is worth noting that females with above average backfat thickness and fatty tissue surface over Longissimus dorsi muscle had a shorter weaning-to-service interval ($p \leq 0.05$).

Table 5 presents the influence of the degree of muscularity determined by ultrasound method on the parameters of reproductive performance in the studied population. The analysis of ultrasound measurements of LD muscle performed during insemination showed a significant influence of depth, width and surface of this muscle on reproductive parameters.

Table 4. An effect of fatness degree determined by ultrasound method on reproductive parameters of sows.

Reproduction Parameter ($\bar{x}$,s)	Degree of Fatness Determined by Aloka SSD-500 Apparatus							
	UP2		UP4		UPT		UTŚ	
	≤22.25 mm	>22.25 mm	≤17.36 mm	>17.36 mm	≤25.81 cm^2	>25.81 cm^2	≤1.74%	>1.74%
Farrowing rate (%)	87.89 ^{b}	91.90 ^{a}	87.99 ^{b}	91.99 ^{a}	87.50 ^{b}	91.56 ^{a}	88.00	90.30
Number of piglets born in total (head)	10.34 ^{b} ± 0.83	11.14 ^{a} ± 0.91	10.30 ^{b} ± 0.85	11.25 ^{a} ± 0.92	10.20 ^{B} ± 0.87	11.30 ^{A} ± 0.94	10.45 ± 0.88	10.99 ± 0.95
Number of alive-born piglets (head)	10.04 ^{b} ± 0.85	10.94 ^{a} ± 0.77	9.97 ^{b} ± 0.79	11.06 ^{a} ± 0.76	9.82 ^{B} ± 0.91	11.08 ^{A} ± 0.74	10.19 ± 0.70	10.74 ± 0.82
Number of dead-born piglets (head)	0.30 ^{b} ± 0.41	0.20 ^{a} ± 0.39	0.33 ^{b} ± 0.35	0.19 ^{a} ± 0.43	0.38 ^{B} ± 0.40	0.22 ^{A} ± 0.38	0.26 ± 0.41	0.25 ± 0.36
Number of weaned piglets (head)	9.58 ^{B} ± 0.77	10.59 ^{A} ± 0.69	9.56 ^{B} ± 0.65	10.76 ^{A} ± 0.68	9.39 ^{B} ± 0.75	10.77 ^{A} ± 0.67	9.75 ± 0.60	10.33 ± 0.63
Mortality of piglets until weaning day (head)	0.46 ^{b} ± 0.53	0.35 ^{a} ± 0.51	0.41 ^{b} ± 0.45	0.30 ^{a} ± 0.48	0.43 ^{b} ± 0.54	0.31 ^{a} ± 0.47	0.44 ± 0.50	0.41 ± 0.43
Total litter weight at 1st day (kg)	14.56 ^{b} ± 2.44	17.50 ^{a} ± 2.13	14.26 ^{b} ± 2.32	17.47 ^{a} ± 2.20	14.14 ^{B} ± 2.47	17.95 ^{A} ± 2.05	14.98 ± 2.19	16.11 ± 2.28
Average piglet weight at 1st day (kg)	1.45 ^{B} ± 0.25	1.60 ^{A} ± 0.20	1.43 ^{B} ± 0.27	1.58 ^{A} ± 0.18	1.44 ^{B} ± 0.23	1.62 ^{A} ± 0.20	1.47 ± 0.25	1.50 ± 0.29
Total litter weight at weaning (kg)	80.57 ^{B} ± 10.90	91.07 ^{A} ± 9.85	80.30 ^{B} ± 10.29	92.86 ^{A} ± 9.81	78.78 ^{B} ± 10.83	92.84 ^{A} ± 9.89	82.39 ± 10.50	87.70 ± 9.51
Average piglet weight at weaning (kg)	8.41 ^{b} ± 0.44	8.60 ^{a} ± 0.48	8.40 ^{B} ± 0.51	8.63 ^{A} ± 0.53	8.39 ^{B} ± 0.52	8.62 ^{A} ± 0.51	8.45 ± 0.47	8.49 ± 0.50
Average body gains of piglets (g)	248.57 ^{b} ± 9.85	250.00 ^{a} ± 9.59	248.93 ^{B} ± 10.36	251.79 ^{A} ± 9.55	248.21 ^{b} ± 9.86	250.00 ^{a} ± 9.66	249.29 ± 10.47	249.64 ± 9.57
Gestation period (days)	114.62 ± 2.25	114.09 ± 2.70	114.55 ± 2.44	114.26 ± 2.39	114.72 ± 2.51	114.10 ± 2.73	114.65 ± 2.44	114.60 ± 2.71
Weaning to service interval (days)	11.12 ^{b} ± 10.33	9.77 ^{a} ± 9.78	11.00 ^{b} ± 11.02	9.65 ^{a} ± 9.99	11.20 ^{b} ± 10.31	9.84 ^{a} ± 9.90	10.60 ± 10.45	10.29 ± 10.25

a, b—in the same line denote statistically significant differences between the degrees of fatness, with $p \leq 0.05$, A, B—in the same line denote statistically significant differences between the degrees of fatness, with $p \leq 0.01$.

Table 5. An effect of muscularity degree determined by ultrasound method on the reproductive parameters of sows.

Reproduction Parameter ($\bar{x}$,s)	Degree of Muscularity Determined by Aloka SSD-500 Apparatus					
	UP4M		USLD		UPLD	
	≤61.32 mm	>61.32 mm	≤136.08 mm	>136.08 mm	≤63.66 cm^2	>63.66 cm^2
Farrowing rate (%)	92.59 ^{a}	88.58 ^{b}	91.29 ^{a}	87.11 ^{b}	92.79 ^{a}	88.78 ^{b}
Number of piglets born in total (head)	10.99 ^{a} ± 0.89	10.15 ^{b} ± 0.85	11.02 ^{A} ± 0.90	9.97 ^{B} ± 0.89	11.14 ^{A} ± 0.91	10.05 ^{B} ± 0.88
Number of alive-born piglets (head)	10.81 ^{a} ± 0.75	9.87 ^{b} ± 0.79	10.82 ^{A} ± 0.71	9.67 ^{B} ± 0.69	10.93 ^{A} ± 0.70	9.73 ^{B} ± 0.78
Number of dead-born piglets (head)	0.18 ^{a} ± 0.34	0.28 ^{b} ± 0.31	0.20 ^{a} ± 0.39	0.30 ^{b} ± 0.42	0.21 ^{a} ± 0.39	0.32 ^{b} ± 0.47
Number of weaned piglets (head)	10.55 ^{a} ± 0.65	9.50 ^{b} ± 0.61	10.52 ^{A} ± 0.65	9.26 ^{B} ± 0.72	10.65 ^{A} ± 0.61	9.34 ^{B} ± 0.67
Mortality of piglets until weaning day (head)	0.26 ^{a} ± 0.41	0.37 ^{b} ± 0.45	0.30 ^{a} ± 0.44	0.41 ^{b} ± 0.47	0.28 ^{a} ± 0.40	0.39 ^{b} ± 0.50
Total litter weight at 1st day (kg)	16.76 ± 2.22	14.21 ± 2.25	16.66 ± 2.09	13.83 ± 2.55	17.27 ^{a} ± 2.20	14.30 ^{b} ± 2.24
Average piglet weight at 1st day (kg)	1.55 ^{a} ± 0.23	1.44 ^{b} ± 0.27	1.54 ^{a} ± 0.19	1.43 ^{b} ± 0.29	1.58 ^{a} ± 0.26	1.47 ^{b} ± 0.30
Total litter weight at weaning (kg)	90.31 ^{A} ± 10.24	79.80 ^{B} ± 10.29	90.37 ^{A} ± 9.80	77.97 ^{B} ± 10.99	91.38 ^{A} ± 9.78	78.55 ^{B} ± 10.28
Average piglet weight at weaning (kg)	8.56 ^{a} ± 0.44	8.40 ^{b} ± 0.52	8.59 ^{a} ± 0.55	8.42 ^{b} ± 0.45	8.58 ^{a} ± 0.53	8.41 ^{b} ± 0.47
Average body gains of piglets (g)	250.36 ^{a} ± 10.35	248.57 ^{b} ± 10.39	251.79 ^{a} ± 9.78	249.64 ^{b} ± 9.90	250.00 ^{a} ± 9.87	247.86 ^{b} ± 10.67
Gestation period (days)	114.20 ± 2.45	114.60 ± 2.57	114.15 ± 2.80	114.71 ± 2.75	114.03 ± 2.77	114.76 ± 2.50
Weaning to service interval (days)	9.81 ^{a} ± 9.94	11.16 ^{b} ± 11.32	10.00 ^{a} ± 9.95	11.36 ^{b} ± 9.97	9.93 ^{a} ± 10.32	11.28 ^{b} ± 10.89

a, b—in the same line denote statistically significant differences between the degrees of muscularity, with $p \leq 0.05$, A, B—in the same line denote statistically significant differences between the degrees of muscularity, with $p \leq 0.01$.

Females with the depth of Longissimus dorsi muscle during insemination more than 61.32 mm, width more than 136.08 mm and area more than 63.66 cm^2, achieved worse reproductive results in the range of farrowing rate ($p \leq 0.05$), number of piglets born in total ($p \leq 0.05$; $p \leq 01$), number of alive-born piglets ($p \leq 0.05$; $p \leq 0.01$), number of dead-born piglets ($p \leq 0.05$), number of weaned piglets ($p \leq 0.05$; $p \leq 0.01$) and piglet mortality until the weaning day ($p \leq 0.05$). It should be noted that gilts and sows in which muscularity was above average gave birth to piglets with low body weight ($p \leq 0.05$). Moreover, piglets from these females also had significantly lower weight at weaning ($p \leq 0.05$) and worse body weight gains ($p \leq 0.05$). The described groups of gilts and sows gave birth to litters of low weight, and statistical analysis confirmed that this was mainly due to the area of LD muscle ($p \leq 0.05$). In the case of the total litter weight at weaning, the lowest values of this parameter were also recorded in more muscular females ($p \leq 0.01$). It should be noted that sows with greater muscularity had a longer weaning-to-service interval ($p \leq 0.05$).

The summarized results of breed effect on reproductive parameters are presented in Table 6.

Table 6. Breed effect on reproduction parameters.

Reproduction Parameter ($\bar{x}$,s)	Breed		
	PL	PLW	PL × PLW
Farrowing rate (%)	90.22	88.70 ^{b}	91.30 ^{a}
Number of piglets born in total (head)	10.20 ± 0.77	9.90 ^{b} ± 0.81	11.00 ^{a} ± 0.72
Number of alive-born piglets (head)	10.00 ± 0.70	9.60 ^{b} ± 0.80	10.80 ^{a} ± 0.75
Number of dead-born piglets (head)	0.20 ± 0.31	0.25 ± 0.37	0.15 ± 0.31
Number of weaned piglets (head)	9.70 ± 0.50	9.55 ^{b} ± 0.70	10.51 ^{a} ± 0.61
Mortality of piglets until weaning day (head)	0.44 ± 0.44	0.32 ± 0.50	0.38 ± 0.40
Total litter weight at 1st day (kg)	14.99 ± 2.30	13.80 ^{b} ± 2.51	16.95 ^{a} ± 2.09
Average piglet weight at 1st day (kg)	1.43 ^{b} ± 0.27	1.41 ^{b} ± 0.29	1.55 ^{a} ± 0.20
Total litter weight at weaning (kg)	83.20 ± 10.15	78.77 ^{b} ± 10.66	89.40 ^{a} ± 9.99
Average piglet weight at weaning (kg)	8.45 ± 0.39	8.30 ± 0.42	8.59 ± 0.49
Average body gains of piglets (g)	249.00 ± 10.50	248.01 ^{b} ± 9.71	250.19 ^{a} ± 9.89
Gestation period (days)	114.76 ± 2.35	114.53 ± 2.19	114.79 ± 2.71
Weaning to service interval (days)	10.70 ± 9.71	10.90 ± 10.01	10.66 ± 9.80

a, b—in the same line denote statistically significant differences between the breeds, with $p \leq 0.05$.

In general, the best results in the range of all studied indices were obtained by PL × PLW females. The differences were statistically significant compared to PLW breed ($p \leq 0.05$), except the number of dead-born piglets, piglet mortality, mean piglet weight at weaning as well as parameters related to farrowing and weaning-to-service interval length. The results obtained for PL breed were insignificant compared to two other genotypes, except for mean piglet weight at birth, which differed statistically from hybrid sows.

4. Discussion

A very dynamic development of ultrasonography has been observed in recent years, and it is used to examine and image the tissues in human and veterinary medicine, or for the purpose of in vivo measurements of animals in order to characterize their performance [7]. Ultrasonography is a non-invasive method of imaging diagnostics, which allows to obtain an image of the cross-section of the examined object and its advantage is the lack of harmful side effects on the organism of both the researcher and the examined person. The ultrasound technique is very precise, very accurate and allows the detection of even small changes in the examined organs and tissues. Increasingly lower costs of ultrasound equipment and high mobility have made ultrasound scanners more and more popular in pig research [7,15].

One of the devices used for the intravital ultrasound examination of animals is Aloka SSD-500. This device has been widely used in farm animal studies. The device gained particular popularity in research on cattle [16–18], as well as on pigs [7,13,19,20]. The spectrum of research conducted with this device was very broad and most often referred to the determination of degree of oocytes development [17], fetus development in different stages [18], and the structure of mammary glands [16]. In addition to its use in the broadly understood diagnostics of reproductive performance, the Aloka SSD-500 is now also used to assess slaughter performance. It is most commonly used to evaluate backfat thickness and Longissimus dorsi muscle width [13,20,21]. Studies with the use of the Aloka ultrasound scanner, aiming to evaluate the degree of fatness and muscularity, allowed even an intravital estimation of the meat content in the carcass [19]. The study conducted by Rempel et al. [22] suggests that, despite the better selection procedures observed currently in animal breeding, there is still a relationship between body condition expressed as fatness or muscularity degree, and reproductive parameters, i.e., litter features. The results of our study in the range of fatness degree are slightly different from that presented by Tyra and Żak [23]. In the study on the domestic pig population, the authors obtained lower results and also noted the highest backfat thickness for the Puławska breed and line 990, 19.20 and 18.30 mm, respectively, and a higher degree of fatness (15.20 mm) was noted for PLW breed compared to PL breed (14.20 mm). However, this was quite the opposite in the case of our study: the difference was almost 1 mm.

In turn, in another study on Polish breeds, Tyra et al. [13] confirmed lower backfat thickness at UP4 compared to UP2 for PL and PLW breeds, which is consistent with the results of our study. The authors showed that the mean backfat thickness measured behind the last rib at the border of thoracic and lumbar vertebrae 8 cm from the middle line of the back was 14.90 mm for PL breed, and 14.50 mm for PLW. The lower values of this parameter obtained by the authors may be associated with lower body weight of the examined animals.

Ultrasonographic examination of fat area under LD muscle was performed by Tyra et al. [20], and the results were below 20.00 cm^2, which was related to the low weight of the examined animals, i.e., 100 kg. The authors demonstrated strong correlation between this parameter and backfat thickness at P2 and P4 measured during carcass dissection. Correlation coefficients were at a similar level for these two analyzed parameters and amounted to about 0.53. Moreover, the authors noted that an increase in LD muscle depth is accompanied by a significant decrease in the fat area above the muscle. Subsequent study of UPT parameter conducted by Tyra et al. [20] in the domestic population using the Aloka SSD-500 apparatus showed inter-breed differences. The authors noted the highest values for Duroc breed (21.3 cm^2). The lowest fat area over the Longissimus dorsi was recorded for Pietrain breed (14.1 cm^2). Maternal breeds, i.e., Polish Landrace and Polish Large White, were characterized by intermediate results for this trait. However, PL breed had a slightly higher fatness degree compared to PLW. The authors calculated that PL gilts had, on average, 20.6 cm^2 of fat area over Longissimus dorsi muscle, while this value for PLW gilts was about 19.4 cm^2.

Intramuscular fat in our study was determined in a similar manner as in the study of Bahelka et al. [24]. The results of these authors confirm that Aloka SSD-500 device with 3.5 MHz UST-5044 head, with a signal amplification of 80%, allows to determine intramuscular fat content with a very high accuracy. Intramuscular fat content determined by means of ultrasound amounted to 2.28%, whereas the result obtained based on chemical analysis was 2.22%.

The results of intramuscular fat measurements obtained in our study are similar to the results obtained by Orzechowska et al. [25], who showed that in the PL breed it was 1.72%, and 1.71% in PLW breed. Moreover, the study confirmed that this parameter value changes with an increase in animals' muscularity. In the case of PL breed, the animals with the highest leanness, above 60%, were characterized by the lowest intramuscular fat level (1.71%). A similar tendency was also observed in PL breed, where the carcasses of the most muscular pigs were rated the highest (class S), and had an intramuscular fat level of 1.68%.

Slightly different results of intramuscular fat measurements were obtained by Tyra and Żak [23], who observed higher intramuscular fat content in Longissimus dorsi muscle in PLW breed (1.84%) compared to PL (1.76%). In our study, PLW gilts and sows also had a slightly higher intramuscular fat content in LD muscle. Moreover, the authors recorded very high results in this parameter in Duroc (2.23%) and Puławska (2.17%) breeds. Jankowiak et al. [26] studied the relationship between carcass fatness and intramuscular fat content and fatty acid profile in pig meat. The study was conducted on Puławska breed and hybrids from PLW and PL crossbreeding. Intramuscular fat content in Puławska breed was high and amounted to 1.87%, while in hybrids a lower result was found (1.72%).

With regards to the muscularity degree examined in this study, similar results in the scope of LD muscle depth were obtained by Knecht et al. [27]. The authors demonstrated that hybrid gilts were characterized by an LD muscle depth of 61.47 mm on average. In the case of our study, the average depth of Longissimus dorsi muscle in gilts from such cross-breeding was about 60 mm. This difference can be explained by higher representativeness of the research group, due to more numerous groups of animals participating in the experiment.

It is worth emphasizing that no inter-breed differences in the case of the UPLD parameter were demonstrated in our study. Orzechowska et al. [25] observed that PL breed is characterized by a larger area of the tenderloin eye compared to PLW breed. LD muscle area in PL breed was 56.50 cm^2, while in PLW it was 55.2 cm^2. In our study, PL sows, which were inseminated for the second and third time, also had a larger area of the tenderloin eye compared to PLW sows. Moreover, the results of the study of the abovementioned authors showed that with leanness exceeding 60%; this parameter was 56.70 cm^2 for PL breed, and 57.40 cm^2 for PLW. The study conducted by Tyra and Żak [23] also showed that PL breed is characterized by a larger surface area of tenderloin eye compared to PLW breed. According to these authors, the area of LD muscle in PL was 53.40 cm^2, whereas in PLW it was 52.70 cm^2. The results obtained are lower than those presented in Table 3 because the animals had a lower body weight (about 100 kg).

Tyra et al. [13] examined the differences in the results of LD muscle surface measurements using ultrasound and dissection methods. In the case of the PL breed, the ultrasound measurement of tenderloin eye area gave the result of 49.9 cm^2, while the dissection measurement result was 52.2 cm^2. The examination of LD muscle area in PLW breed in the case of the ultrasound method gave the result of 51.10 cm^2, whereas after dissection the result was 52.70 cm^2. The differences were therefore insignificant and related to the measurement methodology. Krška et al. [19] also studied tenderloin eye area using the ultrasound method with Aloka SSD-500 apparatus. The authors determined meat content in the carcass on the basis of intravital measurements performed with Piglog-105 probe, SonoMark SM-100 scanner and Aloka SSD-500 apparatus. The highest accuracy was found in the case of Aloka SSD-500, with the leanness estimated using this method amounting to 55.83%, while value determined by dissection was 55.67%.

The results obtained show that an increase in the degree of fatness during insemination may determine the improvement in reproductive indices. De Rensis et al. [28] observed in their study no relationship between backfat thickness and farrowing rate, which is contrary to our results. Matysiak et al. [9] stated that the number of alive-born piglets and the number of piglets at weaning are positively correlated with the backfat thickness on a level of 0.31 and 0.24, which means that females with thicker backfat, similarly to our study, gave birth to more piglets and recorded a higher number of piglets at weaning. Holm et al. [4] point out that when energy reserves are too low (the female has a lower degree of fatness during the first insemination), the development and implantation of embryos in the uterus may not proceed properly, which increases the number of embryos that are resorbed in the uterus and, as a result, contributes to a reduction in the number of alive-born piglets. Bergsma et al. [29] claim that lactation can be a period with maximum energy expenditure, and this may limit reproductive performance, e.g., litter growth. Vanroose et al. [30] and Knecht et al. [6] also pay attention to other factors influencing embryo resorption, such as viral infections, stress, poor nutrition or seasonality of reproduction.

In our study, it was found that the lowest birth weights of piglets were recorded in gilts and sows, which were characterized by lower or equal to the mean population thickness of backfat and the surface of fat tissue over LD muscle during insemination (UP2 $\leq$ 22.25 mm, UP4 $\leq$ 17.36 mm,

UPT ≤ 25.81 cm). It should be emphasized that a lower degree of gilt fatness during insemination may interfere with fetal development and worsen fetal growth, and may also affect the lower weight of piglets at birth, which probably contributed to the poorer reproductive performance indices in this range. Milligan et al. [31] emphasize that a low body weight at birth is accompanied by an increase in the number of dead-born piglets and an increase in mortality during the rearing period, which is associated with a very high physiological effort of the gilt organism.

Immediately after parturition, reserves of energy accumulated earlier are released in the form of adipose tissue. The metabolism of gilts and sows, oriented for a long period of pregnancy to store energy reserves, has problems with switching in a very short period of time to the use of nutrients taken from feed for milk production. Weaker gestational anabolism is likely to result in weaker lactation, which may result in a reduction in the number of weaned piglets, inter alia due to insufficient milk production, piglets may also have a lower weaning weight due to insufficient feed intake [32]. It was also demonstrated in the study by Strathe et al. [32] that backfat thickness and backfat losses during lactation did not affect the number of piglets born in subsequent parity, which is contrary to our findings, however, the differences could be attributed to different rearing strategies, breed or environmental conditions.

The study by Tummaruk et al. [3] confirm that gilts with a high growth rate have more numerous and heavier litters than sows. The authors suggest that this may be related to higher feed intake resulting in a thicker backfat. The gilts with a higher growth rate consume more feed during their growth, thus their health and nutritional status is better, which is then reflected in improved reproductive parameters compared to gilts with a lower growth rate. Moreover, Holm et al. [4] state that the thickness of backfat considered as a source of energy for sows may also play an important role in subsequent breeding cycles. Szulc et al. [33] report that too high a level of female fatness during mating/insemination may lead to hormonal disorders. In the case of these authors' study, it was noted that females with the thickest backfat (>15 mm) gave birth to the lowest number of alive piglets. The authors explained their results by the characteristic transformation of estradiol to estriol, a weaker estrogen that occurs in more fattened females. The opposite effect was observed in our study. Females in which backfat thickness was higher than the average of population at both examined points produced and reared the highest number of piglets characterized by very high weights at birth and weaning. Tummaruk et al. [3] obtained similar results in their study. However, the differences can be due to genetics. It may be supposed that, in the case of a very lean genetics, the benefit of a certain threshold of backfat is positive, like in the current study, but excess backfat might indicate that the animal is too heavy, too old or in heat, and that might be the explanation for the different findings.

Selection aimed at improving fattening and slaughter traits led to a significant increase in the muscularity of the animals, but at the expense of a reduction in fatness level, which led to a deterioration in reproductive capacity. Many authors confirmed a very strong positive correlation between backfat thickness and litter size, e.g., [3,34–36]. Our study confirms the thesis that females with higher backfat thickness at insemination may produce more piglets with higher birth weight. A similar trend can also be observed in the analysis of fat surface over LD muscle. This can be explained by the fact that the beneficial effect of LD muscle could be a direct effect of weight. That higher body weight increases the size of reproductive organs as well as LD muscle.

Kawęcka et al. [37] found negative correlations between gilts' muscularity and their subsequent reproductive performance in three consecutive litters, which means that too much muscularity caused a deterioration in reproductive indices. A similar trend was observed in our study. Decreasing the muscularity level in the range of UP4M, USLD and UPLD parameters resulted in an improvement in reproductive performance indices. It should be also born in mind that fat reserves can be mobilized relatively easily, while muscle store mobilization for energy purposes is quite the opposite [38].

5. Conclusions

The measurements using Aloka SSD-500 demonstrated the highest fat thickness, as well as the highest fat area over LD muscle in hybrid gilts (PL × PLW) in the first parity. The muscularity of pure-bred gilts was lower compared to hybrid gilts. Fatness degree influenced the reproductive parameters of females in which the backfat thickness in UP2 was above 22.25 mm, in UP4 point above 17.36 mm and the fat area over LD muscle was above 25.81 cm^2. These females achieved a better farrowing rate and higher numbers of born piglets. Decreased mortality, higher gains of piglets and higher body weight at weaning were observed, and the weaning to service interval was shortened. Intramuscular fat content did not affect reproductive parameters. Muscularity evaluated by ultrasound affected reproductive performance parameters, except gestation period. Females with too high muscularity were characterized by the lowest reproductive indices.

Author Contributions: Conceptualization, D.K.; methodology, D.K.; software, S.Ś.; investigation, S.Ś.; formal analysis, D.K. and S.Ś.; writing—original draft preparation, S.Ś. and K.C.; writing—review and editing, D.K. and K.C. supervision, D.K. All authors have read and agreed to the published version of the manuscript.

Funding: This research received no external funding.

Conflicts of Interest: The authors declare no conflict of interest.

References and Notes

1. POLSUS. *Results of the Evaluation of Pigs in 2017*; Polish Association of Pig Farmers and Producers: Warsaw, Poland, 2018; pp. 1–46.
2. Rozeboom, D.W.; Pettigrew, J.E.; Mosel, R.L.; Cornelius, S.G.; El Kandelgy, S.M. Influence of gilt age and body composition at first breeding on sow reproductive performance and longevity. *J. Anim. Sci.* **1996**, *74*, 138–150. [CrossRef] [PubMed]
3. Tummaruk, P.; Lundeheim, N.; Einarsson, S.; Dalin, A.M. Effect of birth litter size, birth parity number, growth rate, backfat thickness and age at first mating of gilts on their reproductive performance as sow. *Anim. Reprod. Sci.* **2001**, *66*, 225–235. [CrossRef]
4. Holm, B.; Bakken, M.; Klemetsdal, G.; Vangen, O. Genetic correlations between reproduction and production traits in swine. *J. Anim. Sci.* **2004**, *82*, 3458–3464. [CrossRef] [PubMed]
5. Kim, J.S.; Yang, X.; Pangeni, D.; Baidoo, S.K. Relationship between backfat thickness of sows Turing late gestation and reproductive efficiency at different parities. *Acta Agric. Scand. Sect. A* **2015**, *65*, 1–8. [CrossRef]
6. Knecht, D.; Duziński, K. The effect of parity and date of service on the reproductive performance of polish large white × polish landrace (plw × pl) crossbred sows. *Ann. Anim. Sci.* **2014**, *14*, 69–79. [CrossRef]
7. Knecht, D.; Środoń, S.; Jankowska-Mąkosa, A.; Duziński, K. Appropriate level of gilt fatness and muscularity during insemination can improve the efficiency of piglet production. *Anim. Reprod.* **2017**, *14*, 1138–1146. [CrossRef]
8. Quinton, V.M.; Wilton, J.W.; Robinson, J.A.; Mathur, P.K. Economic weights for sow productivity traits in nucleus pig populations. *Livest. Sci.* **2006**, *99*, 69–77. [CrossRef]
9. Matysiak, B.; Kawęcka, M.; Jacynio, E.; Kołodziej-Skalska, A.; Pietruszka, A. Relationships between test of gilts before day at first mating on their reproduction performance. *Acta Sci. Pol. Zoot.* **2010**, *9*, 29–38.
10. Grela, E.; Skomiał, J. *Nutritional Recommendations and Nutritional Value of Feedingstuffs for Pigs. Standards of pigs Nutrition*; Jan Kielanowski Institute of Physiology and Animal Nutrition, Polish Academy of Sciences: Jabłonna, Poland, 2014.
11. Journal of Laws of 2010, no 116, item 778, Ordinance of the Minister of Agriculture and Rural Development of 28 June 2010 on minimum conditions for maintaining farm animal species other than those for which the standards of protection are laid down in the European Union regulations.
12. Journal of Laws of 26 February 2015, item 266, Act of 15 January 2015 on the protection of animals used for scientific and educational purposes.
13. Tyra, M.; Szyndler-Nędza, M.; Eckert, R. Possibilities of using ultrasonography in breeding work with pigs, Part I—Analysis of ultrasonic, ultrasonographic and dissection measurements of the most numerous breeds of pigs raised in Poland. *Annals Anim. Sci.* **2011**, *11*, 27–40.

14. Schulte, K.J.; Baas, T.J.; Wilson, D.E. An evaluation of equipment and procedures for the prediction of intramuscular fat in live swine. *Anim. Ind. Rep.* **2011**, *657*, 1–5.
15. Mörlein, D.; Rosner, F.; Brand, S.; Jenderka, K.V.; Wicke, M. Non-destructive estimation of the intramuscular fat content of the longissimus muscle of pigs by means of spectral analysis of ultrasound echo signals. *Meat Sci.* **2005**, *69*, 187–199. [CrossRef] [PubMed]
16. Strzetelski, J.; Bilik, K.; Niwińska, B.; Skrzyński, G.; Łuczyńska, E. Ultrasound evaluation of the mammary gland structure in preparturient heifers vs performance of first calves. *J. Anim. Feed Sci.* **2004**, *13*, 5–8. [CrossRef]
17. Adamiak, S.J.; Mackie, K.; Watt, R.G.; Webb, R.; Sinclair, K.D. Impact of nutrition on oocyte quality: Cumulative effects of body composition and diet leading to hyperinsulinemia in cattle. *Biol. Reprod.* **2005**, *73*, 918–926. [CrossRef] [PubMed]
18. McNeill, R.E.; Diskin, M.G.; Sreenan, J.M.; Morris, D.G. Associations between milk progesterone concentration on different days and with embryo survival during the early luteal phase in dairy cows. *Theriogenology* **2006**, *65*, 1435–1441. [CrossRef] [PubMed]
19. Krška, P.; Bahelka, I.; Demo, P.; Peškovičová, D. Meat content in pigs estimated by various methods and compared with objective lean meat content. *Czech J. Anim. Sci.* **2002**, *47*, 206–211.
20. Tyra, M.; Orzechowska, B.; Żak, G. Relationships between ultrasonic and dissection measurements of backfat thickness and m. Longissimus dorsi of pigs using Piglog 105 and Aloka SSD 500. *Ann. Anim. Sci.* **2005**, *5*, 279–286.
21. Kearns, C.F.; McKeever, K.H.; Roegner, V.; Brady, S.M.; Malinowski, K. Adiponectin and leptin are related to fat mass in horses. *Vet. J.* **2006**, *172*, 460–465. [CrossRef]
22. Rempel, L.A.; Vallet, J.L.; Lents, C.A.; Nonneman, D.J. Measurements of body composition during late gestation and lactation in first and second parity sows and its relationship to piglet production and post-weaning reproductive performance. *Livest. Sci.* **2015**, *178*, 289–295. [CrossRef]
23. Tyra, M.; Żak, G. Characteristics of the Polish breeding population of pigs in terms of intramuscular fat (IMF) content of m. Longissimus dorsi. *Ann. Anim. Sci.* **2010**, *10*, 241–248.
24. Bahelka, I.; Oravcová, M.; Peškovičová, D.; Tomka, J.; Hanusová, E.; Lahučký, R.; Demo, P. Comparison of accuracy of intramuscular fat prediction in live pigs using five different ultrasound intensity levels. *Animal* **2009**, *3*, 1205–1211. [CrossRef]
25. Orzechowska, B.; Tyra, M.; Mucha, A.; Żak, G. Quality of carcasses from polish large white and polish landrace pigs with particular consideration of intramuscular fat (IMF) content depending on carcass meat percentage. *Roczniki Nauk. Zoot.* **2012**, *39*, 77–85.
26. Jankowiak, H.; Bocian, M.; Kapelanski, W.; Roslewska, A. The relationship between carcass fatness and intramuscular fat content, and fatty acids profile in pig meat. *Food Sci. Technol. Qual.* **2010**, *17*, 199–208. [CrossRef]
27. Knecht, D.; Środoń, S.; Duziński, K. In vivo evaluation of the fat content and muscularity of gilts with different genotypes using Aloka SSD-500 ultrasound scanner in relation to selected reproductive performance indicators. *Roczn. Nauk. Pol. Tow. Zoot.* **2014**, *10*, 25–35.
28. De Rensis, F.; Gherpelli, M.; Superchi, P.; Kirkwood, R.N. Relationships between backfat depth and plasma leptin during lactation and sow reproductive performance after weaning. *Anim. Reprod. Sci.* **2005**, *90*, 95–100. [CrossRef]
29. Bergsma, R.; Kanis, E.; Verstegen, M.W.A.; van der Peet-Schwering, C.M.C.; Knol, E.F. Lactation efficiency as a result of body composition dynamics and feed intake in sows. *Livest. Sci.* **2009**, *125*, 208–222. [CrossRef]
30. Vanroose, G.; de Kruif, A.; Van Soom, A. Embryonic mortality and embryo–pathogen interactions. *Anim. Reprod. Sci.* **2000**, *60–61*, 131–143. [CrossRef]
31. Milligan, B.N.; Dewey, C.E.; de Grau, A.F. Neonatal-piglet weight variation and its relation to pre-weaning mortality and weight gain on commercial farms. *Prevent. Vet. Med.* **2002**, *56*, 119–127. [CrossRef]
32. Strathe, A.V.; Bruun, T.S.; Hansen, C.F. Sows with high milk production had both a high feed intake and high body mobilization. *Animal* **2017**, *11*, 1913–1921. [CrossRef]
33. Szulc, K.; Knecht, D.; Jankowska-Mąkosa, A.; Skrzypczak, E.; Nowaczewski, S. The influence of fattening and slaughter traits on reproduction in Polish Large White sows. *Ital. J. Anim. Sci.* **2013**, *12*, 16–20. [CrossRef]
34. Karsten, S.; Rohe, R.; Schulze, V.; Looft, H.; Kalm, E. Genetic association between individual feed intake during performance test and reproductions traits in pigs. *Archiv Tierz.* **2000**, *43*, 451–461.

35. Bečková, R.; Daněk, P.; Václavková, E.; Rozkot, M. Influence of growth rate, backfat thickness and meatiness on reproduction efficiency in Landrace gilts. *Czech J. Anim. Sci.* **2005**, *50*, 535–544. [CrossRef]
36. Stalder, K.J.; Saxton, A.M.; Conatser, G.E.; Serenius, T.V. Effect of growth and compositional traits on first parity and lifetime reproductive performance in U.S. Landrace sows. *Livest. Prod. Sci.* **2005**, *97*, 151–159. [CrossRef]
37. Kawęcka, M.; Matysiak, B.; Kamyczek, M.; Delikator, B. Relationships between growth, fatness and meatiness traits in gilts and their subsequent reproductive performance. *Ann. Anim. Sci.* **2009**, *9*, 249–258.
38. Lewis, C.R.; Bunter, K.L. Body development in sows, feed intake and maternal capacity. Part2: Gilt body condition before and after lactation, reproductive performance and correlations with lactation feed intake. *Animal* **2011**, *5*, 1855–1867. [CrossRef]

MDPI

Article

Effect of Different Cross-Fostering Strategies on Growth Performance, Stress Status and Immunoglobulin of Piglets?

Xiaojun Zhang, Meizhi Wang, Tengfei He, Shenfei Long, Yao Guo and Zhaohui Chen *

State Key Laboratory of Animal Nutrition, College of Animal Science and Technology, China Agricultural University, Beijing 100193, China; zhangxiaojun14@163.com (X.Z.); meizhiwang@cau.edu.cn (M.W.); hetengfei@cau.edu.cn (T.H.); longshenfei@cau.edu.cn (S.L.); guoyaocau@163.com (Y.G.)

* Correspondence: chenzhaohui@cau.edu.cn; Tel.: +86-10-6273-2763

Simple Summary: Piglet survival in large litters can be increased if surplus piglets are cross-fostered to smaller litters, exploiting surplus teats in these sows. We aimed (1) to investigate the effect of cross-fostering piglets of different birth weights into new litters on the growth performance of piglets; (2) to determine the effect of cross-fostering piglets of different ages on the growth performance, stress and immunity of these piglets. Cross-fostering on day 2 after birth reduced average daily gain (ADG) in high birth weight (HBW) piglets. Late cross-fostering on day 7 after birth decreased ADG, affected the teat order and increased the cortisol level of piglets. Therefore, these results provide suitable cross-fostering strategies for improving cross-fostering piglets' welfare.

Abstract: The effect of different cross-fostering strategies on the growth performance, stress and immunity of piglets was investigated in this study. In the first experiment, a total of 20 litters (i.e., 20 sows) and 120 piglets were classified into one of six treatments in a 2 × 3 factorial arrangement. The treatments consisted of piglets without or with cross-fostering and different birth weights (low birth weight, LBW; intermediate birth weight, IBW; high birth weight, HBW). The weaning weight (WW) and average daily gain (ADG) of LBW piglets and IBW piglets were not significantly different between the not cross-fostered (NC-F) group and the cross-fostered (C-F) group. There was a higher ($p < 0.05$) ADG in the control piglets compared with the cross-fostered piglets. This effect on ADG was only seen in the HBW piglets. In the second experiment, six sows with a similar body condition and farrowed on the same day were selected. Three female piglets with a birth weight of 0.6–0.85 kg were selected from each litter as experimental piglets. Eighteen piglets were grouped into three treatments: (1) not cross-fostered (NC-F1), (2) cross-fostered at 36–48 h after birth (C-F1), (3) cross-fostered at day 7 after birth (C-F2). The growth performance of NC-F1 and C-F1 piglets was higher than C-F2 piglets ($p < 0.05$), and the suckling positions of NC-F1 and C-F1 piglets on days 8, 12, 16 and 20 were more forward than the C-F2 piglets ($p < 0.05$). Plasma cortisol (COR) concentrations of NC-F1 and C-F1 piglets were lower than C-F2 piglets ($p < 0.05$). A significant negative correlation was observed between BW at day 21 and plasma COR concentration. In conclusion, cross-fostering within 24 h of birth has adverse influences on the ADG of HBW piglets, while it has no negative effect on the ADG of LBW and IBW piglets. Moreover, for IBW piglets, late cross-fostering (i.e., on day 7 after farrowing) has negative impacts on the growth performance and teat order of piglets, and it increases the cortisol level of piglets.

Keywords: cross-fostering; piglets; growth performance; stress status; immunoglobulin

Citation: Zhang, X.; Wang, M.; He, T.; Long, S.; Guo, Y.; Chen, Z. Effect of Different Cross-Fostering Strategies on Growth Performance, Stress Status and Immunoglobulin of Piglets. *Animals* **2021**, *11*, 499. https://doi.org/10.3390/ani11020499

Academic Editors: María-Luz García and María-José Argente

Received: 15 January 2021
Accepted: 11 February 2021
Published: 14 February 2021

1. Introduction

With the improvement of gene breeding technology and management practices, the number of surviving piglets per sow per year is increasing. When piglet numbers exceed available functional teats, piglets could have less milk intake, which could lead to a high risk of mortality [1,2]. Large litter size has negative animal welfare impacts on piglets and sows [2,3]. Therefore, some management strategies should be applied to reduce mortality and improve animal welfare for piglets in commercial farms. Cross-fostering is a management practice that transfers extra piglets from large litters to smaller litters, so that the sows with more functional teats can be put to good use [4–6]. Some authors have studied the effects of cross-fostering on the mortality and growth performance of piglets [7,8]. Other studies show that the age/stage of lactation and body weight/size of piglets affect the efficiency of cross-fostering [9,10].

Good maternal behavior and high milk production of sows have a positive effect on the growth performance of piglets [1,11]. Therefore, sows with good maternity and lactation capacity should be selected to improve the efficiency of cross-fostering. Small piglets with large litter mates spend more time competing for teats [12], thus, fostered piglet size is also an important factor for cross-fostering. Ferrari et al. [13] found that heavy birth weight fostered piglets had a higher weight at days 14 and 20 pre-weaning than light birth weight fostered piglets. Large piglets that were fostered in mixed litters consisting of equal numbers of light-weight and heavy-weight piglets had a greater growth rate than large piglets, but the growth performance of small fostered piglets in uniform litters was significantly increased [14]. However, Souza et al. [15] reported that small piglets had a similar body weight during lactation, regardless of being mixed with piglets of higher weights or not. Similar-sized piglets may face more competition [16], and homogeneous litters with only small piglets may not fully stimulate the breast, resulting in reduced milk production. Therefore, the body weight of small piglets that were grouped with large piglets in mixed litters could be improved because heavier piglets are able to stimulate the teats to remove more milk from the mammary glands [17]. There is still controversy about the effect of cross-fostering on the performance of piglets with different birth weights [1].

Teat order is established by piglets in early lactation [18], and cross-fostering should be completed as early as possible during this time [5]. Teat order was not established when piglets were fostered at 12–24 h after birth and they could absorb more colostrum immunoglobulins [5,19,20]. There was little impact on the weight gain of piglets when they were fostered at 12 h after parturition [12]. Heim et al. [5] showed that cross-fostering at 14–24 h after birth had no negative effect on the growth performance and survival rate of fostered piglets. It was shown that cross-fostering at 48 h after birth had no adverse effect on growth performance, but fostered piglets had a higher risk of death [21]. In practice, some piglets fall behind through lactation so that they may be subjected to late cross-fostering. Late cross-fostering means fostered piglets will face more competition because the teat order has been established. Some studies indicated that late cross-fostering reduced the body weight gain of piglets [22,23]. Robert et al. [24] showed that piglets performed more aggressive behavior when they were fostered on day 7 post-partum. Environmental factors could affect the blood stress hormones of pigs [25,26], but few studies on cross-fostering have focused on the blood stress hormones and immunoglobulins of piglets. At present, most research on cross-fostering time reported that late cross-fostering had an adverse effect on the growth performance, mortality and behavior of piglets, but the effect of cross-fostering piglets of different ages still remains to be investigated.

The purpose of this study was (1) to investigate the effect of cross-fostering piglets of different birth weights into new litters on the growth performance of piglets; (2) to determine the effect of cross-fostering piglets of different ages on the growth performance, stress and immunity of these piglets.

2. Materials and Methods

The experimental protocols used in this experiment were approved by the Institutional Animal Care and Use Committee of China Agricultural University (Beijing, China) (No. AW09089102-6).

2.1. Animals and Experimental Design

This study was conducted on a breeding farm in Guangdong, China, with a herd size of 400 sows, from January to April 2019, and two experiments were conducted. Experimental sows and piglets were purebred Luchuan pigs—this is one of the famous native breeds in China. Luchuan pigs are mainly distributed in southeast China, and they have the characteristics of small size, high productivity, early sexual maturity and good maternity [27]. We analyzed the birth weight of 13,222 piglets from December 2017 to December 2018 and found that the average birth weight of piglets was 0.68 (±0.18) kg. Sows were moved into the farrowing house from day 4 before predicted parturition date and were housed in farrowing crates, which occupied an area of 1.8 m × 0.6 m, and farrowing pens occupied an area of 2.3 m × 1.8 m. The temperature of the farrowing house was maintained at 24–27 °C during the first week after parturition, and 22–24 °C during other experimental periods. All sows were fed two times a day, at 07:30 and 16:30, and received 0.5 kg on day 2 after parturition and then 0.5–1 kg extra per day onwards to a maximum of 5 kg on day 10 after parturition. On day 5 after parturition, male piglets were castrated under analgesia and isoflurane anesthesia. On day 7 after parturition, piglets received piglet feed. Both sows and piglets had free access to water.

In the first experiment, piglets were divided into three different levels, including high birth weight (HBW—>0.8–1.15 kg), intermediate birth weight (IBW—0.6–0.8 kg) and low birth weight (LBW—<0.6–0.35 kg) piglets according to their birth weight. Six piglets (2 HBW piglets, 2 IBW piglets and 2 LBW piglets) were selected from each litter as experimental piglets, and three (1 HBW piglet, 1 IBW piglet and 1 LBW piglet) of the six piglets were fostered piglets and cross-fostered (C-F) at 18–24 h after birth. The rest were not cross-fostered (NC-F) piglets, except for three fostered piglets in the same litter, and they were born on the same day. There were 20 litters (i.e., 20 sows) in the experiment. A total of 120 piglets were classified to one of 6 treatments in a 2 × 3 factorial arrangement. The treatments consisted of piglets without or with cross-fostering (C-F, NC-F) and different birth weights (LBW, IBW, HBW). Each treatment included 20 piglets. The sex of piglets was evenly distributed across fostered and non-fostered experimental piglets (i.e., 8 male HBW fostered piglets and 8 male HBW non-fostered piglets, 7 male IBW fostered piglets and 7 male IBW non-fostered piglets, 4 male LBW fostered piglets and 4 male LBW non-fostered piglets), with 31.7% being males and 69.3% being females. During the experiment, we did not find that sows harmed the piglets and all experimental piglets survived. Twenty sows of parity 3–5 had similar body condition. Each litter involved 12–14 piglets. Piglets were weaned at 23–27 days after farrowing. The average litter size was 11.85 ± 0.24 (mean ± SEM, SEM means standard error of mean) piglets at weaning, and the mortality of piglets was 6.32% during the experiment.

In the second experiment, six sows (parity 3–5) with similar body condition and farrowed on the same day were selected. Each litter included 12–14 piglets. Three female piglets with a birth weight of 0.6–0.85 kg (mean ± SEM, 0.74 ± 0.08 kg) were selected from each litter as experimental piglets. The 18 piglets were grouped to the three treatments: (1) not cross-fostered (NC-F1, n = 6), (2) cross-fostered at 36–48 h after birth (C-F1, n = 6), (3) cross-fostered at 7 day after birth (C-F2, n = 6). A cross-fostered piglet was cross-fostered to another of the experimental litters in exchange for a cross-fostered piglet from this litter. The piglets were weaned at 21 day of age. The average litter size was 12.33 ± 0.42 (mean ± SEM) piglets at weaning, and the mortality of piglets was 7.5% during the experiment.

2.2. Measurements

In the first experiment, experimental piglets were weighted individually at birth and weaning. Average daily gain (ADG) was calculated from birth and weaning.

In the second experiment, experimental piglets were weighted individually at 0, 7, 14 and 21 days of age. ADG was calculated for each time period. Suckling positions of experimental piglets (i.e., 1 represented the first teat pair, 2 represented the second teat pair) were recorded two times at 8:30–9:00 and 15:30–16:00 on days 3, 5, 8, 12, 16 and 20 after parturition when suckling behavior occurred. Blood samples were taken from the experimental piglets' jugular vein at 7:00–9:00 on day 21 after parturition. Blood was collected in a Heparin tube and centrifuged at 3000 r/min for 10 min. Plasma was recovered for the growth hormone (GH), cortisol (COR), alpha-amylase (α-AMY), immunoglobulin A (IgA), immunoglobulin G (IgG), and immunoglobulin M (IgM) assay. GH levels were measured by enzyme immunoassay (Growth hormone EIA kit, Nanjing Jiangcheng Bioengineering Institute, Nanjing, China), and COR levels were measured by enzyme immunoassay (Cortisol EIA kit, DRG International, Springfield, NJ, USA). A-AMY levels were measured by the iodine-starch colorimetric method (Amylase EIA kit, Nanjing Jiangcheng Bioengineering Institute, Nanjing, China). IgA, IgG and IgM levels were measured by turbidimetric inhibition immuno assay (Automatic biochemical analyzer, HITACHI, Tokyo, Japan).

2.3. Statistical Analyses

Data analysis was carried out using the statistical software JMP 14.1 (SAS Institute Inc., Cary, NC, USA). All data were tested for normality by the Shapiro–Wilk test. Significant differences were considered at the 95% confidence level ($p < 0.05$). In the first experiment, birth weight (BW), weaning weight (WW) and ADG were analyzed using the linear mixed model, and the sow number is included as a random factor. In the second experiment, body weight (BW1), ADG and suckling position were analyzed as repeated measures by the linear mixed model, with the fixed effect of cross-fostering time, day, and interaction between these two factors [5]. One-way analysis of variance (ANOVA) was used to test the cross-fostering time effects on ADG from 0 day to 21 day, COR, α-AMY, IgA, IgG and IgM of piglets. Correlations between COR, α-AMY, IgA, IgG, IgM and growth performance were analyzed using the Pearson correlation coefficient.

3. Results

3.1. Birth Weight, Weaning Weight and Average Daily Gain

Significant birth weight class (BWC) effects were exhibited in BW, WW and ADG ($p < 0.05$), and there were no significant effects of treatment and interactions between treatment and BWC in BW, WW and ADG. For LBW piglets and IBW piglets, the WW and ADG were not significantly different between the NC-F group and the C-F group. The ADG of NC-F was significantly higher compared with C-F in the HBW piglets ($p < 0.05$) (Table 1).

Table 1. The effect of different piglets' birth weight through cross-fostering on the BW, WW and ADG of piglets.

Item	Treatment	Birth Weight Class (BWC)			SEM	*p*-Values		
		LBW	IBW	HBW		Treatment	BWC	Treatment × BWC
BW (kg)	NC-F	0.46	0.71	0.94	0.02	0.88	<0.01	0.93
	C-F	0.47	0.70	0.94				
WW (kg)	NC-F	3.78	4.46	5.06	0.06	0.19	<0.01	0.53
	C-F	3.68	4.31	4.80				
ADG (g)	NC-F	128.69	145.52	159.70 ^{a}	1.72	0.07	<0.01	0.48
	C-F	124.76	139.59	149.58 ^{b}				

BW—birth weight; WW—weaning weight; ADG—average daily gain; NC-F—not cross-fostered; C-F—cross-fostered; LBW—low birth weight; IBW—intermediate birth weight; HBW—high birth weight; $N_{(NC\text{-}F,\ LBW)} = 20$, $N_{(C\text{-}F,\ LBW)} = 20$, $N_{(NC\text{-}F,\ IBW)} = 20$, $N_{(C\text{-}F,\ IBW)} = 20$, $N_{(NC\text{-}F,\ HBW)} = 20$, $N_{(C\text{-}F,\ HBW)} = 20$, N means the number of piglets per group; SEM—standard error of mean; $^{a,\ b}$—Values with different superscripts differ significantly ($p < 0.05$) among different treatments (columns).

3.2. Growth Performance

There were significant effects of treatment and interactions between treatment and day on the body weight (BW1) and ADG ($p < 0.05$). On days 14 and 21, the BW1 of not cross-fostered (NC-F1) and cross-fostered at 36–48 h after birth (C-F1) piglets was higher than cross-fostered at 7 day after birth (C-F2) piglets ($p < 0.05$). The NC-F1 and C-F1 piglets had higher ADG between 7 day of age and 14 day of age and between birth and weaning compared with the C-F2 piglets ($p = 0.01$) (Table 2).

Table 2. Effect of different cross-fostering times on the growth performance of piglets.

Item	Day	Treatment			SEM	*p*-Values		
		NC-F1	C-F1	C-F2		Treatment	Day	Treatment × Day
BW1 (kg)	0	0.73	0.77	0.73	0.12	0.04	<0.01	0.02
	7	1.51	1.47	1.47				
	14	2.69 ^{a}	2.61 ^{a}	1.99 ^{b}				
	21	3.58 ^{a}	3.50 ^{a}	2.73 ^{b}				
ADG (g)	0–7	111.07	99.52	104.76	5.82	0.01	0.06	0.03
	7–14	169.52 ^{a}	163.33 ^{b}	75.17 ^{c}				
	14–21	126.43	127.02	105.07				
ADG (g)	0–21	135.67 ^{a}	129.96 ^{a}	95.00 ^{b}	6.46	0.01	-	-

BW1—body weight; ADG—average daily gain; NC-F1—not cross-fostered, N $_{(NC-F1)}$ =6, N means the number of piglets per group; C-F1—cross-fostered at 36–48 h after birth, N $_{(C-F1)}$ = 6; C-F2—cross-fostered at 7 day after birth, N $_{(C-F2)}$ = 6; SEM—standard error of mean; $^{a, b, c}$—Values with different superscripts differ significantly ($p < 0.05$) among different treatments (rows).

3.3. Suckling Positions

A significant treatment ($p = 0.04$) effect was exhibited and significant interactions between treatment and day ($p = 0.01$) were observed in suckling position. On days 3 and 5, there was no difference in the suckling position. The suckling positions of NC-F1 and C-F1 piglets on days 8, 12, 16 and 20 were more forward than the C-F2 piglets ($p < 0.05$) (Table 3).

Table 3. Effect of different cross-fostering times on the suckling position of piglets.

Item	Day	Treatment			SEM	*p*-Values		
		NC-F1	C-F1	C-F2		Treatment	Day	Treatment × Day
Suckling position	3	3.00	3.08	3.08	0.11	0.04	0.20	0.01
	5	3.33	3.33	3.25				
	8	3.16 ^{a}	3.25 ^{a}	4.33 ^{b}				
	12	2.75 ^{a}	3.08 ^{a}	4.92 ^{b}				
	16	2.92 ^{a}	3.12 ^{a}	4.92 ^{b}				
	20	3.08 ^{a}	3.08 ^{a}	5.00 ^{b}				

NC-F1—not cross-fostered, N $_{(NC-F1)}$ = 6, N means the number of piglets per group; C-F1—cross-fostered at 36–48 h after birth, N $_{(C-F1)}$ = 6; C-F2—cross-fostered at 7 day after birth, N $_{(C-F2)}$ = 6; SEM—standard error of mean; $^{a, b}$—Values with different superscripts differ significantly ($p < 0.05$) among different treatments (rows) within the same day.

3.4. Plasma Serum Parameters

Plasma cortisol concentrations of NC-F1 and C-F1 piglets were less than C-F2 piglets ($p < 0.05$) (Table 4). There were no significant effects of different treatments on plasma GH, α-AMY, IgA, IgG and IgM concentrations ($p > 0.05$) (Table 4).

Table 4. Effect of different cross-fostering times on the plasma parameters of piglets.

Plasma Parameters	NC-F1	C-F1	C-F2	SEM	*p*-Value
GH (ng/mL)	1.99	1.90	1.87	0.13	0.93
COR (ng/mL)	14.08 [a]	15.52 [a]	17.44 [b]	0.56	0.03
α-AMY (U/dL)	153.12	148.71	150.85	5.90	0.96
IgA (g/L)	1.04	1.01	0.92	0.03	0.36
IgG (g/L)	9.31	9.52	9.46	0.16	0.87
IgM (g/L)	0.86	0.83	0.77	0.03	0.50

GH—growth hormone; COR—cortisol; α-AMY—alpha-amylase; IgA—immunoglobulin A; IgG—immunoglobulin G; IgM—immunoglobulin M; NC-F1—not cross-fostered, $N_{(NC\text{-}F1)} = 6$, N means the number of piglets per group; C-F1—cross-fostered at 36–48 h after birth, $N_{(C\text{-}F1)} = 6$; C-F2—cross-fostered at 7 day after birth, $N_{(C\text{-}F2)} = 6$; SEM—standard error of mean; [a, b]—Values with different superscripts differ significantly ($p < 0.05$) among different treatments (rows).

3.5. Correlations between Growth Performance and Plasma Serum Parameters

A significant positive correlation was observed between BW1 at day 21 and plasma GH concentration (r = 0.527, $p < 0.05$, r means correlation coefficient), as well as between ADG from day 0 to day 21 and plasma GH concentration (r = 0.488, $p < 0.05$); a significant negative correlation was observed between BW1 at day 21 and plasma COR concentration (r = −0.734, $p < 0.05$), as well as between ADG from day 0 to day 21 and plasma COR concentration (r = −0.736, $p < 0.05$). There was no significant correlation between BW1 at day 21 and plasma α-AMY, IgA, IgG, and IgM ($p > 0.05$), as well as between ADG from day 0 to day 21 and plasma α-AMY, IgA, IgG, and IgM ($p > 0.05$) (Table 5).

Table 5. Correlation analysis between growth performance and plasma parameters.

Growth Performance	GH	COR	α-AMY	IgA	IgG	IgM
BW1 at day 21	0.527 *	−0.734 *	−0.197	0.132	0.108	0.216
ADG from day 0 to day 21	0.488 *	−0.736 *	−0.211	0.134	0.024	0.133

BW1—body weight; ADG—average daily gain; GH—growth hormone; COR—cortisol; α-AMY—alpha-amylase; IgA—immunoglobulin A; IgG—immunoglobulin G; IgM—immunoglobulin M; NC-F1—not cross-fostered, N = 6; C-F1—cross-fostered at 36–48 h after birth, N = 6; C-F2—cross-fostered at 7 day after birth, N = 6; * $p < 0.05$, (−)denotes a negative association.

4. Discussion

A higher concentration of immunoglobulins in sow colostrum within the 12 h period post farrowing was found than at other times [28], and they were better absorbed through the intestinal barrier of piglets during this time [29,30]. In order to ensure that piglets sucked enough colostrum, piglets were cross-fostered more than 12 h after birth in our experiments. Alexopoulos et al. [1] also advised that piglets should stay with their birth sow for at least 12 h. Cross-fostering within 24 h of birth had no effect on the BW1 of non-fostered piglets and fostered piglets [4,5,31]—there was a lack of difference in the WW and ADG of LBW and IBW piglets between treatments in the current study. However, we found that cross-fostering had an adverse effect on the ADG of HBW fostered piglets, which could be due to the fact that HBW fostered piglets fought more and sucked less milk than HBW non-fostered piglets [30]. Piglet size was an important factor for growth and survival [1]. Souza et al. [15] found that LBW fostered piglets missed more nursing episodes when they were mixed with high BW fostered piglets, and HBW piglets vigorously stimulate the udder and suck more milk [32,33]; this may lead to a lower growth performance of LBW piglets than HBW piglets [14,34]. In the present study, we also found that the WW and ADG of LBW piglets (including C-F and NC-F piglets) were lower than HBW piglets.

The age of cross-fostering was an important factor for the growth performance and behavior of fostered piglets. Many authors advised that cross-fostering should be implemented soon after farrowing [35–37]. However, some piglets with low growth performance appear in litters during lactation (i.e., over 3 day after parturition), and they may also need to be fostered. Kooij et al. [38] and Wattanaphansak et al. [39] found that cross-fostering on days 2–3 after farrowing had no significant effect on growth performance, whereas Robert et al. [24] and Calderón et al. [8] found that cross-fostering in the first week after farrowing impaired the growth performance of piglets. Our results showed that piglets which were cross-fostered on day 7 after birth had lower BW1 and ADG compared with NC-F1 and C-F1 piglets, which was in line with the above research. In addition, we found that the ADG of C-F2 piglets in the second week after farrowing was lower than other experimental periods. Some studies have reported that the teat order of piglets was developed in the first few days after birth and was relatively stable after one week of birth [40,41], and the stability of teat order had a positive effect on the growth performance of piglets [42]. On days 3 and 5 post-partum, there was lack of difference in the suckling position of piglets among treatments, and when piglets were cross-fostered on day 7 post-partum, the suckling position of C-F2 piglets was more backward compared with the other two treatments in the present study. Huting et al. [14] reported that the BW1 of piglets suckling the anterior teats was higher than piglets suckling the posterior teats. The results demonstrated that cross-fostering on day 7 after farrowing damaged the stability of teat order, and this may lead to the lower growth performance of C-F2 piglets.

In terms of the stress of piglets, plasma COR and α-AMY were selected to evaluate the stress level of piglets [43,44]. Differences were observed such that the COR concentrations of C-F2 piglets were higher than NC-F1 and C-F1 piglets, which indicated that cross-fostering on day 7 after farrowing increased the stress level of piglets. Similar results have been found that the stress level of piglets was increased when unfamiliar piglets were mixed together [45]. Horrell et al. [46] reported that mother-offspring bonds formed early in lactation (i.e., within 3 days after parturition). In fostered piglets removed from their mother to a new environment on day 7 after farrowing, the maternal bond was destroyed, and fostered piglets fought with non-littermates [9,47]. Hence, it is possible that cross-fostering on day 7 after farrowing increased the stress level of fostered piglets. In addition, we found a significant negative correlation between plasma COR concentration and growth performance, which indicated that stress had an adverse effect on the growth performance of piglets. There was no difference in the immunoglobulins level of piglets among treatments.

5. Conclusions

Cross-fostering within 24 h of birth has adverse influences on the average daily gain of high birth weight piglets, while it has no negative effect on the average daily gain of low birth weight and intermediate birth weight piglets. Moreover, for intermediate birth weight piglets, late cross-fostering (i.e., on day 7 after farrowing) has negative impacts on the growth performance and teat order of piglets, and it increases the stress level of piglets. We suggest that low birth weight and intermediate birth weight piglets might be more suitable to cross foster as early as possible (i.e., 12–48 h after farrowing) when cross-fostering strategy is implemented.

Author Contributions: Conceptualization, Z.C. and X.Z.; Performed the experiments, X.Z.; Data curation, X.Z. and M.W.; Writing—original draft preparation, X.Z.; Writing—review and editing, M.W., T.H., S.L., Y.G. and Z.C.; Funding acquisition, M.W. and Z.C.; All authors have read and agreed to the published version of the manuscript.

Funding: This work is supported by the funding of the National Key Research and Development Program of China (2018YFD0501200).

Institutional Review Board Statement: The study was conducted according to the guidelines of the Declaration of Helsinki, and approved by the Institutional Animal Care and Use Committee of China Agricultural University (Beijing, China) (No. AW09089102-6).

Acknowledgments: The support of the Guangdong Yihao Foodstuff Co., Ltd is greatly appreciated.

Conflicts of Interest: The authors declare no potential conflict of interest. The funders had no role in the design of the study; in the collection, analyses, or interpretation of data; in the writing of the manuscript, or in the decision to publish the results.

References

1. Alexopoulos, J.G.; Lines, D.S.; Hallett, S.; Plush, K.J. A Review of Success Factors for Piglet Fostering in Lactation. *Animals* **2018**, *8*, 38. [CrossRef]
2. Rutherford, K.; Baxter, E.; D'Eath, R.; Turner, S.; Arnott, G.; Roehe, R.; Ask, B.; Sandøe, P.; Moustsen, V.; Thorup, F.; et al. The welfare implications of large litter size in the domestic pig I: Biological factors. *Anim. Welf.* **2013**, *22*, 199–218. [CrossRef]
3. Prunier, A.; Heinonen, M.; Quesnel, H. High physiological demands in intensively raised pigs: Impact on health and welfare. *Animal* **2010**, *4*, 886–898. [CrossRef]
4. Milligan, B.N.; Fraser, D.; Kramer, D.L. The effect of littermate weight on survival, weight gain, and suckling behavior of low-birth-weight piglets in cross fostered litters. *J. Swine Health Prod.* **2001**, *99*, 161–166.
5. Heim, G.; Mellagi, A.; Bierhals, T.; De Souza, L.; De Fries, H.; Piuco, P.; Seidel, E.; Bernardi, M.; Wentz, I.; Bortolozzo, F. Effects of cross-fostering within 24h after birth on pre-weaning behaviour, growth performance and survival rate of biological and adopted piglets. *Livest. Sci.* **2012**, *150*, 121–127. [CrossRef]
6. Schmitt, O.; Baxter, E.M.; Boyle, L.A.; O'Driscoll, K. Nurse sow strategies in the domestic pig: I. Consequences for selected measures of sow welfare. *Animal* **2019**, *13*, 580–589. [CrossRef] [PubMed]
7. Bierhals, T.; Magnabosco, D.; Ribeiro, R.; Perin, J.; Da Cruz, R.; Bernardi, M.; Wentz, I.; Bortolozzo, F. Influence of pig weight classification at cross-fostering on the performance of the primiparous sow and the adopted litter. *Livest. Sci.* **2012**, *146*, 115–122. [CrossRef]
8. Calderón, D.J.A.; García, M.E.; Alessia, D. Cross-Fostering implications for pig mortality, welfare and performance. *Front. Vet. Sci.* **2018**, *5*, 1–10. [CrossRef] [PubMed]
9. Baxter, E.M.; Rutherford, K.M.D.; D'Eath, R.B.; Arnott, G.; Turner, S.P.; Sandoe, P.; Moustsen, V.A.; Thorup, F.; Edwards, S.A.; Lawrence, A.B. The welfare implications of large litter size in the domestic pig II: Management factors. *Anim. Welf.* **2013**, *22*, 219–238. [CrossRef]
10. Pajžlar, L.; Skok, J. Cross-fostering into smaller or older litter makes piglets integration difficult: Suckling stability-based rationale. *Appl. Anim. Behav. Sci.* **2019**, *220*, 104856. [CrossRef]
11. Illmann, G.; Chaloupková, H.; Melišová, M. Impact of sow prepartum behavior on maternal behavior, piglet body weight gain, and mortality in farrowing pens and crates1. *J. Anim. Sci.* **2016**, *94*, 3978–3986. [CrossRef]
12. Deen, M.; Bilkei, G. Cross fostering of low-birthweight piglets. *Livest. Prod. Sci.* **2004**, *90*, 279–284. [CrossRef]
13. Ferrari, C.; Sbardella, P.; Bernardi, M.; Coutinho, M.; Vaz, I.; Wentz, I.; Bortolozzo, F. Effect of birth weight and colostrum intake on mortality and performance of piglets after cross-fostering in sows of different parities. *Prev. Veter. Med.* **2014**, *114*, 259–266. [CrossRef]
14. Huting, A.M.S.; Almond, K.; Wellock, I.; Kyriazakis, I. What is good for small piglets might not be good for big piglets: The consequences of cross-fostering and creep feed provision on performance to slaughter1,2. *J. Anim. Sci.* **2017**, *95*, 4926–4944. [CrossRef] [PubMed]
15. Souza, L.; Fries, H.; Heim, G.; Faccin, J.; Hernig, L.; Marimon, B.; Bernardi, M.; Bortolozzo, F.; Wentz, I. Behaviour and growth performance of low-birth-weight piglets cross-fostered in multiparous sows with piglets of higher birth weights. *Arq. Bras. Med. Vet. Zootec.* **2014**, *66*, 510–518. [CrossRef]
16. Arnott, G.; Elwood, R.W. Assessment of fighting ability in animal contests. *Anim. Behav.* **2009**, *77*, 991–1004. [CrossRef]
17. King, R.; Mullan, B.; Dunshea, F.; Dove, H. The influence of piglet body weight on milk production of sows. *Livest. Prod. Sci.* **1997**, *47*, 169–174. [CrossRef]
18. Skok, J.; Škorjanc, D. Group suckling cohesion as a prelude to the formation of teat order in piglets. *Appl. Anim. Behav. Sci.* **2014**, *154*, 15–21. [CrossRef]
19. Bandrick, M.; Pieters, M.; Pijoan, C.; Molitor, T.W. Passive transfer of maternal *Mycoplasma hyopneumoniae*-specific cellular immunityto piglets. *Clin. Vaccine Immunol.* **2008**, *15*, 540–543. [CrossRef]
20. Pieters, M.; Bandrick, M.; Pijoan, C.; Baidoo, S.; Molitor, T. The effect of cross-fostering on the transfer of Mycoplasma hyopneumoniae maternal immunity from the sow to the offspring. *Clin. Vaccine Immunol.* **2008**, *15*, 540–543.
21. Neal, S.M.; Irvin, K.M. The effects of crossfostering pigs on survival and growth. *J. Anim. Sci.* **1991**, *69*, 41–46. [CrossRef]
22. Straw, B.E.; Bürgi, E.J.; Dewey, C.E. Effects of extensive cross fostering on performance of pigs on a farm. *J. Am. Vet. Med. A* **1998**, *212*, 855–856.
23. Giroux, S.; Robert, S.; Martineau, G.P. The effects of cross-fostering on growth rate and post-weaning behavior of segregated early-weaned piglets. *Can. J. Anim. Sci.* **2000**, *80*, 533–538. [CrossRef]
24. Robert, S.; Martineau, G.P. Effects of repeated cross-fosterings on preweaning behavior and growth performance of piglets and on maternal behavior of sows. *J. Anim. Sci.* **2001**, *79*, 88–93. [CrossRef] [PubMed]
25. Cronin, G.; Van Amerongen, G. The effects of modifying the farrowing environment on sow behaviour and survival and growth of piglets. *Appl. Anim. Behav. Sci.* **1991**, *30*, 287–298. [CrossRef]
26. Jarvis, S.; Vegt, B.J.V.D.; Lawrence, A.B. The effect of parity and environmental restriction on behavioural and physiological responses of pre–parturient pigs. *Appl. Anim. Behav. Sci.* **2001**, *71*, 203–216. [CrossRef]
27. Ran, M.-L.; He, J.; Tan, J.-Y.; Yang, A.-Q.; Li, Z.; Chen, B. The complete sequence of the mitochondrial genome of Luchuan pig (*Sus scrofa*). *Mitochondrial DNA* **2014**, *27*, 1–2. [CrossRef] [PubMed]

28. Klobasa, F.; Werhahn, E.; Butler, J.E. Regulation of humoral immunity in the piglet by immunoglobulin of maternal origin. *Res. Vet. Sci.* **1981**, *32*, 195–206. [CrossRef]
29. Tuboly, S.; Bernath, S.; Glavits, R.K.; Medveczky, I. Intestinal absorption of colostral lymphoid cells in newborn piglets. *Vet. Immunol. Immunop.* **1988**, *20*, 75–85. [CrossRef]
30. Bandrick, M.; Pieters, M.; Pijoan, C.; Baidoo, S.K.; Molitor, T.W. Effect of cross-fostering on transfer of maternal immunity to Mycoplasma hyopneumoniae to piglets. *Veter. Rec.* **2011**, *168*, 100. [CrossRef] [PubMed]
31. English, J.G.H.; Bilkeit, G. The effect of litter size and littermate weight on pre-weaning performance of low-birth-weight piglets that have been cross-fostered. *Anim. Sci.* **2004**, *79*, 439–443. [CrossRef]
32. Algers, B.; Madej, A.; Rojanasthien, S.; Uvns-Moberg, K. Quantitative relationships between suckling-induced teat stimultion and the release of prolactin, gastrin, somatostatin, insulin, glucagon and vasoactive intestinal polypeptide in sows. *Vet. Res. Commun.* **1991**, *15*, 395–407. [CrossRef] [PubMed]
33. English, P.R. Ten basic principles of fostering piglets. *Pig Prog.* **1998**, *4*, 39–41.
34. Milligan, B.N.; Fraser, D.; Kramer, D.L. Birth weight variation in the domestic pig: Effects on offspring survival, weight gain and suckling behaviour. *Appl. Anim. Behav. Sci.* **2001**, *73*, 179–191. [CrossRef]
35. Kingston, N.G. Farrowing house management. *Pig. Vet. J.* **1989**, *22*, 62–74.
36. Vaillancourt, J.P.; Tubbs, R.C. Preweaning mortality. *Vet. Clin. N. Am. Food A* **1992**, *8*, 685–706. [CrossRef]
37. Price, E.O.; Hutson, G.D.; Price, M.I.; Borgwardt, R. Fostering in swine as affected by age of offspring1. *J. Anim. Sci.* **1994**, *72*, 1697–1701. [CrossRef] [PubMed]
38. Kooij, E.V.E.-V.D.; Kuijpers, A.; Van Eerdenburg, F.; Tielen, M. Coping characteristics and performance in fattening pigs. *Livest. Prod. Sci.* **2003**, *84*, 31–38. [CrossRef]
39. Wattanaphansak, S.; Luengyosluechakul, S.; Larriestra, A.; Deen, J. The impact of cross-fostering on swine production. *Thai J. Vet. Med.* **2002**, *32*, 101–106.
40. Puppe, B.; Tuchscherer, A. Developmental and territorial aspects of suckling behaviour in the domestic pig (*Sus scrofa f. domestica*). *J. Zool.* **1999**, *249*, 307–313. [CrossRef]
41. Skok, J.; Škorjanc, D. Formation of teat order and estimation of piglets' distribution along the mammary complex using mid-domain effect (MDE) model. *Appl. Anim. Behav. Sci.* **2013**, *144*, 39–45. [CrossRef]
42. Horrell, I.; Bennett, J. Disruption of teat preferences and retardation of growth following cross-fostering of 1-week-old pigs. *Anim. Sci.* **1981**, *33*, 99–106. [CrossRef]
43. Petrowski, K.; Wintermann, G.-B.; Schaarschmidt, M.; Bornstein, S.R.; Kirschbaum, C. Blunted salivary and plasma cortisol response in patients with panic disorder under psychosocial stress. *Int. J. Psychophysiol.* **2013**, *88*, 35–39. [CrossRef]
44. Kang, Y. Psychological stress-induced changes in salivary alpha-amylase and adrenergic activity. *Nurs. Health Sci.* **2011**, *12*, 477–484. [CrossRef]
45. Arey, D.; Edwards, S. Factors influencing aggression between sows after mixing and the consequences for welfare and production. *Livest. Prod. Sci.* **1998**, *56*, 61–70. [CrossRef]
46. Horrell, I.; Hodgson, J. The bases of sow-piglet identification. 2. Cues used by piglets to identify their dam and home pen. *Appl. Anim. Behav. Sci.* **1992**, *33*, 329–343. [CrossRef]
47. D'Eath, R.B. Socialising piglets before weaning improves social hierarchy formation when pigs are mixed post-weaning. *Appl. Anim. Behav. Sci.* **2005**, *93*, 199–211. [CrossRef]

Article

Correlated Response to Selection for Litter Size Residual Variability in Rabbits' Body Condition?

Iván Agea [1], María de la Luz García [1], Agustín Blasco [2], Peter Massányi [3], Marcela Capcarová [3] and María-José Argente [1,*]

1 Departamento de Tecnología Agroalimentaria, Escuela Politécnica Superior de Orihuela, Universidad Miguel Hernández de Elche, Ctra. de Beniel km 3.2, 03312 Orihuela, Spain; iagea@umh.es (I.A.); mariluz.garcia@umh.es (M.d.l.L.G.)

2 Institute for Animal Science and Technology, Universitat Politècnica de València, P.O. Box 22012, 46022 València, Spain; ablasco@dca.upv.es

3 Department of Animal Physiology, Faculty of Biotechnology and Food Sciences, Slovak University of Agriculture, 949 76 Nitra, Slovakia; massanyip@gmail.com (P.M.); marcela.capcarova@uniag.sk (M.C.)

* Correspondence: mj.argente@umh.es; Tel.: +34-966-749-708

Received: 11 November 2020; Accepted: 17 December 2020; Published: 21 December 2020

Simple Summary: Selection for decreasing litter size residual variance has been proposed as an indirect way to select for resilience. Resilience has been directly related to welfare. A good body condition and efficient body fat mobilization have been associated with an optimal level of animal welfare. Two rabbit lines have been divergently selected for litter size residual variability. The low line selected for decreasing litter size variance more efficiently managed the body fat from mating to weaning in the second productive cycle in females compared to the high line, which could be related to the lower culling rate reported previously in the low line. Therefore, body condition can be used as a useful biomarker of resilience.

Abstract: A divergent selection experiment for residual variance of litter size at birth was carried out in rabbits during twelve generations. Residual variance of litter size was estimated as the within-doe variance of litter size after pre-correction for year and season as well as parity and lactation status effects. The aim of this work was to study the correlated response to selection for litter size residual variability in body condition from mating to weaning. Body condition is related directly to an animal's fat deposits. Perirenal fat is the main fat deposit in rabbits. Individual body weight (IBW) and perirenal fat thickness (PFT) were used to measure body condition at second mating, delivery, 10 days after delivery, and weaning. Litter size of the first three parities was analyzed. Both lines decreased body condition between mating to delivery; however, the decrease in body condition at delivery was lower in the low line, despite this line having higher litter size at birth (+0.54 kits, $p = 0.93$). The increment of body condition between delivery and early lactation was slightly higher in the low line. On the other hand, body condition affected success of females' receptivity and fertility at the third mating, e.g., receptive females showed a higher IBW and PFT than unreceptive ones (+129 g and +0.28 mm, respectively), and fertile females had a higher IBW and PFT than unfertile ones (+82 g and +0.28 mm, respectively). In conclusion, the does selected for reducing litter size variability showed a better deal with situations of high-energy demand, such as delivery and lactation, than those selected for increasing litter size variability, which would agree with the better health and welfare condition in the low line.

Keywords: body condition; fertility; litter size variability; rabbits; selection

---

1. Introduction

Animal welfare is a priority in livestock production for ethical reasons and also because poor animal welfare is associated with low production, poor health, and larger culling rate [1]. Resilience is defined as the ability of an animal to maintain or quickly recover its performance in spite of environmental perturbations [2,3], thus it is directly related to welfare. The ability of an animal to efficiently mobilize its fat reserves can be essential for it to maintain, or quickly return to its production level. Body condition has been traditionally employed to measure the mobilization of fat reserves in livestock animals (Schröder and Staufenbiel [4] in cattle; Maes et al. [5] in pigs; Pascual et al. [6] in rabbits). Body condition has been commonly used as a welfare indicator, due to its relations with fertility success and prevention of diseases (Barletta et al. [7] in cattle; van Staaveren et al. [8] in pigs; Sánchez et al. [9] in rabbits). Therefore, body condition may be connected to resilience, and monitoring it may be useful in resilience assessments.

Recently, residual variance has been proposed as a measure of resilience [10,11]. A direct divergent selection experiment for residual variance in litter size has been performed successfully in rabbits at the Universidad Miguel Hernández de Elche [12]. The high and low lines showed a remarkable difference in residual variance of litter size (4.5% of the mean of the base population). There were also differences in sensitivity to stress and diseases, which lowered the culling rate in the low line [11]. In this regard, the more homogeneous line coped better with environmental stressors such as infections and acute stress than the heterogeneous line which showed higher resilience [11,13].

In an early experiment with the first generations of selection, García et al. [14] found that the low line had a favorable correlated response to selection in body condition and fat reserve mobilization at birth. The objective of this work was to study the correlated response to selection for litter size residual variability in the development of body condition from mating to weaning.

2. Materials and Methods

2.1. Ethics Statement

All experimental procedures were approved by the Miguel Hernández University of Elche Research Ethics Committee, according to Council Directives 98/58/EC and 2010/63/EU (reference number 2017/VSC/PEA/00212).

2.2. Experimental Animals

Animals came from the twelfth generation of a divergent selection experiment for residual variance of litter size (see more details in Blasco et al. [12]). A total of 121 females of the low line (homogeneous) and 124 females of the high line (heterogeneous) were used to estimate the response to selection and correlated responses in litter size at first, second and third parity, and correlated responses in individual body weight at 4 weeks and 9 weeks old. A subset of 100 primiparous females from the low line and 74 primiparous females from the high line were used to measure the development of body condition in the second reproductive cycle and to study the body condition effect on doe's receptivity and fertility.

All animals were kept on a farm at the Miguel Hernández University of Elche (Spain). Rabbits were fed a standard commercial diet (17% crude protein, 16% fiber, 3.5% fat, Nutricun Elite Gra®, De Heus Nutrición Animal, La Coruña). Food and water were provided ad libitum. Females were housed in individual cages (37.5 cm × 33 cm × 90 cm) under a constant photoperiod of 16 h continuous light (8 h continuous darkness and controlled ventilation throughout the experiment). The experiment took place from December to August. They were first mated at 18 weeks of age and at 10 days after parturition thereafter. Gestation was checked by abdominal palpation 12 d after mating. Litters were not standardized and weaning was at 28 d after delivery.

2.3. Traits

Individual body weight at 4 weeks and 9 weeks old and litter size at birth were recorded. Residual variance of litter size was estimated for all females of the twelfth generation considering all parties, after pre-correcting litter size for the effects of year and season and parity and lactation status. Individual body weight and perirenal fat thickness were recorded at four different physiological stages: second mating, delivery, 10 days after delivery and weaning. Perirenal fat thickness was measured by ultrasound imaging to evaluate body fat reserves as described by Pascual et al. [15] using Toshiba NemioMX SSA-590 ultrasound equipment (Toshiba, Tokyo, Japan). Receptivity and fertility were recorder at third mating (i.e., 10 days after second delivery). Receptivity (acceptance or rejection of the male at mating) was defined as a binary trait as was fertility (pregnant or non-pregnant females at palpation).

2.4. Statistical Analysis

2.4.1. Correlated Response to Selection for Residual Variance

Models included a different set of effects depending on the trait. The following models were used:

residual variance of litter size at birth had only the effect of line (two levels, high and low line); individual body weight at 4 weeks and 9 weeks old, litter size at first parity had the effects of line and season;

litter size at second and third parity had the effects of line, season and lactation status (two levels: lactating and non-lactating female at mating);

individual body weight and perirenal fat thickness had the effects of line-time (eight levels: low line at mating, high line at mating, low line at delivery, high line at delivery, low line 10 days after delivery, high line 10 days after delivery, low line at weaning, and high line at weaning), season, lactation status (two levels: lactating and non-lactating female when recording data) and the dam permanent effect.

All analyses were performed using Bayesian methodology [16]. Bounded uniform priors were used for all effects with the exception of the dam permanent effect, considered normally distributed with mean 0 and variance σ_p^2. Residuals were a priori normally distributed with mean 0 and variance σ_e^2 and uncorrelated with the dam effects. The priors for the variances were also bounded uniform. Features of the marginal posterior distributions for all unknowns were estimated using Gibbs sampling. The Rabbit program developed by the Institute for Animal Science and Technology (Valencia, Spain) was used for all procedures. We used a chain of 60,000 samples, with a burn-in period of 10,000. Only one out of every 10 samples were saved for inferences. Convergence was tested using the Z criterion of Geweke and Monte Carlo sampling errors were computed using time-series procedures.

2.4.2. Effect of Body Condition on Receptivity and Fertility

We analyzed the difference on body condition at third mating (i.e., at 10 days after second delivery) between receptive and non-receptive does, using a model with the effects of line, season, lactation status, and receptivity with two levels (acceptance or rejection of the male at first attempt). In order to study the difference on body condition at mating between fertile and unfertile does, we used a model with the effects of line, season, lactation status and fertility with two levels (pregnant or non-pregnant female at palpation).

A probit regression was performed to assess the effect of individual body weight and perirenal fat thickness at third mating on probability of successful receptivity and fertility using the former models. The probit procedure of the statistical package SAS was used for this analysis (SAS Institute, 2019, Cary, CA, USA).

3. Results

3.1. Correlated Response to Selection for Residual Variance

Table 1 shows the features of marginal posterior distributions of the differences between lines for litter size residual variance, litter size at first, second and third parity, and individual body weight at 4 weeks and 9 weeks old. The probability of these differences being greater than zero if $D_{L-H} > 0$ or lower than zero if $D_{L-H} < 0$ is shown. In a Bayesian context there are no significance levels; instead, we offer the actual probability of the differences. As the environmental effects are the same for both lines, the differences between lines ($D_{L\text{-}H}$) are genetic differences, so they estimate the response and correlated responses to selection. The low line showed a lower litter size variability than the high line (−1.45 kits^2, p =1.00), and a higher litter size in the first parities (+0.42 kits, p = 0.90 in first parity; +0.54 kits, p = 0.93 in second parity; +0.66 kits, p = 0.94 in third parity). The low line had similar body weight to the high line at 4 weeks and 9 weeks old.

Table 1. Features of the marginal posterior distribution of the differences for litter size residual variance at birth (Ve), litter size at first (LS1), second (LS2) and third (LS3) parity, and individual body weight at 4 weeks (IB4w) and at 9 weeks old (IB9w) in rabbits.

	L (n = 121)	H (n = 124)	$D_{L\text{-}H}$	$HPD_{95\%}$	p
Ve, kits^2	2.78	4.23	−1.45	−2.22, −0.67	1.00
LS1, kits	7.54	7.12	0.42	0.26, 1.04	0.90
LS2, kits	8.31	7.77	0.54	−0.19, 1.29	0.93
LS3, kits	8.93	8.27	0.66	−0.19, 1.48	0.94
IB4w, g	732	754	−22	−93.2, 43.9	0.74
IB9w, g	1836	1823	13	−70.9, 95.3	0.61

n: number of data. L: mean of the low line. H: mean of the high line. $D_{L\text{-}H}$: differences between the low and the high line. $HPD_{95\%}$: highest posterior density region at 95%. p: probability of the difference being >0 when $D_{L\text{-}H} > 0$ or being <0 when $D_{L\text{-}H} < 0$.

Figure 1 displays the development of body condition from second mating to weaning in the high and the low line. Individual body weight and perirenal fat thickness showed a reduction from mating to delivery in both lines. However, this reduction was lesser in the low line than in the high line. Both lines exhibited a recovery of body reserves from delivery to 10 days after delivery, but the increment was slightly higher in the low line. Body condition showed a decrease from 10 days after delivery to weaning, but the decrease was slightly higher in the low line. We notice that although number of kits at birth was higher in the low line than the high one (8.31 kits vs. 7.77 kits respectively, p = 0.93) perirenal fat thickness was higher in the low line than the high one in the critical moments of delivery (7.71 mm versus 7.44 mm, p = 0.99) and 10 days after delivery (8.17 mm versus 7.90 mm, p = 0.99).

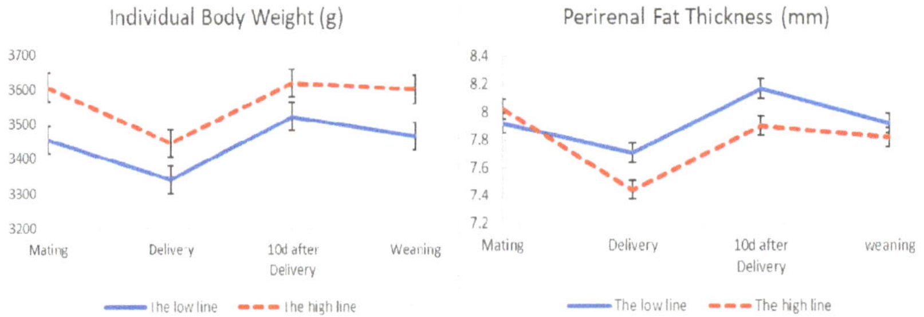

Figure 1. Development of individual body weight and perirenal fat thickness at second mating, delivery, 10 days after delivery and weaning in the low and high lines. The bars show standard deviation.

3.2. *Effect of Body Condition on Receptivity and Fertility*

Table 2 shows that receptive females had higher individual body weight and perirenal fat thickness than unreceptive females (+129 g, p = 0.97 for body weight; +0.28 mm, p = 0.96 for perirenal fat thickness). Individual body weight and perirenal fat thickness were higher in fertile females compared to unfertile females (+82 g, p = 0.94 for body weight; +0.28 mm, p = 0.99 for perirenal fat thickness).

The probabilities of acceptance of mating and pregnancy were not affected by line and season. However, non-lactating females always showed a higher probability for accepting the male and becoming pregnant than lactating females (Figures 2 and 3). For a body weight between 2900 and 4400 g, the probability of acceptance of the male ranged from 75% to 100% in non-lactating does and from 65% to 95% in lactating does. For a perirenal fat thickness between 6.0 and 10.0 mm, the probability of acceptance of the male ranged from 80% to 95% in non-lactating does and from 60% to 95% in lactating does. For the same range of weights, the probability of pregnancy extended from 60% to 95% in non-lactating does and from 30% to 60% in lactating does. For the same range of perirenal fat thickness, the probability of pregnancy ranged from 50% to 95% in non-lactating does and from 20% to 80% in lactating does.

Table 2. Features of the marginal posterior distribution of the differences for individual body weight (IBW10d) and perirenal fat thickness (PFT10d) at 10 days after delivery for receptivity and fertility.

	Receptive	**Non-Receptive**	**D**	**HPD$_{95\%}$**	p
IBW10d (g)	3581	3452	129	6.59, 260	0.97
PFT10d (mm)	8.09	7.81	0.28	−0.03, 0.57	0.96
	Fertile	**Infertile**			
IBW10d (g)	3593	3511	82	−20, 182	0.94
PFT10d (mm)	8.15	7.87	0.28	0.04, 0.52	0.99

D: differences between receptive and non-receptive does or fertile and unfertile does. HPD$_{95\%}$: highest posterior density region at 95%. p: probability of the difference being >0 when D > 0 or being <0 when D < 0.

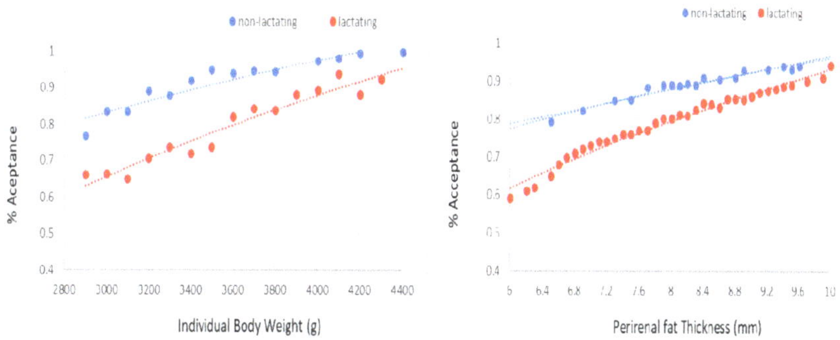

Figure 2. Probability of acceptance of the male at mating on individual body weight and perirenal fat thickness at 10 day after delivery (i.e., third mating in lactating and non-lactating does).

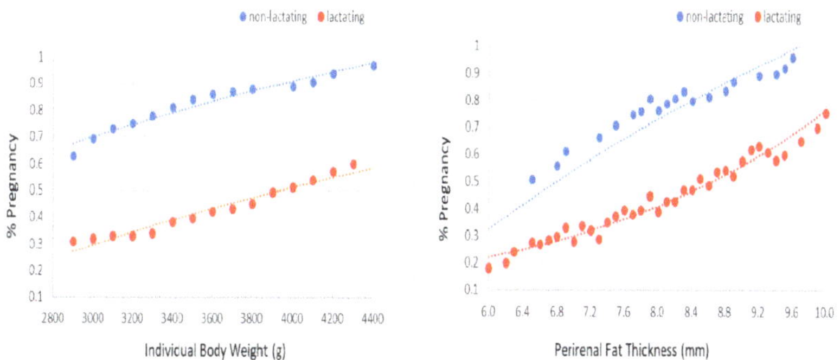

Figure 3. Probability of pregnancy at third gestation on individual body weight and perirenal fat thickness at 10 day after delivery (i.e., third mating, in lactating and non-lactating does).

4. Discussion

4.1. Correlated Response to Selection for Residual Variance

We have found that, as in former generations [12], selection to reduce litter size residual variance produces females with more uniform litters. Uniformity in litter size and body weight has been related to immune response and resistance to diseases (see review [17]). Additionally, we have found that selection for reducing residual variance of litter size increases litter size without affecting the individual weight neither at birth nor at weaning [18].

In relation to development of body condition, we have observed that both lines decrease the body condition from mating to delivery. This is due to the negative energy balance during the last week of gestation, as a consequence of the growing fetuses and the decreasing feed intake in the mother [19]. However, in agreement with previous results from an early experiment in those lines [14], the decrease in body condition at delivery is lower in females from the low line, despite that fact that this line is gestating on average more fetuses.

Immediately after delivery, milk production is low and feed intake is sufficient for covering the nutritional needs for both maintenance and lactation [20]; therefore, body fat reserves are recovered [21]. In accordance with Theilgaard et al. [21], the low and high line increase their body condition between delivery and early lactation; however, the increment is higher in the low line than the high line. A low body condition and high fat mobilization have been related to a high risk of dying or being culled [14,22]; thus, a higher body condition at delivery and a larger fat deposition between delivery from 10 days after delivery in the low line would agree with the lower involuntary elimination rate reported in this line by Argente et al. [11]. In current rabbit production systems, does are mated between 10 and 12 days post-delivery, arriving at the end of weaning with lactation and gestation overlapping [23]. The high energetic needs for milk production and development of fetuses are not entirely compensated with doe's increasing feed intake at the end of weaning (review by Castellini [24]). Therefore, there is an important increase in the mobilization of dam's body reserves, which leads them to lose body condition (review by Castellini et al. [25]). In this sense, we also observed a decrease in body condition between 10 days after delivery and weaning in both lines, although the decrease is slightly higher in the low line due to a large number of kits at weaning [18].

We see that selection to reduce residual variance of litter size has a favorable correlated response in body condition and fat mobilization in the dam, playing an important role in coping to environmental challenges.

4.2. Effect of Body Condition on Receptivity and Fertility

As previously commented, current rabbit breeding programs are based on an interval between delivery and artificial insemination or mating between 10 and 12 days. Therefore, females have to simultaneously allocate their fatness resources for both maintenance and milk production [26], and poor body condition at mating can limit mating success [27]. Several studies have reported a negative effect of lactation on fertilization rate [20,28]. We stress that our study quantifies for the first time the negative effect of lactating on receptivity and fertility. Non-lactating females have from 10% to 20% more probability to accept to mating than lactating females. The effect is even more relevant for fertility. In this regard, probability of becoming pregnant was from 30% to 35% higher in non-lactating females than in lactating females. No differences in receptivity and fertility were found between lines.

These findings support that lactation mobilizes a large amount of doe's fat reserves and has an important effect on receptivity and fertility. Therefore, females must arrive to mate with a good body fatness level which will allow them to have a long and successful reproductive lifespan. The low line showed a greater perirenal fat thickness than the high line at mating (8.17 mm versus 7.90 mm). However, this difference was not enough to result in relevant differences between lines in receptivity and fertility.

5. Conclusions

Selection for litter size variability showed a correlated response between body condition and fat mobilization. The does selected for litter size homogeneity did better in situations with high-energy demand such as delivery and lactation, compared to those selected for increasing litter size variability. This means the animals in the homogenous line had better health and welfare levels.

Author Contributions: Conceptualization M.-J.A. and M.d.l.L.G.; data curation, I.A., M.-J.A. and M.d.l.L.G.; funding acquisition, A.B. and M.-J.A.; formal analysis I.A.; writing—original draft preparation I.A.; writing—review and editing, I.A., M.d.l.L.G., A.B., P.M., M.C., M.-J.A. All authors have read and agreed to the published version of the manuscript.

Funding: This research was supported by Project AGL2017-86083-C2-2-P, funding by Ministerio de Ciencia e Innovación (MIC)-Agencia Estatal de Investigación (AEI) and el Fondo Europeo de Desarrollo Regional (FEDER).

Conflicts of Interest: The authors declare that there are no conflicts of interest to this publication.

References

1. Passillé, D.A.M.; Rushen, J. Preslaughter handling|Welfare Including Housing Conditions. In *Encyclopedia of Meat Sciences*, 2nd ed.; Dikeman, M., Devine, C., Eds.; Academic Press: Cambridge, MA, USA, 2014; pp. 102–107.
2. Colditz, I.G.; Hine, B.C. Resilience in farm animals: Biology, management, breeding and implications for animal welfare. *Anim. Prod. Sci.* **2016**, *56*, 1961–1983. [CrossRef]
3. Berghof, T.V.L.; Poppe, M.; Mulder, H.A. Opportunities to improve resilience in animal breeding programs. *Front. Genet.* **2019**, *9*, 692. [CrossRef] [PubMed]
4. Schröder, U.J.; Staufenbiel, R. Invited Review: Methods to Determine Body Fat Reserves in the Dairy Cow with Special Regard to Ultrasonographic Measurement of Backfat Thickness. *J. Dairy Sci.* **2006**, *89*, 1–14. [CrossRef]
5. Maes, D.G.; Janssens, G.P.; Delputte, P.; Lammertyn, A.; de Kruif, A. Backfat measurements in sows from three commercial pig herds: Relationship with reproductive efficiency and correlation with visual body condition scores. *Livestock Prod. Sci.* **2004**, *91*, 57–67. [CrossRef]
6. Pascual, J.J.; Castella, C.; Cervera, C.; Blas, E.; Fernández-Carmona, J. The use of ultrasound measurement of perirenal fat thinckness to estimate changes in body condition of young females rabbits. *Anim. Sci.* **2000**, *70*, 435–442. [CrossRef]
7. Barletta, R.V.; Maturana Filho, M.; Carvalho, P.D.; Del Valle, T.A.; Netto, A.S.; Rennó, F.P.; Mingoti, R.D.; Gandra, J.R.; Mourão, G.B.; Fricke, P.M.; et al. Association of changes among body condition score during the transition period with NEFA and BHBA concentrations, milk production, fertility, and health of Holstein cows. *Theriogenology* **2017**, *104*, 30–36. [CrossRef]
8. van Staaveren, N.; Doyle, B.; Manzanilla, E.G.; Calderón Díaz, J.A.; Hanlon, A.; Boyle, L.A. Validation of carcass lesions as indicators for on-farm health and welfare of pigs. *J. Anim. Sci.* **2017**, *95*, 1528–1536. [CrossRef]
9. Sánchez, J.P.; de la Fuente, L.F.; Rosell, J.M. Health and body condition of lactating females on rabbit farms. *J. Anim. Sci.* **2012**, *90*, 2353–2361. [CrossRef]
10. Mulder, H.A.; Rashidi, H. Selection on resilience improves disease resistance and tolerance to infections. *J. Anim. Sci.* **2017**, *95*, 3346–3358. [CrossRef]
11. Argente, M.J.; García, M.L.; Zbyňovká, K.; Petruška, P.; Capcarová, M.; Blasco, A. Correlated response to selection for litter size environmental variability in rabbit's resilience. *Animal* **2019**, *13*, 2348–2355. [CrossRef]
12. Blasco, A.; Martínez-Álvaro, M.; García, M.L.; Ibáñez-Escriche, N.; Argente, M.J. Selection for environmental variance of litter size in rabbit. *Genet. Sel. Evol.* **2017**, *49*, 48–56. [CrossRef] [PubMed]
13. Beloumi, D.; Blasco, A.; Muelas, R.; Santacreu, M.A.; García, M.L.; Argente, M.J. Inflammatory correlated response in two lines of rabbit selected divergently for litter size environmental variability. *Animals* **2020**, *10*, 1540. [CrossRef] [PubMed]
14. García, M.L.; Blasco, A.; García, M.E.; Argente, M.J. Correlated response in body condition and energy mobilisation in rabbits selected for litter size variability. *Animal* **2019**, *13*, 784–789. [CrossRef] [PubMed]
15. Pascual, J.J.; Blanco, J.; Piquer, O.; Quevedo, F.; Cervera, C. Ultrasound measurements of perirenal fat thickness to estimate the body condition of reproducing rabbit does in different physiological status. *World Rabbit Sci.* **2004**, *12*, 7–22. [CrossRef]

16. Blasco, A. *Bayesian Data Analysis for Animal Scientists*; Springer: New York, NY, USA, 2017.
17. Lung, L.H.D.S.; Carvalheiro, R.; Neves, H.H.D.R.; Mulder, H.A. Genetics and genomics of uniformity and resilience in livestock and aquaculture species: A review. *J. Anim. Breed Genet.* **2020**, *137*, 263–280. [CrossRef]
18. Agea, I.; García, M.L.; Blasco, A.; Argente, M.J. Litter survival differences between divergently selected lines for environmental sensitivity in rabbits. *Animals* **2019**, *9*, 603. [CrossRef]
19. Fortun-Lamothe, L. Energy balance and reproductive performance in rabbit does. *Anim. Reprod. Sci.* **2006**, *93*, 1–15. [CrossRef]
20. Feugier, A.; Fortun-Lamothe, L. Extensive reproductive rhythm and early weaning improve body condition and fertility of rabbit does. *Anim. Res.* **2006**, *55*, 459–470. [CrossRef]
21. Theilgaard, P.; Baselga, M.; Blas, E.; Friggens, N.C.; Cercera, C.; Pascual, J.J. Differences in productive robustness in rabbits selected for reproductive longevity or litter size. *Animal* **2009**, *3*, 637–646. [CrossRef]
22. Theilgaard, P.; Sánchez, J.P.; Pascual, J.J.; Friggens, N.C.; Baselga, M. Effect of body fatness and selection for prolificacy on survival of rabbit does assessed using a cryopreserved control population. *Livest. Sci.* **2006**, *103*, 65–73. [CrossRef]
23. Arias-Alvarez, M.; Garcia-Garcia, R.M.; Rebollar, P.G.; Revuelta, L.; Millan, P.; Lorenzo, P.L. Influence of metabolic status on oocyte quality and follicular characteristics at different postpartum periods in primiparous rabbit does. *Theriogenology* **2009**, *72*, 612–623. [CrossRef] [PubMed]
24. Castellini, C. Reproductive activity and welfare of rabbit does. *Ital. J. Anim. Sci.* **2007**, *6*, 743–747. [CrossRef]
25. Castellini, C.; Dal Bosco, A.; Arias-Alvarez, M.; Lorenzo, P.L.; Cardinali, R.; Rebollar, P.G. The main factors affecting the reproductive performance of rabbit does: A review. *Anim. Reprod. Sci.* **2010**, *122*, 174–182. [CrossRef] [PubMed]
26. Xiccato, G.; Bernardini, M.; Castellini, C.; Dalle Zotte, A.; Queaque, P.I.; Trocino, A. Effect of post-weaning feeding on the performance and energy balance of female rabbits at different physiological states. *J. Anim. Sci.* **1999**, *77*, 416–426. [CrossRef]
27. Cardinali, R.; Dal Bosco, A.; Bonanno, A.; Di Grigoli, A.; Rebollar, P.G.; Lorenzo, P.L.; Castellini, C. Connection between body condition score, chemical chracteristics of body and reproductive traits of rabbit does. *Livest. Sci.* **2008**, *116*, 209–215. [CrossRef]
28. Castellini, C.; Dal Bosco, A.; Cardinali, R. Long term effect of post-weaning rhythm on the body fat and performance of rabbit doe. *Reprod. Nutr. Dev.* **2006**, *46*, 195–204. [CrossRef]

Article

Relationship between Prenatal Characteristics and Body Condition and Endocrine Profile in Rabbits

María-Luz García [1,*], Raquel Muelas [1], María-José Argente [1] and Rosa Peiró [2]

1 Departamento de Tecnología Agroalimentaria, Universidad Miguel Hernández de Elche, Ctra de Beniel km 3.2, 03312 Orihuela, Spain; raquel.muelas@umh.es (R.M.); mj.argente@umh.es (M.-J.A.)
2 Instituto de Conservación y Mejora de la Agrodiversidad Valenciana, Universitat Politècnica de València, P.O. Box 22012, 46071 Valencia, Spain; ropeibar@btc.upv.es
* Correspondence: mariluz.garcia@umh.es

Simple Summary: Litter size is an essential trait in rabbit production, and it depends on ovulation rate and embryonic and foetal survival. The period between 8 and 18 d of gestation is critical for foetal survival, as the placenta controls foetal nutrition during this period. Ovulation rate and foetal survival at 12 d of gestation are affected by body condition and metabolic and hormonal profile. Higher foetal survival is related to a higher number of vessels arriving at the implantation site, and may be due to higher available space for the foetus.

Abstract: This study evaluated the relationship between prenatal characteristics and body condition and endocrine profile. A total of 25 non-lactating multiparous females were used. Body condition, measured as body weight and perirenal fat thickness, non-esterified fatty acids (NEFA), leptin, progesterone and 17β-estradiol were recorded at mating and 12 d of gestation. Ovulation rate, number of foetuses, ovary and foetal weight, length and weight of uterine horn, available space per foetus and maternal and foetal placental morphometry were recorded at 12 d of gestation. Ovulation rate showed a positive linear relationship with number of foetuses, ovary weight and NEFA. A negative linear relationship between ovulation rate and perirenal fat thickness and leptin was obtained. Ovulation rate was maximum when body weight and 17β-estradiol were 4.4 kg and 22.7 pg/mL, respectively. Foetal weight showed a positive relationship with perirenal fat thickness and a negative relationship with leptin. An increase in progesterone and NEFA concentration was related to a positive linear increase in number of foetuses and in uterine horn weight. Space available per foetus was affected both by the number of vessels that reach the implantation site and by position of the foetus in the uterine horn. In conclusion, body condition during mating and early gestation should be maintained within an optimal range to ensure the best prenatal characteristics. While 17β-estradiol, NEFA and leptin affected the ovulation rate, progesterone and NEFA affected foetal development. The number of vessels that reach the implantation site determines early foetal survival.

Keywords: estradiol; foetuses; leptin; NEFA; progesterone; perirenal fat thickness; placenta; ovulation rate

Citation: García, M.-L.; Muelas, R.; Argente, M.-J.; Peiró, R. Relationship between Prenatal Characteristics and Body Condition and Endocrine Profile in Rabbits. *Animals* **2021**, *11*, 95. https://doi.org/10.3390/ani11010095

Received: 28 November 2020
Accepted: 4 January 2021
Published: 6 January 2021

1. Introduction

A high ovulation rate and high litter size are characteristics of females in the rabbit industry [1]. Ovulation rate reaches higher values than litter size, approximately 20% to 40% of the ovulated ova do not reach gestational term (see review by [2]). Most of these losses occur mainly up to 18 d of gestation [3,4]. The early foetal period, between 8 and 18 d of gestation, is critical for foetal survival, as the placenta controls foetal nutrition during this period [5]. Thus, early foetal survival seems to be associated with the placental development, foetal available space and vascular supply [6,7].

Body condition is traditionally employed to measure the mobilization of fat reserves [8]. Furthermore, an optimal body condition of rabbit females is an important issue considered to improve the effectiveness of reproductive performance [9]. Body reserve status is reflected by changes in some metabolic parameters, such as non-esterified fatty acids (NEFA), and leptin concentrations [10,11]. Briefly, NEFA permits follicle growth, ovulation and development of the *corpus luteum* [12]. Leptin concentration is also related to the reproductive function in rabbit females [12,13] since it is implicated in steroidogenesis [14], ovulation [15], and pregnancy and lactation [16]. Specifically, leptin may act as the critical link between adipose tissue and the reproductive system, indicating whether adequate energy reserves are present for normal reproductive function [17]. Other hormones such as estradiol and progesterone are essential for ovulation and the maintenance of pregnancy [18,19].

A detailed understanding of how ovulation rate and early foetal development and survival is affected by body condition and metabolic and hormonal profile could improve the productivity of rabbit females. Therefore, the objective of this work was to study the relationship between ovulation rate and early foetal characteristics and body condition, NEFA, leptin, 17β -estradiol and progesterone.

2. Materials and Methods

2.1. Ethics Statement

All experimental procedures involving animals were approved.

2.2. Experiment Animals

A total of 25 non-lactating multiparous rabbit females were used. Females belonged to a cross population of two lines selected divergently by uterine capacity [20]. Both lines were derived from the V line [21]. The females were held on the experimental farm at the Universidad Miguel Hernández de Elche (Spain). All animals were reared in individual cages and fed ad libitum with a commercial diet (crude protein, 17.5%; crude fiber, 15.5%; ether extract, 5.4%; ash, 8.1%) during their reproductive life. The photoperiod was 16 h light: 8 h dark.

Females that had finished the forth lactation were mated and blood samples were collected from the central ear artery early in the morning, before feeding, to prevent the effect of feeding [9]. Tubes containing EDTA were used. At 12 d after mating, blood samples were also collected after positive abdominal palpation and then females were euthanized by intravenous administration of sodium thiopental in a dose of 50 mg/kg of body weight (Thiobarbital, B. Braun Medical S.A., Barcelona, Spain). The entire reproductive tract was immediately removed in order to measure reproductive traits. Plasma was obtained after centrifugation at 3000× g for 15 min at 4 °C and stored at −20 °C until the metabolite and hormones assays were performed.

2.3. Reproductive Traits

Total ovulation rate was estimated as the number of *corpora lutea*. The ovaries were weighted. The implantation sites were considered when foetus, and maternal and foetal placenta, were presented. The number of foetuses in each uterine horn was recorded. Foetuses were classified into live foetuses if normal development was observed, or dead foetuses if they were not developed. The number of blood vessels arriving at the implantation sites and position of each foetus in the uterine horn were counted [6]. The uterine positions were: oviduct (the first foetus nearest the ovarian end), middle (foetus in middle of the uterine horn) and cervix (the last foetus in the uterine horn from the ovarian end). All foetuses with their foetal and maternal placental were removed from the uterine horn and were weighted. The empty uterine horn was weighted and its length was measured. The length of each maternal placenta and the distance between maternal placentas or to the end of the uterine horn were measured. Perimeter and area of foetal and maternal placenta were calculated using the AUTOCAD program.

2.4. Metabolite and Hormonal Assays

Non-esterified fatty acid (NEFA, mmol/L) concentrations were analyzed in duplicate, using an in vitro enzymatic colorimetric method (NEFA-C®, Wako Chemicals GmbH, Neuss, Germany). NEFA in samples was converted to Acyl-CoA by the action of Acyl-CoA synthetase, under the coexistence with coenzyme A. Obtained Acyl-CoA was oxidized and yielded hydrogen peroxide by the action of Acyl-CoA oxidase. In the presence of peroxidase, the hydrogen peroxide formed yields a blue purple pigment. NEFA concentration was obtained by measuring absorbance of the blue purple colour.

Duplicate aliquots of plasma for the sample tube were assayed. The leptin concentrations were measured by RIA antibody using the multi-species leptin kit (XL-85K, Linco Research Inc.®, St. Charles, MO, USA). The detection limit was 1.0 to 50.0 ng/mL Human Equivalents (HE). The 17β-estradiol and progesterone concentrations were assayed using a commercial 125I RIA kit (07-238102 and 07-270102, respectively; ICN Pharmaceuticals Inc.®, Diagnostic Division, Costa Mesa, CA, USA). The detection range was 10 to 3000 pg/mL and 0.15 to 80.00 ng/mL, respectively. Intra and inter-assay coefficients of variations were <5% for all hormones.

2.5. Body Condition

Body weight and perirenal fat thickness were recorded at mating and 12 d of gestation. Perirenal fat thickness was measured by ultrasound imaging as described by [8], using Justvision 200 SSA-320A Toshiba ultrasound equipment.

2.6. Traits

2.6.1. At Mating

Variables measured on each female were body weight, perirenal fat thickness, NEFA, 17β-estradiol, progesterone and leptin.

2.6.2. At 12 d of Gestation

Variables measured on each female were body weight, perirenal fat thickness, NEFA, 17β-estradiol, progesterone, leptin, ovulation rate, number of foetuses and uterine weight and length. Total foetal weight, and total foetal and maternal placenta weight per female were calculated.

Variables measured on each uterine horn were weight and length of uterine horn, ovulation rate and ovary weight per ovary and number of foetus per uterine horn.

Variables measured on each foetus were individual foetal weight, foetal and maternal placenta weight, perimeter and area, and maternal placenta length. The available space per foetus was calculated as the length of its maternal placenta plus one-half the total distance to their two adjacent maternal placentas. For extreme foetuses, available space per foetus was the length from the tip of the uterine horn to the maternal placenta plus the length of its maternal placenta and one-half the distance to adjacent maternal placenta [22].

2.7. Statistical Analyses

2.7.1. Differences between Mating and 12 d of Gestation

Body weight, perirenal fat thickness, NEFA, 17β-estradiol, progesterone and leptin were analysed with a model that included fixed effect of moment (mating and 12 d of gestation) and random effect of female. MIXED procedure of SAS was used [23].

2.7.2. Relationship between Traits at Mating

In order to study the relationships between ovulation rate and body weight, perirenal fat thickness, NEFA, 17β-estradiol, progesterone and leptin, the model included the linear and quadratic regression coefficients. If the quadratic regression coefficient was not significant, the linear relationship was tested. In addition, the relationship between ovulation rate and number of foetuses was analyzed. The GLM procedure of SAS was used for these analyses.

The model used for ovulation rate and ovary weight per ovary included the random effect of the female. A MIXED procedure of SAS was used for these analyses including the lineal and quadratic regression coefficients. If the quadratic regression coefficient was not significant, the linear relationship was tested.

2.7.3. Relationship between Traits at 12 d of Gestation

In order to study the relationships between uterine weight and length, total foetal weight and total foetal and maternal placenta weight with perirenal fat thickness, NEFA, 17β-estradiol, progesterone and leptin, the model included the linear and quadratic regression coefficient. If the quadratic regression coefficient was not significant, the linear relationship was tested. The GLM procedure of SAS was used for these analyses.

The model used for traits of the uterine horn included the random effect of the female, and the linear and quadratic regression coefficients. If the quadratic regression coefficient was not significant, the linear relationship was tested. For foetal traits, a random effect of the uterine horn was also included. A MIXED procedure of SAS was used for these analyses.

2.7.4. Blood Supply and Uterine Position

The number of live and dead foetuses according to the number of vessels reaching the implantation site with four levels (1, 2, 3 or more than 3 vessels) and the foetal position in the uterine horn (oviduct, central or cervix) was analysed using Chi-square test.

Traits measured in each foetus were analysed with the model:

$$Y_{ijklm} = \mu + V_i + P_j + m_{ijk} + h_{ijkl} + b_1 \times NF_{ijklm} + e_{ijklm};$$

where V_i is the number of blood vessels reaching the implantation site of the foetus effect with four levels previously described, P_j is the foetal position in the uterine horn effect with three levels previously described, m_{ijk} is the random effect of the female, h_{ijkl} is the random effect of the uterine horn, b_1 is the regression coefficient of the covariate number of foetuses in each uterine horn (NF_{ijklm}) and e_{ijklm} is the residual term.

The MIXED procedure of SAS statistical package was used for the analyses.

3. Results

3.1. *Differences between Mating and 12 d of Gestation*

Table 1 shows descriptive statistics of the traits. Body condition differed between mating and 12 d of gestation (Table 2). Body weight and perirenal fat thickness increased 4.6% and 5.7%, respectively and NEFA decreased 20%. Levels of 17β-estradiol and progesterone were similar but leptin increased 22% between mating and 12 d of gestation.

3.2. *Relationships at Mating*

Only significant relationships are shown in the figures. Ovulation rate showed a positive linear relationship with total number of foetuses (Figure 1a). Each extra *corpus luteum* was associated with an increase of 0.72 foetuses. Ovary weight also increased linearly with the ovulation rate (Figure 1b). There was a quadratic relationship between body weight and ovulation rate (Figure 1c). The maximum ovulation rate was reached with 4.4 kg of body weight. There was a linear and negative relationship between ovulation rate and perirenal fat thickness (Figure 1d).

Figure 2 shows the relationship between ovulation rate with NEFA, 17β-estradiol and leptin. The relationship was positive linear with NEFA (Figure 2a), and negative linear with leptin (Figure 2c). The 17β-estradiol had a significant quadratic relationship (Figure 2b). The equation predicted a maximum of 15.6 *corpora lutea* when the 17β-estradiol was 22.7 pg/mL.

Table 1. Summary statistics of the traits.

Trait	N	Average	Minimum	Maximum	Standard Deviation
At mating					
Body Weight (Kg)	25	4.10	3.65	5.25	0.41
Perirenal Fat Thickness (mm)	25	7.68	6.04	9.16	0.92
Non-esterified fatty acids (mmol/L)	25	0.34	0.08	0.70	0.15
17β -estradiol (pg/mL)	25	20.01	4.20	39.23	10.73
Progesterone (ng/mL)	25	30.82	9.67	72.25	18.47
Leptin (ng/mL)	25	3.89	1.28	8.75	2.11
At 12 d of gestation					
Body Weight (Kg)	25	4.28	3.94	5.13	0.30
Perirenal Fat Thickness (mm)	25	7.97	6.24	9.95	1.08
Non-esterified fatty acids (mmol/L)	25	0.28	0.11	0.47	0.10
17β -estradiol (pg/mL)	25	22.61	12.24	33.23	5.05
Progesterone (ng/mL)	25	30.31	9.18	62.19	15.14
Leptin (ng/mL)	25	4.48	1.52	8.40	1.68
Traits per female					
Ovulation Rate	25	13.31	8.00	19.00	2.54
Number of Foetuses	25	10.32	3.00	18.00	3.59
Uterine Weight (g)	25	51.25	22.45	69.64	12.98
Uterine Length (cm)	25	27.31	12.70	34.04	5.28
Total Foetal Weight (g)	25	0.85	0.09	2.31	0.58
Total Foetal Placenta Weight (g)	25	3.28	0.43	7.17	1.89
Total Maternal Placenta Weight (g)	25	14.24	2.89	27.07	5.47
Traits per uterine horn					
Ovary Weight (g)	50	0.60	0.36	0.94	0.14
Ovulation Rate	50	6.56	2.00	13.00	2.22
Number of Foetuses	50	5.34	1.00	11.00	2.44
Tract Weight (g)	50	25.36	8.14	47.98	9.14
Tract Lenght (cm)	50	14.24	8.41	21.60	2.79
Traits per foetus					
Foetal Weight (g)	261	0.13	0.02	0.46	0.08
Foetal Placenta Weight (g)	261	0.35	0.03	0.90	0.18
Foetal Placenta Permieter (cm)	261	4.50	1.36	6.53	1.00
Foetal Placenta Area (cm^2)	261	1.54	0.19	5.78	0.85
Maternal Placenta Weight (g)	261	1.34	0.15	3.10	0.46
Maternal Placenta Perimeter (cm)	261	7.85	3.41	13.15	1.20
Maternal Placenta Area (cm)	261	3.79	1.45	7.16	0.96
Maternal Placenta Length (cm)	261	1.60	0.23	2.26	0.34
Aviable space per foetus (cm)	261	2.82	0.50	10.56	1.16

Table 2. Least square means and standard error of body weight, perirenal fat thickness, 17β -estradiol, NEFA, progesterone and leptin at mating and 12 d of gestation.

Trait	Mating	12 d of Gestation
Body Weight (Kg)	4.09 ± 0.04^{a}	4.28 ± 0.04^{b}
Perirenal Fat Thickness (mm)	7.68 ± 0.10^{a}	8.12 ± 0.10^{b}
Non-esterified fatty acids (mmol/L)	0.35 ± 0.02^{b}	0.28 ± 0.01^{a}
17β-estradiol (pg/mL)	19.72 ± 1.61	22.93 ± 1.54
Progesterone (ng/mL)	31.71 ± 3.42	30.69 ± 2.55
Leptin (ng/mL)	3.85 ± 0.11^{a}	4.69 ± 0.14^{b}

^{a,b} Different superscripts on the same line differ at $p < 0.05$.

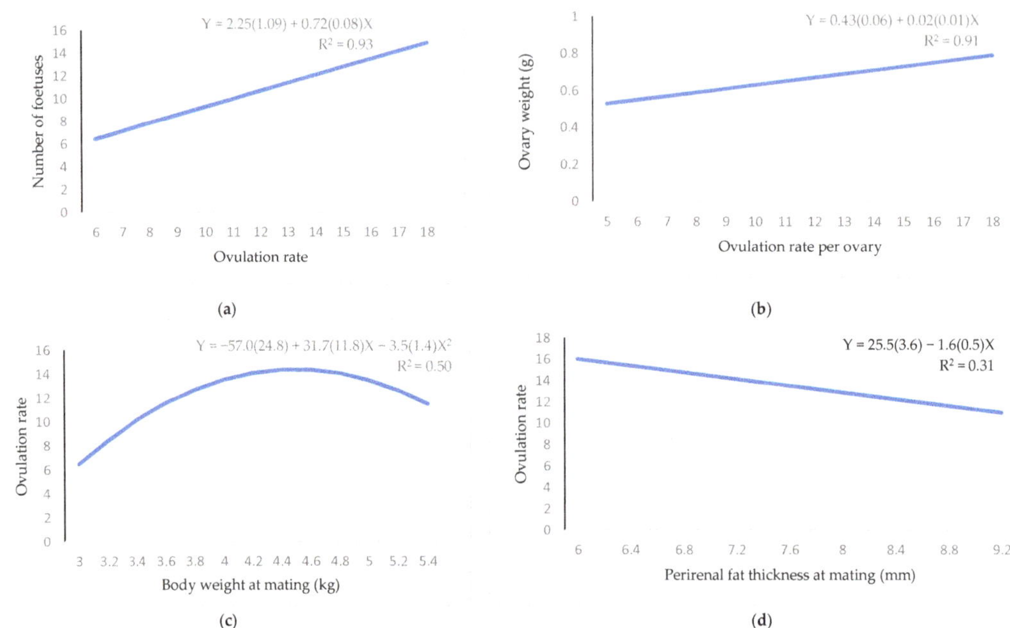

Figure 1. Regression equation (standard error between brackets) at mating for the relationship between: (**a**) ovulation rate and number of foetuses; (**b**) ovulation rate per ovary and its ovary weight; (**c**) body weight and ovulation rate; (**d**) perirenal fat thickness and ovulation rate.

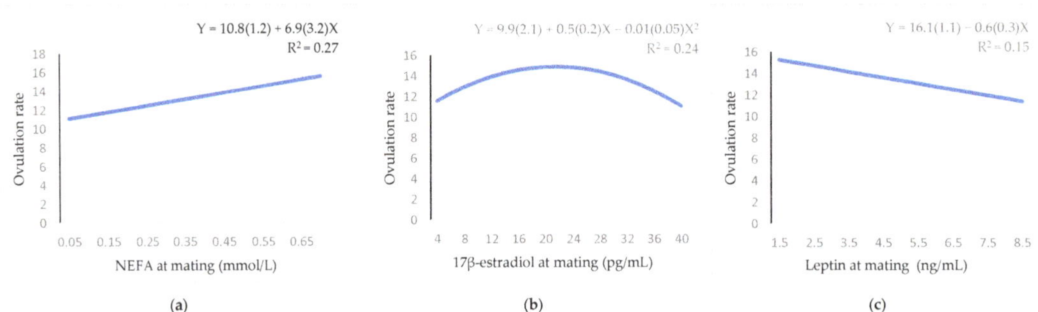

Figure 2. Regression equation (standard error between brackets) at mating for the relationship between ovulation rate: (**a**) NEFA; (**b**) 17β-estradiol; (**c**) leptin.

3.3. Relationships at 12 d of Gestation

Total foetal weight had a positive relationship with perirenal fat thickness (Figure 3a) and a negative relationship with leptin (Figure 3b). Both the total foetal weight and maternal placenta weight showed a positive linear relationship with progesterone (Figure 3c,d). The number of foetuses (Figure 4a) and weight and length of the uterine horn (Figure 4b,c) showed a positive linear relationship with progesterone.

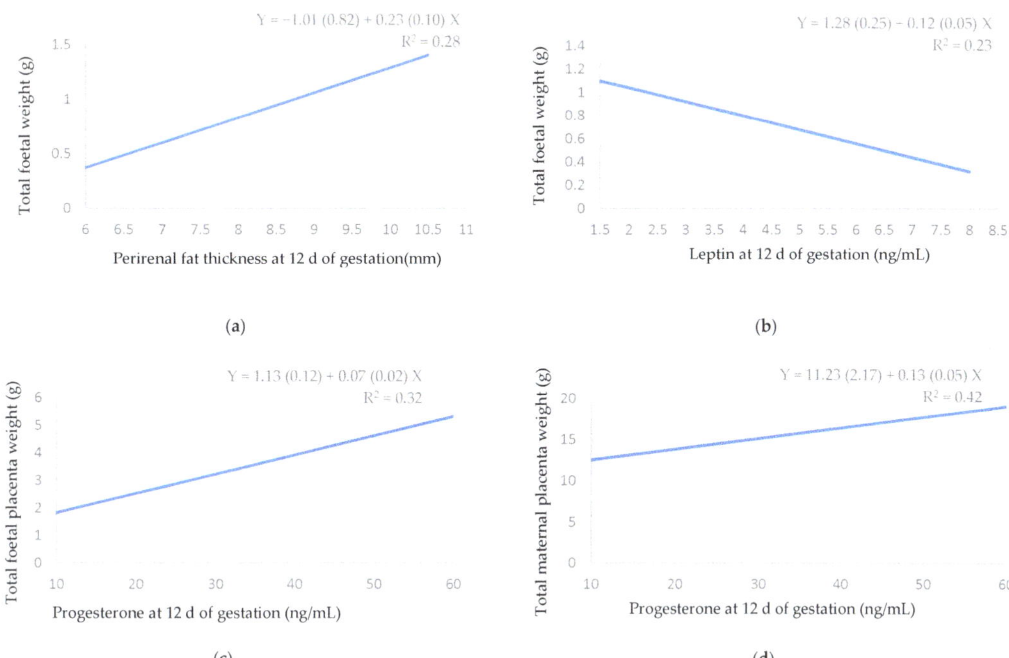

Figure 3. Regression equation (standard error between brackets) at 12 d of gestation for the relationship between: (**a**) perirenal fat thickness and total foetal weight; (**b**) leptin and total foetal weight; (**c**) progesterone and total foetal placenta weight; (**d**) progesterone and total maternal placenta weight.

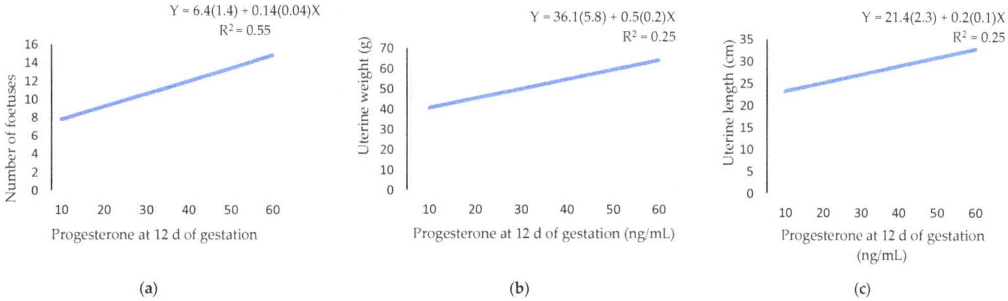

Figure 4. Regression equation (standard error between brackets) at 12 d of gestation for the relationship between progesterone and: (**a**) number of foetuses; (**b**) uterine weight; (**c**) uterine length.

Figure 5 shows the positive linear relationship between NEFA and uterine weight (Figure 5a), number of foetuses (Figure 5b) and progesterone (Figure 5c). Thus, each extra 0.1 mmol/L of NEFA was related to an increase of 4.52 g of uterine weight, 1.33 foetuses and 4.59 ng/mL of progesterone.

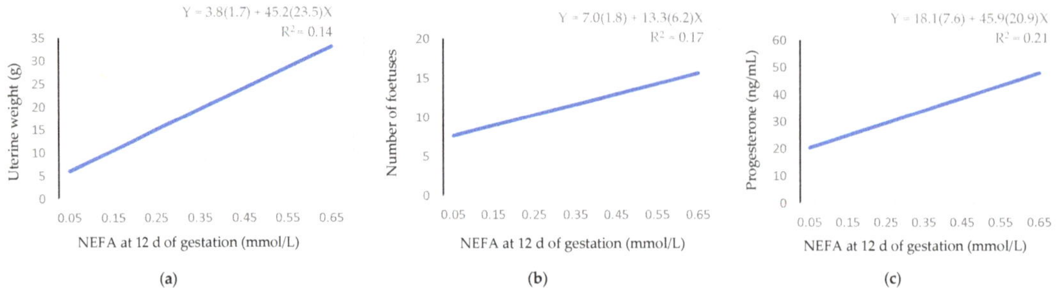

Figure 5. Regression equation (standard error between brackets) at 12 d of gestation for the relationship between NEFA and: (**a**) uterine weight; (**b**) number of foetuses; (**c**) progesterone.

3.4. Uterine Position, Blood Supply and Foetal Development

Number of foetuses showed a quadratic relationship with tract weight (Figure 6a). The maximum weight was 42.5 g when 13.5 foetuses were implanted. A positive linear relationship was seen between number of foetuses and tract length (Figure 6b), but it was a negative linear relationship with maternal placenta length and area (Figure 7a,b, respectively). A convex curve was observed when the relationship between number of foetuses and available space per foetus was studied (Figure 7c). If nine embryos were implanted, the space was minimum (2.2 cm).

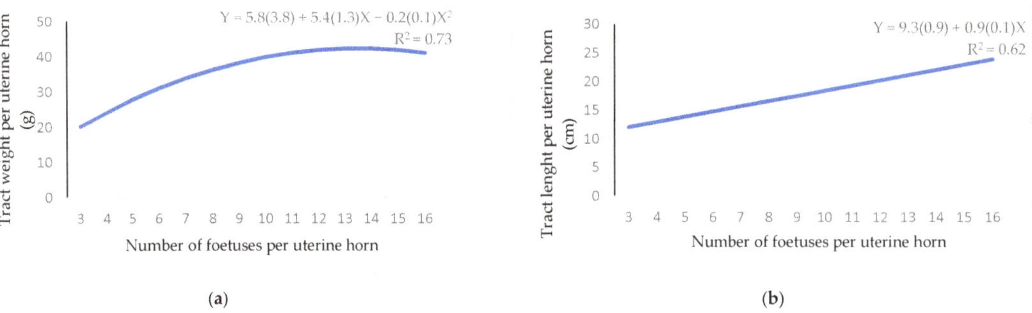

Figure 6. Regression equation (standard error between brackets) at 12 d of gestation for the relationship between number of foetuses per uterine horn and: (**a**) tract weight; (**b**) tract length.

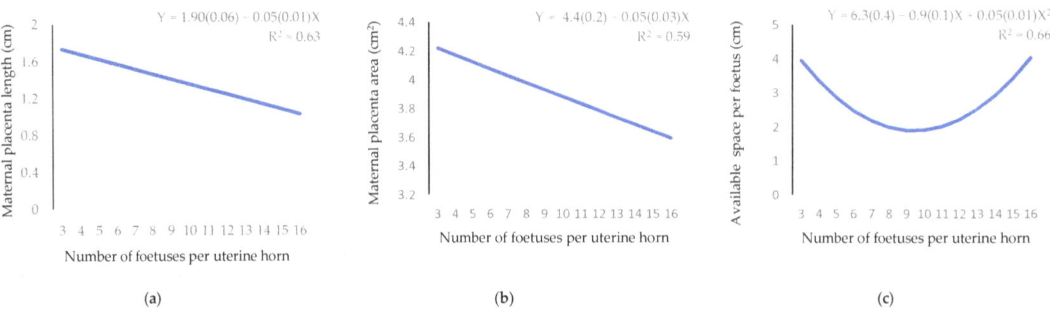

Figure 7. Regression equation (standard error between brackets) at 12 d of gestation for the relationship between number of foetuses per uterine horn and: (**a**) maternal placenta length; (**b**) maternal placenta area; (**c**) available space per foetus.

Table 3 shows that foetuses with a poor blood supply had a higher probability of death. There were no differences in the percentages of dead foetuses between the different positions within the uterine horn (oviduct, middle or cervix).

Foetal weight was not affected by number of vessels and position in the uterine horn (Table 4). Available space per foetus was lower with one vessel than with more than two vessels. Available space per foetus was lower in the middle of the uterine horn than in the oviduct and cervix.

Foetal placenta weight was higher with three or more vessels reaching the implantation site than with one vessel. The highest foetal placenta perimeter and area were shown when more than three vessels reached the implantation site (4.97 cm and 2.07 cm^2, respectively). Regarding the position in the uterine horn, foetal placenta weight was 6% higher in the middle than in the oviduct and cervix. Neither the perimeter nor the area was affected by the position.

A maternal placenta with three or more vessels showed higher weight than with less than three vessels. Length of maternal placenta was higher for two or more vessels, and this length was higher in the oviduct than in the middle, but similar in the cervix (Table 5).

Table 3. Percentage of live and dead foetuses according to the number of vessels reaching the implantation site and the position of the foetus in the uterine horn.

Trait	Number of Vessels				Position		
	1	2	3	>3	Oviduct	Middle	Cervix
Live foetuses (%)	83	98	99	98	92	97	98
Dead foetuses (%)	17	2	1	2	8	3	2
	χ^2 = 19.24 p = 0.0002				χ^2 = 3.45 p = 0.18		

Table 4. Least square means and standard errors of foetal weight, available uterine space per foetus and foetal placenta morphometry according to the number of vessels reaching each implantation site and the position of the foetus in the uterine horn.

Effect	Level	Foetal		Foetal Placenta		
		Weight (g)	Available Space (cm)	Weight (g)	Perimeter (cm)	Area (cm^2)
Number of vessels	1	0.127 ± 0.02	2.77 ± 0.14 ^{a}	0.306 ± 0.03 ^{a}	4.30 ± 0.20 ^{a}	1.52 ± 0.15 ^{a}
	2	0.128 ± 0.01	3.03 ± 0.09 ^{ab}	0.355 ± 0.02 ^{ab}	4.38 ± 0.11 ^{a}	1.48 ± 0.09 ^{a}
	3	0.136 ± 0.01	3.16 ± 0.10 ^{b}	0.371 ± 0.02 ^{b}	4.63 ± 0.13 ^{a}	1.50 ± 0.11 ^{a}
	>3	0.116 ± 0.01	3.21 ± 0.14 ^{b}	0.378 ± 0.03 ^{b}	4.97 ± 0.16 ^{b}	2.07 ± 0.17 ^{b}
Position	Oviduct	0.123 ± 0.01	3.58 ± 0.11 ^{b}	0.358 ± 0.03 ^{a}	4.58 ± 0.16	1.69 ± 0.14
	Middle	0.127 ± 0.01	2.77 ± 0.07 ^{a}	0.339 ± 0.01 ^{b}	4.48 ± 0.08	1.59 ± 0.08
	Cervix	0.131 ± 0.01	3.19 ± 0.13 ^{b}	0.361 ± 0.03 ^{a}	4.66 ± 0.16	1.65 ± 0.15

^{a,b} Rows within the same column with a different superscripts on the same line differ at $p < 0.05$.

Table 5. Least square means and standard errors of the maternal placenta morphometry according to the number of vessels reaching each implantation point and the position of the foetus in the uterine horn.

Effect	Level	Placenta			
		Weight (g)	Perimeter (cm)	Area (cm^2)	Length (cm)
Number of vessels	1	1.26 ± 0.07 ^{a}	7.58 ± 0.22	3.79 ± 0.17	1.47 ± 0.05 ^{a}
	2	1.29 ± 0.04 ^{a}	7.70 ± 0.13	3.68 ± 0.10	1.60 ± 0.03 ^{b}
	3	1.42 ± 0.06 ^{b}	7.85 ± 0.15	3.77 ± 0.12	1.68 ± 0.04 ^{b}
	>3	1.43 ± 0.06 ^{b}	8.03 ± 0.20	3.95 ± 0.16	1.66 ± 0.05 ^{b}
Position	Oviduct	1.31 ± 0.07	7.68 ± 0.18	3.72 ± 0.15	1.70 ± 0.04 ^{a}
	Middle	1.33 ± 0.04	7.91 ± 0.11	3.82 ± 0.08	1.60 ± 0.02 ^{b}
	Cervix	1.42 ± 0.07	7.79 ± 0.19	3.86 ± 0.15	1.65 ± 0.04 ^{ab}

^{a,b} Rows within the same column with a different superscript indicate significant differences ($p < 0.05$).

4. Discussion

4.1. Relationships at Mating

Body condition, measured as body weight and perirenal fat thickness, is a common tool for assessing the energy status of rabbit females. Moreover, NEFA is used to measure energy mobilization [9]. While body condition increases from mating to 12 d of gestation, NEFA level decreases. Leptin is also higher during gestation due to the role it plays in foetal development. Therefore, gestation induces a hormonal and metabolic adaptation necessary to fulfil the energy requirements of both females and foetuses [24].

Concentration of 17β-estradiol and progesterone at mating is similar to 12 d of gestation. Similar results were found by Fortun et al. [25]. 17β-estradiol is essential for ovulation and normal luteal function in the pregnant females [18]. A peak of progesterone has been shown at mating with a similar concentration during gestation [26].

4.2. Relationships at Mating

Ovulation rate is affected by body condition and metabolic and hormone profile. We have found a quadratic relationship between body weight at mating and ovulation rate. Thus, depleted body weight or being overweight have a negative effect on ovulation rate. A similar relationship was found for sexual receptivity and fertility [27]. Our results demonstrated that a high perirenal fat thickness produces a lower ovulation rate and this could account for the lower litter size at birth found in females with high perirenal fat thickness [28].

The ovulation rate is influenced by levels of NEFA, leptin and 17β-estradiol at mating but it is not affected by progesterone. NEFA acts at the ovarian level by modifying endocrine, paracrine, and autocrine regulation, which permit follicle growth, ovulation and development of the *corpus luteum* [12]. The ovulation rate is positively related to NEFA concentration. A higher number of oocytes would increase the energy demand, which was reflected in the higher NEFA concentration [29] and consequently lower leptin [11]. Leptin plays a dual role in regulating reproduction [30]. On the one hand, a minimum threshold level of leptin is necessary to ensure normal reproduction [31]. On the other hand, elevated leptin levels negatively influence normal ovarian function and oocyte quality [32,33]. Our results confirm the negative relationship between leptin concentration and ovulation rate.

Measurements of 17β-estradiol levels have been used to assess follicular growth [34]. Thus, high levels of 17β-estradiol were related to a high population of antral follicles [35] that would ultimately imply a higher ovulation rate. But the quadratic relationship between both traits would indicate that an excess of 17β-estradiol levels would decrease the ovulation rate.

4.3. Relationships at 12 d of Gestation

An optimal body condition of females during gestation improves birth weight and litter uniformity [36]. It seems that perirenal fat thickness plays an important role in early foetal development. While body weight at 12 days of gestation does not show any relationship with the early foetal characteristics, the perirenal fat thickness is positively related to foetal weight. Similar results have been found in pigs [37].

The levels of progesterone and NEFA present a linear and positive relationship with the prenatal characteristics. High progesterone concentrations are necessary for high prenatal survival [38]. Our results would indicate that it could be due to the fact that progesterone increases not only length and weight of uterine horn but also placenta weights. Moreover, a high uterine weight and number of foetuses implies a greater energy expenditure that entails increasing the levels of NEFA in plasma during gestation [11,24]. We have found that leptin decreases proportionally with foetal weight, which is consistent which this NEFA increment.

The 17β-estradiol is essential for normal luteal function in the pregnant females [18] and therefore to maintain the gestation. However, no relationship has been found between 17β-estradiol and the number of foetuses or the uterine and placental characteristics.

4.4. Uterine Position, Blood Supply and Foetal Development

Female rabbits have a duplex uterus, i.e., constituted by two separated fully functional uterine horns and cervices opening into a sole vagina. Therefore, embryo inter-horn migration is not possible [39]. This anatomical characteristic implies that foetal growth depends on their number, irrigation and position in the uterine horn [40,41].

The number of foetuses shows a convex curve with uterine horn weight but the curve is concave for space available per foetus. A reduction in the available uterine space, could increase the number of dead foetuses [21]. Each embryo requires a certain minimum space of uterus to attach, survive, and develop, as previously indicated in pigs [42] and rabbits at 25 d of gestation [6]. Maternal placenta length and area are the traits negatively affected by the increase in the number of foetuses.

At 12d of gestation, two or more vessels reaching implantation point guarantee that more than 98% of foetuses survive. Nevertheless, number of vessels is increased to three or more for a 90 or 95% of survival at 18 d of gestation [43]. The higher foetal placenta weight, perimeter and area, and the higher placenta maternal weight and length found in foetuses with a higher number of vessels could increase foetal survival at the middle of gestation. However, foetal weight is the most important parameter to guarantee foetal survival at the end of gestation [6].

The foetuses in the middle of the uterine horn had a lower availability space and lower length of maternal placenta than those near the oviduct or cervix because their littermates flanked them on both sides. However, this condition does not affect their survival. These results were confirmed at 18 d and 25 d of gestation [6,43]. It seems that the degree of irrigation is a more determining factor than the position of the foetus in the uterus to guarantee its survival.

5. Conclusions

In conclusion, body condition during mating and early gestation should be maintained within an optimal range to ensure the best prenatal performance. While 17β-estradiol, NEFA and leptin is related to ovulation rate, progesterone, NEFA and leptin levels affect early foetal development. The number of vessels that reach the implantation site determines foetal and placental development and therefore early foetal survival.

Author Contributions: Conceptualization, M.-L.G., and R.P.; methodology, M.-L.G. and R.M.; formal analysis, M.-L.G. and M.-J.A.; data curation, M.-L.G. and R.M.; writing—original draft preparation, M.-L.G.; writing—review and editing, M.-L.G., M.-J.A., R.M.; funding acquisition, M.-L.G. and R.P. All authors have read and agreed to the published version of the manuscript.

Funding: This research was funded by Generalitat Valenciana, grant number GVPRE/2008/145.

Institutional Review Board Statement: The study was conducted according to the guidelines of the Council Directives 98/58/EC and 2010/63/EU, and approved by the University Miguel Hernández of Elche Research Ethics Committee (Reference number DTA-MJA-002-08 approved on 24 November 2008).

Informed Consent Statement: Not applicable.

Data Availability Statement: The data presented in this study are available on request from the corresponding author.

Conflicts of Interest: The authors declare no conflict of interest.

References

1. Baselga, M. Genetic improvement of meat rabbits. Programmes and diffusion. In Proceedings of the 8th World Rabbit Congress, Puebla, Mexico, 7–10 September 2004.
2. Blasco, A.; Bidanel, J.P.; Bolet, G.; Haley, C.S.; Santacreu, M.A. The genetics of prenatal survival of pigs and rabbits. *Livest. Prod. Sci.* **1993**, *37*, 1–21. [CrossRef]
3. Adams, C.E. Studies on prenatal mortality in the rabbit, *Oryctolagus cuniculus*: The amount and distribution of loss before and after implantation. *J. Endocrinol.* **1960**, *19*, 325–344. [CrossRef] [PubMed]

4. Laborda, P.; Mocé, M.L.; Blasco, A.; Santacreu, M.A. Selection for ovulation rate in rabbits: Genetic parameters and correlated responses on survival rates. *J. Anim. Sci.* **2012**, *90*, 439–446. [CrossRef] [PubMed]
5. Adams, C.E. Prenatal mortality in the rabbit *Oryctolagus cuniculus*. *J. Reprod. Fertil.* **1960**, *1*, 36–44. [CrossRef] [PubMed]
6. Argente, M.J.; Santacreu, M.A.; Climent, A.; Blasco, A. Relationships between uterine and fetal traits in rabbit selected on uterine capacity. *J. Anim. Sci.* **2003**, *81*, 1265–1273. [CrossRef] [PubMed]
7. Akkuş, T.; Erdoğan, G. Ultrasonic evaluation of feto-placental tissues at different intrauterine locations in rabbit. *Theriogenology* **2019**, *138*, 16–23. [CrossRef]
8. Pascual, J.J.; Castella, F.; Cervera, C.; Blas, E.; Fernández-Carmona, J. The use of ultrasound measurement of perirenal fat thickness to estimate changes in body condition of young female rabbits. *Anim. Sci.* **2000**, *70*, 435–442. [CrossRef]
9. Calle, E.W.; García, M.L.; Blasco, A.; Argente, M.J. Relationship between body condition and energy mobilization in rabbit does. *World Rabbit Sci.* **2017**, *25*, 37–41. [CrossRef]
10. Jorritsma, R.; Wensing, T.; Kruip, T.; Vos, P.; Noordhuizen, J. Metabolic changes in early lactation and impaired reproductive performance in dairy cows. *Endocrinology* **2003**, *143*, 1922–1931. [CrossRef] [PubMed]
11. Menchetti, L.; Andoni, E.; Barbato, O.; Canali, C.; Quattrone, A.; Vigo, D.; Cadini, M.; Curone, G.; Brecchia, G. Energy homeostasis in rabbit does during pregnancy and pseudopregnancy. *Anim. Reprod. Sci.* **2020**, *208*, 106505. [CrossRef]
12. Fortun-Lamothe, L. Energy balance and reproductive performance in rabbit does. *Anim. Reprod. Sci.* **2003**, *93*, 1–15.
13. Zerani, M.; Boiti, C.; Zampini, D.; Brecchia, G.; Dall'Aglio, C.; Ceccarelli, P.; Gobbetti, A. Ob receptor in rabbit ovary and leptin in vitro regulation of corpora lutea. *J. Endocrinol.* **2004**, *183*, 279–288. [CrossRef] [PubMed]
14. Brannian, J.D.; Zha, Y.; McElroy, M. Leptin inhibits gonadotrophin-stimulated granulosa cell production by antagonizing insulin action. *Hum. Reprod.* **1999**, *14*, 1445–1448. [CrossRef]
15. Ryan, N.K.; Woodhouse, C.M.; Van Der Hoeck, K.H.; Gilchrist, R.B.; Armstrong, D.T.; Norman, R.J. Expression of leptin and its receptor in the murine ovary: Possible role in the regulation of oocyte maturation. *Biol. Reprod.* **2002**, *66*, 1548–1554. [CrossRef] [PubMed]
16. Mukherjea, R.; Castonguat, T.W.; Douglas, L.W.; Moser-Veillon, P. Elevated leptin concentrations in pregnancy and lactation: Possible role as a modulator of substrate utilization. *Life Sci.* **1999**, *65*, 1183–1193. [CrossRef]
17. Moschos, K.; Chan, J.L.; Mantzoros, C.S. Leptin and reproduction: A review. *Fertil. Steril.* **2002**, *77*, 433–444. [CrossRef]
18. Gadsby, J.E.; Keyes, P.L.; Bil, C.H. Control of corpus luteum function in the pregnant rabbit: Role of estrogen and lack of a direct luteotropic role of the placenta. *Endocrinology* **1983**, *113*, 2255. [CrossRef]
19. McCarthy, S.M.; Foote, R.H.; Maurer, R.R. Embryo mortality and altered uterine luminal proteins in progesterone-treated rabbits. *Fertil. Steril* **1977**, *28*, 101. [CrossRef]
20. Argente, M.J.; Merchán, M.; Peiró, R.; García, M.L.; Santacreu, M.A.; Folch, J.M.; Blasco, A. Candidate gene analysis for reproductive traits in two lines of rabbits divergently selected for uterine capacity. *J. Anim. Sci.* **2010**, *88*, 828–836. [CrossRef]
21. Garcia, M.L.; Baselga, M. Estimation of genetic response to selection in litter size of rabbits using a cryopreserved control population. *Livest. Prod. Sci.* **2000**, *74*, 45–53. [CrossRef]
22. Argente, M.J.; Santacreu, M.A.; Climent, A.; Blasco, A. Effect of intra uterine crowding on available uterine space per fetus in rabbits. *Livest. Sci.* **2008**, *114*, 211–219. [CrossRef]
23. SAS. *Statistical Analysis System User's Guide: Version 9.4*, 2nd ed.; Statistical Analysis Systems: Cary, NC, USA, 2020.
24. Menchetti, L.; Brecchia, G.; Canali, C.; Cardinali, R.; Polisca, A.; Zerani, M.; Boiti, C. Food restriction during pregnancy in rabbits: Effects on hormones and metabolites involved in energy homeostasis and metabolic programming. *Res. Vet. Sci.* **2015**, *98*, 7–12. [CrossRef]
25. Fortun, L.; Prunier, A.; Lebas, F. Effects of lactaction of fetal survival and development in rabbit does mated shortly after parturition. *J. Anim. Sci.* **1993**, *71*, 1882. [CrossRef] [PubMed]
26. Spilman, C.H.; Wilks, J.W. Peripheral plasma progesterone during egg transport in the rabbit. *Proc. Soc. Exp. Biol. Med.* **1976**, *151*, 726. [CrossRef] [PubMed]
27. Cardinali, R.; Dal Bosco, A.; Bonnano, A.; Di Grigoli, A.; Rebollar, P.G.; Lorenzo, P.L.; Castellini, C. Connection between body condition score, chemical characteristics of body and reproductive traits of rabbit does. *Livest. Sci.* **2008**, *116*, 209–215. [CrossRef]
28. Martínez-Paredes, E.; Ródenas, L.; Pascual, J.J.; Savietto, D. Early development and reproductive lifespan of rabbit females: Implications of growth rate, rearing diet and body condition at first mating. *Animal* **2018**, *12*, 2347–2355. [CrossRef]
29. Rebollar, P.G.; Pereda, P.G.; Schwarz, B.F.; Millán, P.; Lorenzo, P.L.; Nicodemus, N. Effect of feed restriction or feeding high-fibre diet during the rearing period on body composition, serum parameters and productive performance of rabbit. *Anim. Feed Sci. Technol.* **2011**, *163*, 67–76. [CrossRef]
30. Caprio, M.; Fabbrini, E.; Isidori, A.M.; Aversa, A.; Fabbri, A. Leptin in reproduction. *Trends Endocrinol. Metab.* **2001**, *21*, 65–72. [CrossRef]
31. Martínez-Paredes, F.; Ródenas, L.; Martínez-Vallespín, B.; Cervera, C.; Blas, E.; Brecchia, G.; Boiti, C.; Pascual, J.J. Effects of feeding programme on the performance and energy balance of nulliparous rabbit does. *Animal* **2012**, *6*, 1086–1095. [CrossRef]
32. Smith, G.D.; Jackson, L.M.; Foster, D.L. Leptin *regulation* of reproductive function and fertility. *Theriogenology* **2002**, *57*, 73–86. [CrossRef]

33. Arias-Álvarez, M.; García-García, R.M.; Torres-Rovira, L.; Gónzalez-Bulnes, A.; Rebollar, P.G.; Lorenzo, P.L. Influence of leptin on in vitro maturation and steroidogenic secretion of cumulus-oocyte complexes through JAK2/STAT3 and MEK1/2 pathways in the rabbit model. *Reproduction* **2010**, *139*, 523–532. [CrossRef] [PubMed]
34. Wallach, E.E.; Noriega, C. Effects of local steroids on follicular development and atresia in the rabbit. *Fertil. Steril.* **1970**, *21*, 253–267. [CrossRef]
35. Garcia-Garcia, R.M.; Arias-Alvarez, M.; Rebollar, P.G.; Revuelta, L.; Lorenzo, P.L. Influence of different reproductive rhythms on serum estradiol and testosterone levels, features of follicular population and atresia rate, and oocyte maturation in controlled suckling rabbits. *Anim. Reprod. Sci.* **2009**, *114*, 423–433. [CrossRef] [PubMed]
36. Pascual, J.J.; Cervera, C.; Baselga, M. Genetic selection and nutritive resources allocation in reproductive rabbit does. In Proceedings of the 10th World Rabbit Congress, Sharm El-Sheikh, Egypt, 3–6 September 2012.
37. Zhou, Y.; Xu, T.; Cai, A.; Wu, Y.; Wei, H.; Jiang, S.; Peng, J. Excessive backfat of sows at 109 d of gestation induces lipotoxic placental environment and is associated with declining reproductive performance. *J. Anim. Sci.* **2018**, *96*, 250–257. [CrossRef] [PubMed]
38. Graham, J.D.; Clarke, C.L. Physiological Action of Progesterone in Target Tissues. *Endocr. Rev.* **1997**, *18*, 4.
39. García, M.L. Embryo manipulation techniques in the rabbit. In *New Insights into Theriogenology*, 1st ed.; Payan-Carreira, R., Ed.; IntechOpen: London, UK, 2018; Chapter 7; pp. 113–133.
40. Bautista, A.; Rödel, H.G.; Monclús, R.; Juárez-Romero, M.; Cruz-Sánchez, E.; Martínez-Gómez, M.; Hudson, R. Intrauterine position as a predictor of postnatal growth and survival in the rabbit. *Physiol. Behav.* **2015**, *138*, 101–106. [CrossRef]
41. Szendrö, Z.; Cullere, M.; Atkári, T.; Dalle Zotte, A. The birth weight of rabbits: Influence factors and effect on behavioural, productive and reproductive traits: A review. *Livest. Sci.* **2019**, *230*, 103841. [CrossRef]
42. Chen, Z.Y.; Dziuk, P.J. Influence of initial length of uterus per embryo and gestation stage on prenatal survival, development, and sex ratio. *J. Anim. Sci.* **1993**, *71*, 1895–1901. [CrossRef]
43. Argente, M.J.; Santacreu, M.A.; Climent, A.; Blasco, A. Influence of available uterine space per fetus on fetal development and prenatal survival in rabbits selected for uterine capacity. *Livest. Sci.* **2006**, *102*, 83–91. [CrossRef]

MDPI

Article

Effect of Postbiotic Based on Lactic Acid Bacteria on Semen Quality and Health of Male Rabbits

Jesús V. Díaz Cano [1], María-José Argente [2] and María-Luz García [2,*]

[1] Pentabiol S.L., Polígono Noain-Esquiroz s/n, 31110 Pamplona, Spain; jesus@pentabiol.es
[2] Centro de Investigación e Innovación Agroalimentaria y Agroambiental (CIAGRO-UMH), Miguel Hernández University, 03312 Orihuela, Spain; mj.argente@umh.es
* Correspondence: mariluz.garcia@umh.es

Simple Summary: Postbiotics, especially those derived from metabolites of *Lactobacillus*, have been proposed as an alternative to the use of antibiotics for prevention and treatment of some diseases. This study was performed in rabbits due to their economic importance as a livestock species in Mediterranean countries, as well as being an experimental model in biomedicine. In this work, the use of a diet enriched with a postbiotic based on lactic acid bacteria is proposed to improve the seminal characteristics of rabbits and their health.

Abstract: The aim of this study was to evaluate the effect of lactic acid bacteria-based postbiotic supplementation on semen characteristics and hematological and biochemical profiles in rabbits. A total of 28 males were randomly allocated into two groups. Males received a Control diet and Enriched diet supplemented with postbiotic for 15 weeks (4 weeks of adaptation period and 11 weeks of experimental period). Body weight, feed intake and semen characteristics were recorded weekly. Hematological profile was recorded at the beginning and end of the experiment and biochemical profile at 0, 5, 10 and 15 weeks. Bayesian methodology was used for the statistical analysis. Feed intake was higher in Control diet (125.2 g) than in the Enriched diet (118.6 g, $p = 1.00$). The percentages of abnormal spermatozoa were higher in Control diet than in Enriched diet (30% and 22%; $p = 0.93$) and the acrosome integrity percentage was lower (97% and 96%; $p = 0.87$). The hematological profile was within the range for healthy rabbits. The plasmatic level of alanine aminotransferase was higher in Control diet than Enriched diet at 5 and 10 weeks ($p = 0.93$ and $p = 0.94$, respectively) and alkaline phosphatase was similar in Control diet throughout the experiment, but decreased in Enriched diet ($p = 0.97$). No difference was found in kidney parameters (uric nitrogen and creatinine). Enriched diet showed higher total protein and globulin than Control diet ($p = 0.99$). Phosphorus was lower ($p = 0.92$) in Control diet than in Enriched diet. In conclusion, the addition of the postbiotic based on lactic acid bacteria seems to improve the quality of the semen and the liver profile in rabbits.

Keywords: fermented food; hepatic profile; lactic acid bacteria; postbiotic; rabbit; semen profile

Citation: Díaz Cano, J.V.; Argente, M.-J.; García, M.-L. Effect of Postbiotic Based on Lactic Acid Bacteria on Semen Quality and Health of Male Rabbits. *Animals* **2021**, *11*, 1007. https://doi.org/10.3390/ani11041007

Academic Editor: Francesco Gai

Received: 23 February 2021
Accepted: 31 March 2021
Published: 3 April 2021

1. Introduction

Probiotics are live microorganisms which, when administered in adequate amounts, confer health benefits on the host [1]. Probiotic microorganisms are primarily lactic acid-producing bacteria of the genus *Lactobacillus* [2]. These probiotics can regulate the balance of gut microbes, promote the growth and productivity of animals and improve host resistance to diseases [3]. To this end, they have been extensively used in dairy cattle [4], beef cattle [5], pigs [6], hens [7] and rabbits [8]. Postbiotics are defined as soluble products or metabolites secreted by probiotics that have physiological benefits to the host [9]. Postbiotics consist of a wide range of effector molecules [10] and they are capable of reducing the gut pH and, in turn, inhibiting the proliferation of opportunistic pathogens in the feed and gut microbiota [10,11]. Postbiotics, especially those derived from metabolites of

Lactobacillus, have been proposed as an alternative to the use of antibiotics, not only in humans, but also in monogastrics [12]. Currently, the application of postbiotics in human food, animal feed and pharmaceutical industries is increasing and postbiotic products derived from *Lactobacillus* species are commercially available for prevention or treatment of some diseases [10].

Rabbit is a livestock species reared either for the production of meat, hair or skin or as an experimental reference for other species, such as pigs or humans [13]. In rabbit meat production, artificial insemination is being widely used in intensive production farms [14]. The success of artificial insemination programs in rabbits depends to a great extent on both male health and reproductive performance [15]. Thus, the productivity, welfare and health of males should be improved by handling or feeding. Unlike other monogastric animals, data on the use of postbiotics in rabbits are quite scarce [12]. The aim of this study is to study the effect of supplementation with a postbiotic based on lactic acid bacteria on semen characteristics and hematological and biochemical profiles in male rabbits.

2. Materials and Methods

2.1. Ethics Statement

All experimental procedures were approved by the Miguel Hernández University of Elche Research Ethics Committee, according to Council Directives 98/58/EC and 2010/63/EU (reference number 2019/VSC/PEA/0163).

2.2. Product Description

The fermented food product tested was the result of a specific process of fermentation of a substrate and a combination of specific lactic acid bacteria and yeast. The substrate was a plant-based food product primarily composed of soya, alfalfa and wheat, along with other minor components. The fermented food product contained the phyla Firmicutes (38.7%), Proteobacteria (26.7%), Bacteroidetes (18.3%), Actinobacteria (14.5%) and Saccharibacteria (1.8%). At genus level, *Lactobacillus* was the predominant, accounting for more than 6% of identified species [16].

2.3. Animals

A total of 28 rabbit males aged between 9 and 12 months were used [17]. The animals were kept on an experimental farm at the Universidad Miguel Hernández de Elche (Spain). All animals were reared in individual cages (37.5 × 33 × 90 cm) during the entire experiment. The photoperiod was 16 h light:8 h dark.

2.4. Diets

Two diets were used. The control diet presented the following composition: 17% crude protein, 15% crude fiber, 9% crude ash, 3.6% crude fat, 1.2% calcium, 0.6% phosphorus and 0.3% sodium. The enriched diet presented the same composition supplemented with 2.0 kg of a fermented food product in a ton of feed.

2.5. Experimental Design

Animals were randomly divided into two groups of 14 males each; one group received the Control diet and the other the Enriched diet. Animals had a 4-week adaptation period to the feed. The experimental procedure lasted 11 weeks. Animal body weight and feed intake were recorded weekly.

2.6. Semen Collection and Evaluation

Two ejaculates per male were collected each week on a single day using an artificial vagina, with a minimum of 30 min between ejaculate collections. After the adaptation period, semen evaluations were performed for 11 weeks. If gel was present, it was removed. Only ejaculates exhibiting a white color were classified as normal and were evaluated. Ejaculates were diluted (dilution 1:5) with TRIS–citrate-glucose extender. Percentages of motile sperm were evaluated subjectively (from 0 to 5) under a microscope at a magnification of 400× with a thermostatic plate set at 37 °C.

An aliquot from each ejaculate (0.1 mL) was fixed with 0.9 mL of 2% glutaraldehyde solution in DPBS. The sperm concentration was determined using a Thoma-Zeiss cell counting chamber (Marienfield, Germany). A total of 100 spermatozoa were evaluated at a magnification of 400× with a differential interface contrast microscope (Normarski contrast). Spermatozoa were classified as normal or abnormal. The percentage of abnormal spermatozoa was calculated. Abnormalities were referred to tail, head and middle piece. Their percentages were calculated. Presence of cytoplasmic droplets and status of the acrosome (intact or damaged) in the normal spermatozoa were evaluated and their percentages were calculated.

2.7. *Blood Collection and Biochemical and Haematological Parameters*

Following the blood sampling procedure described in [18], blood samples were collected into a tube with tripotassium ethylenediaminetetraacetic acid (K3-EDTA) at weeks 0 and 15. Hematological parameters such as white blood leukocyte count (WBC, 103/μL) and percentage of lymphocytes, neutrophils, monocytes, basophils and eosinophils were determined with the Abacus Junior Vet hematology analyzer (Diatron, Austria).

Blood samples were collected into a lithium heparin tube at weeks 0, 5, 10 and 15. After centrifugation at 4000 rpm for 15 min, the concentrations of total bilirubin (TBIL, μmol/L), alkaline phosphatase (ALP, U/L), albumin (ALB, g/L), alanine aminotransferase (ALT, U/L), total protein (TP, g/L), globulin (GLOB, g/L), glucose (GLU, mmol/L), creatinine (CRE, μmol/L), uric nitrogen (BUN, mmol/L), amylase (AMY, U/L), calcium (Ca^{2+}, mmol/L), potassium (K+, mmol/L), sodium (Na+, mmol/L) and phosphorus (FOS, mmol/L) were assessed. These biochemical parameters were determined with VETSCAN Comprehensive Diagnostic Profile Rotors (Diatron, Austria).

2.8. *Statistical Analyses*

2.8.1. Survival, Body Weight and Feed Intake

A Kaplan–Meier plot was used for the survival analyses (GraphPad Prism 9.0.0)

Body weight and feed intake were analyzed using the following model:

$$Yijkl = \mu + Wi + Dj + Wi \times Dj + mijk + eijkl,$$

where Wi is the week effect (i = 15), Dj is the diet effect (j = 2; Control diet and Enriched diet); Wi × Dj is the interaction between week and diet, mijk is the random effect of the male and eijkl is the residual term. The body weight was also included as covariate for feed intake

2.8.2. Seminal Parameters

The percentage of normal ejaculates was analyzed using Chi-square test. Seminal parameters were analyzed using the following model:

$$Yijklm = \mu + Oi + Wj + Dk + mijkl + eijklm,$$

where Oi is the collection order effect (i = 2; first and second), Wj is the week effect (j = 11), Dk is the diet effect (k = 2; Control diet and Enriched diet), mijkl is the random effect of the male and eijklm is the residual term.

2.8.3. Hematological and Biochemical Traits

Data were analyzed using the following model:

$$Yijkl = \mu + Wi + Dj + Wi \times Dj + mijk + eijkl,$$

where Wi is the week effect (i = 2, week 0 and 15 for haematological traits; i = 4, week 0, 5, 10 and 15 for biochemical traits), Dj is the diet effect (j = 2; Control diet and Enriched diet); Wi × Dj is the interaction effect, mijk is the random effect of the male and eijkl is the residual term.

Residuals and male effects were assumed to be independently normally distributed with the same variance. A Bayesian analysis was used, with bounded flat priors for all unknown parameters. Marginal posterior distributions were estimated for all unknowns using Gibbs sampling. Marginal posterior distributions of the differences between lines were computed with the Rabbit software program developed by the Institute for Animal Science and Technology (Valencia, Spain), using Monte Carlo Markov chains of 60,000 iterations, with a burn-in period of 10,000, and only 1 out of every 10 samples was saved for inferences. Convergence was tested using the Z criterion of Geweke and Monte Carlo sampling errors were computed using time-series procedures.

Results are presented with Bayesian methodology. We provide the difference between diets ($D_{D\text{-}E}$) and the precision of our estimation, finding the shortest interval with 95% probability of containing the true value, which can be asymmetric around the estimation. This is called the highest posterior density interval at 95% probability. We also calculate the actual probability of the difference between the Control diet and Enriched diet $|D_{D\text{-}E}|$ being higher than zero. We consider that there is enough evidence for the Control and Enriched diets being different when the probability of this difference in absolute value $|D_{D\text{-}E}|$ is more than 90%.

3. Results

3.1. Survival, Body Weight and Feed Intake

Males fed with Enriched diet displayed a similar survival rate to those on Control diet (Figure 1a). Survival rate was 78.6% for Enriched diet and 73.3% for Control diet (Chi-square = 0.07; P value = 79%; data not shown in tables).

In general, body weight was 3514 g in Control diet and 3433 g in Enriched diet (p = 0.85, Table 1). Feed intake was 5% higher with the Control diet (125.2 g) than with the Enriched diet (118.6 g; p = 1.00). This difference was not due to a higher body weight of Control diet, as when the body weight was included as a covariate, the difference between diets was maintained. The evolution of the body weight and feed intake each week is shown in Figure 1b.

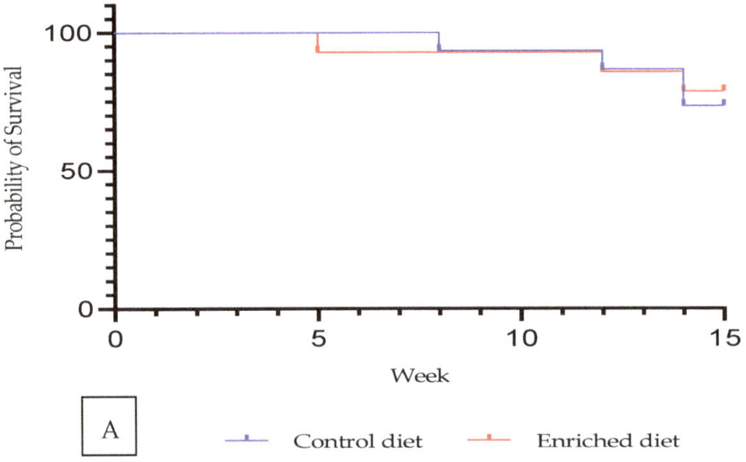

Figure 1. *Cont.*

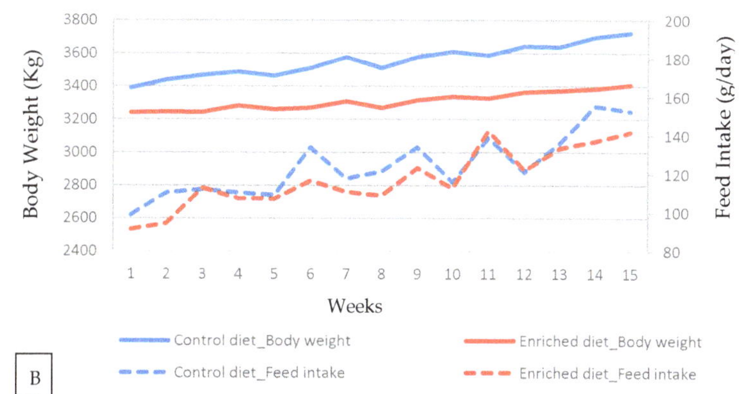

Figure 1. Control and Enriched diet: (**A**) Kaplan–Meier plot. (**B**) Evolution of body weight and feed intake.

Table 1. Effect of diet on body weight and feed intake in male rabbits.

	D	E	$D_{D\text{-}E}$	$HPD_{95\%}$	P
Body weight (g)	3514	3443	71	−66, 202	0.85
Feed intake (g/day)	125.2	118.6	6.6	2.0, 10.7	1.00
Feed intake (g/day) *	125.3	118.3	7.0	2.7, 11.4	1.00

D: Median of the Control diet; E: Median of the Enriched diet; $D_{D\text{-}E}$: Difference between the Control and Enriched diet; $HPD_{95\%}$: Highest posterior density region at 95%; P: Probability of the difference being >0. * Body weight as covariate.

3.2. Sperm Quality

Both diets showed similar percentages of eliminated ejaculates due to low macroscopic quality (12% in the Control diet and 14% in the Enriched diet; Chi-square = 0.58; p = 45%; data not shown in tables).

Volume, motility and production were similar in both diets (Table 2). Enriched diet showed a lower percentage of abnormal spermatozoa than Control diet (22% and 30%, respectively; p = 0.93). This difference was due to the lower percentage of tail abnormalities (16% and 24%, respectively; p = 0.90). Similar percentages of head and middle piece abnormalities were found in both diets (4% and 2%, respectively).

Table 2. Effect of diet on sperm quality in male rabbits.

	D	E	$D_{D\text{-}E}$	$HPD_{95\%}$	P
Volume (mL)	1.09	1.13	0.04	−0.27, 0.18	0.64
Motility	3.72	3.75	−0.03	−0.07, 0.62	0.53
Production (10^6 spz)	266.2	269.1	−3.3	−75,7, 63.1	0.54
Abnormal spz (%)					
Total (%)	30	22	8	−2, 18	0.93
Head (%)	4	4	0	−3, 2	0.64
Tail (%)	24	16	8	−5, 18	0.90
Middle piece (%)	2	2	0	−1, 1	0.62
Cytoplasmic droplet (%)	12	10	2	−5, 8	0.69
Acrosome integrity (%)	96	97	−1	−3, 1	0.87

D: Median of the Control diet; E: Median of the Enriched diet; $D_{D\text{-}E}$: Difference between Control and Enriched diet; $HPD_{95\%}$: Highest posterior density region at 95%; P: Probability of the difference being >0 when $D_{D\text{-}E} > 0$ or being <0 when $D_{D\text{-}E} < 0$.

A similar cytoplasmic droplet was shown for both diets ($p = 0.69$). Acrosome integrity was higher in Enriched than Control diet (97% and 96 % respectively; $p = 0.87$).

3.3. Hematological and Biochemical Parameters

Figure 2 shows the hematological parameters for diets at the beginning and end of the experiment. Lymphocytes increased by 15% and 20% in the Control diet ($p = 0.90$) and in the Enriched diet ($p = 0.93$). Monocytes increased for the Control diet ($p = 0.97$), but they did not vary in the Enriched diet. Neutrophils decreased in the Control diet ($p = 0.90$) and in the Enriched diet ($p = 0.99$). Eosinophils and basophils increased from week 0 to 15 in both Diets ($p = 1.00$ and $p = 0.91$, respectively). WBC did not vary between diets or throughout the experiment, ranging between 8.4 and 9.6 $\times$ $10^3/\mu L$ (data not shown in tables).

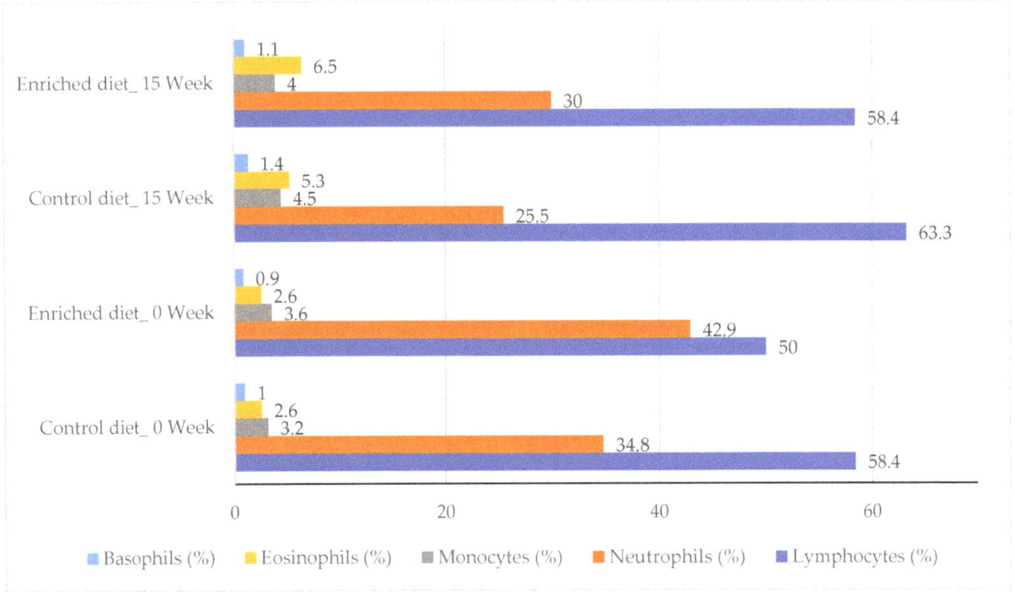

Figure 2. Percentage of lymphocytes, neutrophils, monocytes, eosinophils and basophils for Control and Enriched diets at 0 and 15 weeks.

Alanine aminotransferase is shown for Control and Enriched diets at 0, 5, 10 and 15 weeks in Figure 3a. Alanine aminotransferase was higher in the Control diet than in the Enriched diet at 5 weeks ($p = 0.93$) and at 10 weeks ($p = 0.94$). Both diets decreased the levels of alanine aminotransferase, but this decrease was lower in Control diet (5.6 U/L; from 50.2 to 44.6 U/L) than in Enriched diet (6.0 U/L; from 43.5 to 37.5 U/L; $p = 0.95$; results not shown in Figure). Alkaline phosphatase was similar for both diets and throughout the entire control period (Figure 3b). Nevertheless, while the difference between 0 and 15 weeks was similar in Control diet (39.6 and 35.5 U/L, respectively; $p = 0.62$), the alkaline phosphatase exhibited relevant reduction in Enriched diet (42.7 and 35.5 U/L, respectively; $p = 0.97$). Amylase tends to be higher in Control diet than in Enriched diet, showing differences at week 10 ($p = 0.95$; Figure 3c). Glucose was similar for both diets and ranged from 5.6 to 6.5 mmol/L (Figure 3d).

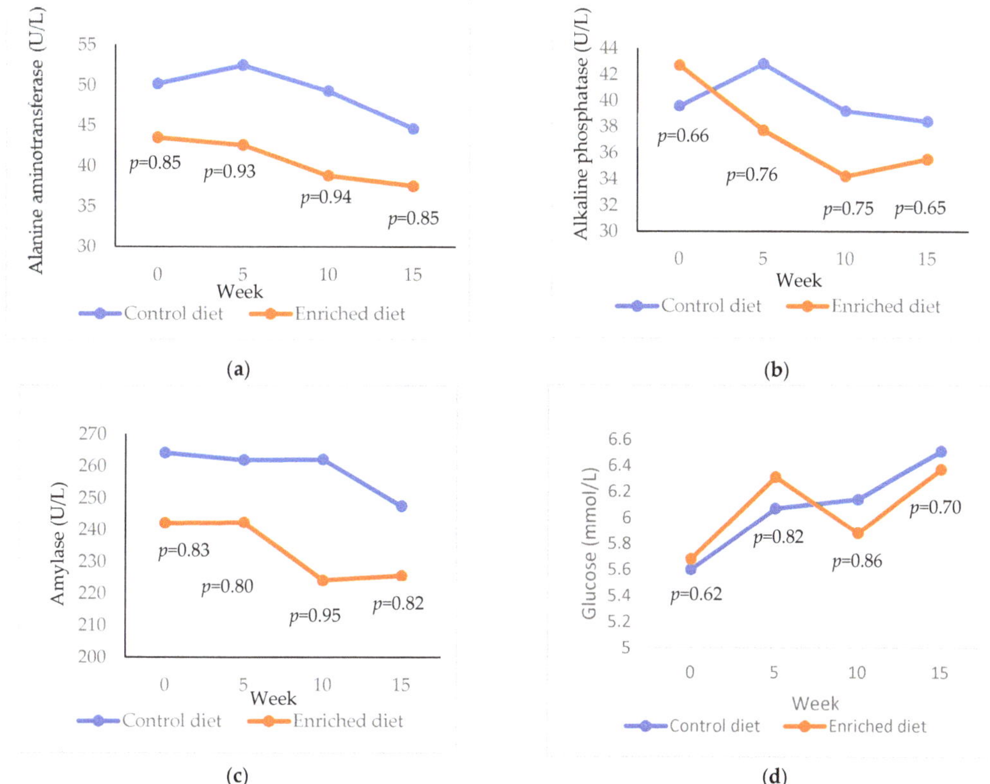

Figure 3. Evolution of (**a**) alanine aminotransferase; (**b**) alkaline phosphatase; (**c**) amylase; (**d**) glucose in males fed with Control and Enriched diet. p is the probability of the difference being >0 when the difference between the diets was >0 or being <0 when this difference was <0.

Enriched diet showed a higher total protein than Control diet after the adaptation period (+ 2.68 g/L; p = 0.99; Figure 4a) and was maintained until week 10 (+3.09 g/L; p = 0.99). However, after feeding Enriched diet for 15 weeks, the total protein was similar to Control diet. Control diet showed a lower globulin concentration than the Enriched diet at both 5 (p = 0.98; Figure 4b) and 10 weeks (p = 0.99). Albumin was higher at the start of the experiment in the Control diet (22.9 g/L; Figure 4c) than in the Enriched diet (21.9 g/L; p = 0.94). Both diets presented similar albumin from 5 to 15 weeks.

Control diet showed higher creatinine values than the Enriched diet (p = 0.92) at week 0, but the values were similar at weeks 5, 10 and 15 (Figure 5a). Both diets decreased creatinine during the experiment (−20.8 µmol/L in Control diet, p = 1.00; −30.5 µmol/L in Enriched diet, p = 1.00). Regarding uric nitrogen, a similar concentration was shown for both diets (Figure 5b) and uric nitrogen increased during the experiment (+0.7 mmol/L in both lines; p = 0.99). Total bilirubin was similar in both diets (Figure 5c) and decreased during the experiment (−0.3 µmol/L in Control diet, p = 0.92; −0.4 µmol/L in Enriched diet, p = 0.96).

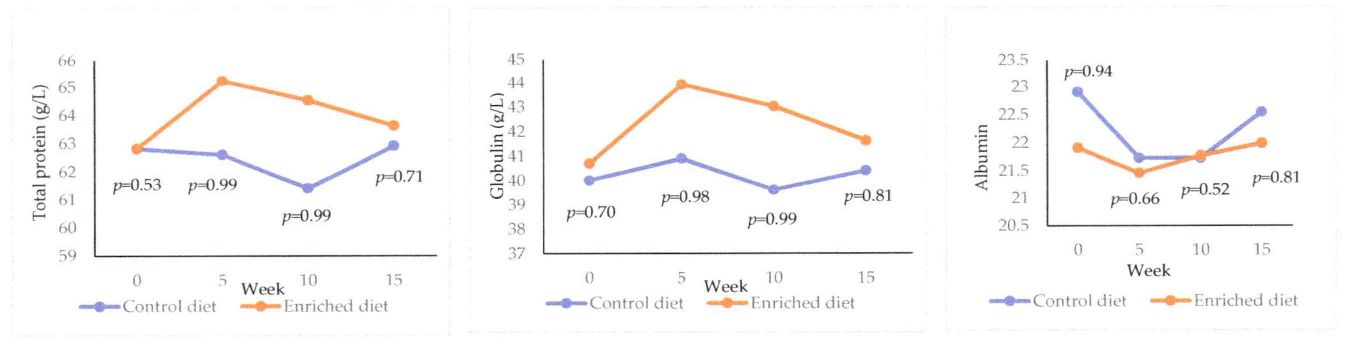

(**a**) (**b**) (**c**)

Figure 4. Evolution of (**a**) total protein; (**b**) globulin; (**c**) albumin in males fed with Control and Enriched diet. p is the probability of the difference being >0 when the difference between the diets was >0 or being <0 when this difference was <0.

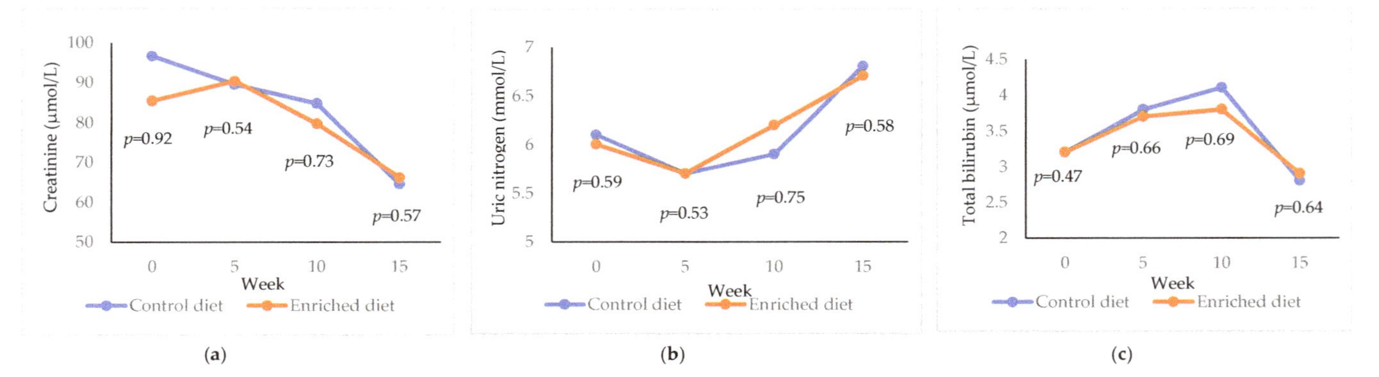

(**a**) (**b**) (**c**)

Figure 5. Evolution of (**a**) creatinine; (**b**) uric nitrogen; (**c**) total bilirubin in males fed with Control and Enriched diet. p is the probability of the difference being >0 when the difference between the diets was >0 or being <0 when this difference was <0.

The results for calcium, phosphorus, potassium and sodium are presented in Figure 6. Calcium was higher in Control diet both at 0 weeks ($p = 0.93$) and at 15 weeks ($p = 0.97$) and phosphorus was lower at 4 weeks ($p = 0.90$) and 15 weeks ($p = 0.92$). Potassium and sodium were similar for the two diets throughout the experimentation period.

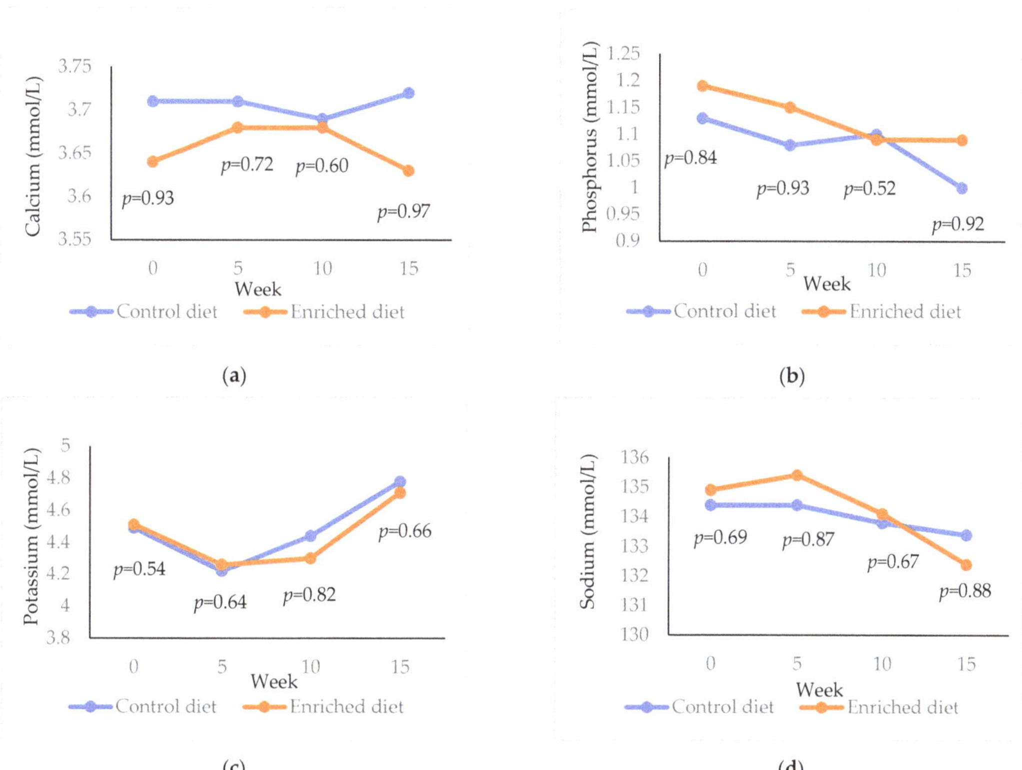

Figure 6. Evolution of (**a**) calcium; (**b**) phosphorous; (**c**) potassium; (**d**) sodium in males fed with Control and Enriched diet. p is the probability of the difference being >0 when the difference between the diets was >0 or being <0 when this difference was <0.

4. Discussion

There is increasing evidence of the role of postbiotics as health promoter. The beneficial effects of postbiotics are mediated through an interaction between the microbial products and the host [10]. In this study, we tested the effectiveness of a postbiotic formulated with a fermented food product in terms of semen quality and health status of male rabbits. Postbiotics have recently demonstrated the ability to improve welfare and health in diabetic rats [16] and dairy heifer calves [19,20].

Food intake was lower with the postbiotic than with control diet from the second week. Nevertheless, survival was not affected. When this diet has been applied to dairy heifer calves, there was also a decrease in consumption from week 5 of intake [19].

Many studies have been carried out to improve the seminal quality in rabbits by supplementing feed with probiotics [21,22]. To the best of our knowledge, no information has been found regarding postbiotics. In our experiment, a slight improvement in the acrosome integrity and spermatozoa with normal tail was obtained in the Enriched diet, although an increase in motility was not achieved.

Hematological parameters provide valuable information on the health status of the animal. In the present study, the hematological profiles were within the range for healthy rabbits both at the beginning and end of the experiment and for both diets [18,23]. Levels of albumin, alkaline phosphatase, alanine aminotransferase, total bilirubin, total protein, globulin, glucose, creatinine, uric nitrogen and amylase are within the wide range of values reported in rabbits [18,24,25].

Alanine aminotransferase and alkaline phosphatase are markers of hepatic diseases [26,27] and alkaline phosphatase is also related to other disorders such as increased bone deposits, intestinal damage, hyperthyroidism and generalized tissue damage [28]. Males fed with postbiotic diet showed lower alanine aminotransferase and alkaline phosphatase concentration, thus the liver profile was improving. The benefit of the postbiotic on liver function has also been demonstrated in rats [16]. Alanine aminotransferase decreased in meat rabbits fed with lactic acid bacteria additives [29]. Moreover, a negative correlation between these biomarkers in plasma and semen quality, mainly the motility and the acrosomal damage, has been reported in rabbits [30] and in goats [31]. As previously mentioned, the improvement in acrosome and tail would agree with this result.

Elevated glucose levels are generally due to various stress factors [32]. Several studies have reported the hypoglycemic effect of probiotic and fermented products [33,34]. Our result indicates that amylase tends to be lower with the postbiotic. This effect is not immediate, but it occurs after consuming the diet for 10 weeks. Although the glucose levels were not modified in this study, they were attenuated with the fermented food product in rats due to changes in the gut microbiota composition [16].

Principal plasma proteins are albumin and globulin [35]. Globulin can be considered as a good indicator of immunity response [36]. The fermented product increased total protein by 2.5% and globulin by 5.2%, whereas the albumin concentration was similar in both diets. Thus, it could be indicated that postbiotic improves immunity to infectious agents. Similar results have been obtained in calves supplemented with this postbiotic [21]. It has been found that postbiotics from *Lactobacillus plantarum* also confer anti-inflammatory responses, as observed in a study in porcine intestinal epithelial cell lines [37]. Regarding seminal quality, higher levels of albumin decrease sperm abnormality and increase acrosomal integrity, whereas these parameters are not affected by total protein and globulin [38]. So, the increase in globulins and total proteins does not seem to have a direct effect on the improvement of sperm quality.

We measured uric nitrogen and creatinine as biomarkers of kidney function status. The results indicate that kidney function was not affected by the use of the postbiotic, as both biomarkers evolved in a similar way during the experiment for the Control and Enriched diet. Uric nitrogen could serve as substrate for reactive oxygen species and thus protect important biomolecules against oxidative damage of the spermatozoa [39]. Nevertheless, no significant changes in concentration of uric nitrogen either in semen plasma or blood are provided with supplemented diets [40].

Little information is available on supplementation of blood minerals in response to postbiotics. Minerals act as structural and functional cofactors in metal-containing enzymes [41]. In addition, phosphorus is part of the ATP molecule, which is the major energy source for cellular function [42]. The postbiotic increased phosphorous levels in rabbit blood. This finding is supported in rabbits fed with probiotics and an improvement in the metabolic state of the rabbits could be expected [41]. It is well known that increased concentrations of phosphorous are associated with increased fertility of males [40]. Nevertheless, the results regarding calcium are not conclusive. The postbiotic equalizes the calcium levels of the animals with those of the Control diet, although the calcium decreased to the initial values in the last week of treatment.

5. Conclusions

In conclusion, postbiotics based on lactic acid bacteria improve the health status of rabbit males, especially with respect to the liver function. Sperm quality was also improved, specifically the quality of the tail and acrosome of the spermatozoid. The improvement in postbiotic intake should be investigated, as it could affect results obtained in the long term.

Author Contributions: Conceptualization, M.-L.G. and J.V.D.C.; formal analysis, M.-L.G. and M.-J.A.; writing—original draft preparation, M.-L.G.; writing—review and editing, M.-L.G., M.-J.A., J.V.D.C. All authors have read and agreed to the published version of the manuscript.

Funding: This research was funded by PENTABIOL.

Institutional Review Board Statement: The study was conducted according to the guidelines of the Council Directives 98/58/EC and 2010/63/EU and approved by the University Miguel Hernández of Elche Research Ethics Committee (reference number 2019/VSC/PEA/0163 approved on 5 September 2019).

Informed Consent Statement: Not applicable.

Data Availability Statement: The data presented in this study are available on request from the corresponding author.

Conflicts of Interest: The authors declare no conflict of interest.

References

1. Hill, C.; Guarner, F.; Reid, G.; Gibson, G.R.; Merenstein, D.J.; Pot, B.; Morelli, L.; Canani, R.B.; Flint, H.J.; Salminen, S.; et al. The international scientific association for probiotics and prebiotics consensus on the scope and appropriate use of the term probiotic. *Nat. Rev. Gastroenterol. Hepatol.* **2014**, *11*, 506–514. [CrossRef] [PubMed]
2. Sarowska, J.; Choroszy-Król, I.; Regulska-Ilow, B.; Frej-Mądrzak, M.; Jama-Kmiecik, A. The therapeutic effect of probiotic bacteria on gastrointestinal diseases (Review). *Adv. Clin. Exp. Med.* **2013**, *22*, 759–766. [PubMed]
3. Bron, P.A.; Tomita, S.; Mercenier, A.; Kleerebezem, M. Cell surface-associated compounds of probiotic lactobacilli sustain the strain-specificity dogma. *Curr. Opin. Microbiol.* **2013**, *16*, 262–269. [CrossRef]
4. Nader-Macías, M.E.F.; Otero, M.C.; Espeche, M.C.; Maldonado, N.C. Advances in the design of probiotic products for the prevention of major diseases in dairy cattle. *J. Ind. Microbiol. Biotechnol.* **2008**, *35*, 1387–1395. [CrossRef] [PubMed]
5. Kelsey, A.J.; Colpoys, J.D. Effects of dietary probiotics on beef cattle performance and stress. *J. Vet. Behav.* **2018**, *27*, 8–14. [CrossRef]
6. Dowarah, R.; Verma, A.K.; Agarwal, N. The use of Lactobacillus as an alternative of antibiotic growth promoters in pigs: A review. *Anim. Nutr.* **2017**, *3*, 1–6. [CrossRef] [PubMed]
7. Zhang, J.L.; Xie, Q.M.; Ji, J.; Yang, W.H.; Wu, Y.B.; Li, C.; Ma, J.Y.; Bi, Y.Z. Different combinations of probiotics improve the production performance, egg quality, and immune response of layer hens. *Poult. Sci. J.* **2012**, *91*, 2755–2760. [CrossRef]
8. Bhatt, R.S.; Agrawal, A.R.; Sahoo, A. Effect of probiotic supplementation on growth performance, nutrient utilization and carcass characteristics of growing Chinchilla rabbits. *J. Appl. Anim. Res.* **2017**, *45*, 304–309. [CrossRef]
9. Aguilar-Toalá, J.E.; Garcia-Varela, R.; Garcia, H.S.; Mata-Haro, V.; González-Córdova, A.F.; Vallejo-Cordoba, B.; Hernádez-Mendoza, A. Postbiotics: An evolving term within the functional foods field. *Trends Food. Sci. Technol.* **2018**, *75*, 105–114. [CrossRef]
10. Teame, T.; Wang, A.; Xie, M.; Zhang, Z.; Yang, Y.; Ding, Q.; Gao, C.; Olsen, R.E.; Ran, C.; Zhou, Z. Paraprobiotics and postbiotics of probiotic *Lactobacilli*, their positive effects on the host and action mechanisms: A Review. *Front. Nutr.* **2020**, *7*, 570344. [CrossRef]
11. Humam, A.M.; Loh, T.C.; Foo, H.L.; Samsudin, A.A.; Mustapha, N.M.; Zulkifli, I.; Izzuddin, W.I. Effects of feeding different postbiotics produced by *Lactobacillus plantarum* on growth performance, carcass yield, intestinal morphology, gut microbiota composition, immune status, and growth gene expression in broilers under heat stress. *Animals* **2019**, *9*, 644. [CrossRef] [PubMed]
12. Saettone, V.; Biasato, I.; Radice, E.; Schiavone, A.; Bergero, D.; Meineri, G. State-of-the-art of the nutritional alternatives to the use of antibiotics in humans and monogastric animals. *Animals* **2020**, *10*, 2199. [CrossRef]
13. García, M.L.; Argente, M.J. The genetic improvement in meat rabbits. In *Lagomorpha Characteristics*; Intechopen: London, UK, 2020. [CrossRef]
14. Catellini, C.; Dal Bosco, A.; Arias-Álvarez, M.; Lorenzo, P.L.; Cardinali, R.; Garcia Rebollar, P. The main factors affecting the reproductive performance of rabbit does: A review. *Anim. Reprod. Sci.* **2010**, *122*, 174–182. [CrossRef] [PubMed]
15. García-Tomás, M.; Sánchez, J.P.; Rafel, O.; Ramón, J.; Piles, M. Variability, repeatability and phenotypic relationships of several characteristics of production and semen quality in rabbit. *Anim. Reprod. Sci.* **2006**, *93*, 88–100. [CrossRef]
16. Cabello-Olmo, M.; Oneca, M.; Torre, P.; Sainz, N.; Moreno-Aliaga, M.J.; Guruceaga, E.; Díaz, J.V.; Encio, I.J.; Barajas, M.; Araña, M. A fermented food product containing lactic acid bacteria protects ZDF rats from the development of type 2 diabetes. *Nutrients* **2019**, *11*, 2530. [CrossRef]

17. Argente, M.J.; García, M.L.; Zbyˇnovká, K.; Petruška, P.; Capcarová, M.; Blasco, A. Correlated response to selection for litter size environmental variability in rabbit's resilience. *Animal* **2019**, *13*, 2348–2355. [CrossRef]
18. Beloumi, D.; Blasco, A.; Muelas, R.; Santacreu, M.A.; García, M.L.; Argente, M.J. Inflammatory correlated response in two lines of rabbit selected divergently for litter size environmental variability. *Animal* **2020**, *10*, 1540. [CrossRef] [PubMed]
19. Rovai, M.; Guifarro, L.; Anderson, J.; Salama, A.A.K. Effects of long-term postbiotic supplementation on dairy heifer calves: Performance and metabolic indicators. Abstracts of the 2019 American Dairy Science Association. Annual Meeting 23–26 June 2019 Cincinnati, Ohio. *J. Dairy Sci.* **2019**, *102* (Suppl. 1), 218–219.
20. Rovai, M.; Guifarro, L.; Salama, A.A.K. Effects of long-term postbiotic supplementation on dairy heifer calves: Health status and wound healing after dehorning. Abstracts of the 2019 American Dairy Science Association. Annual Meeting 23–26 June 2019 Cincinnati, Ohio. *J. Dairy Sci.* **2019**, *102* (Suppl. 1), 221.
21. Attia, Y.A.; Ibrahim, K. Semen quality, testosterone, seminal plasma biochemical and antioxidant profiles of rabbit bucks fed diets supplemented with different concentrations of soybean lecithin. *Animal* **2012**, *6*, 824–833. [CrossRef]
22. Helal, F.I.S.; El- Badawi, A.Y.; Abou-Ward, G.A.; El-Naggar, S.; Hassan, A.A.; Basyoney, M.M.; Morad, A.A.A. Semen quality parameters of adult male NZW rabbits fed diets added with two different types of probiotics. *Egypt. J. Nutr. Feed.* **2018**, *21*, 125–132. [CrossRef]
23. Massányi, M.; Kohút, L.; Argente, M.J.; Halo, M.; Kováčik, A.; Kováčiková, E.; Ondruška, L.; Formicki, G.; Massányi, P. The effect of different sample collection methods on rabbit blood parameters. *Saudi J. Biol. Sci.* **2020**, *27*, 3157–3160. [CrossRef]
24. Leineweber, C.; Müller, E.; Marschang, R.E. Blood reference intervals for rabbits (*Oryctolagus cuniculus*) from routine diagnostic samples. Best immung von Blutreferenzwerten für Kaninchen (Oryctolagus cuniculus) aus Routine diagnostic proben. *Tierarztl Prax Ausg K Kleintiere Heimtiere* **2018**, *46*, 393–398. [CrossRef] [PubMed]
25. Washington, I.M.; Van Hoosier, G. Clinical Biochemistry and Hematology. In *The Laboratory Rabbit, Guinea Pig, Hamster, and Other Rodents*; Suckow, M.A., Stevens, K.A., Wilson, R.P., Eds.; Elsevier: Amsterdam, The Netherlands, 2012; pp. 57–116.
26. Dontas, I.A.; Marinou, K.A.; Iliopoulos, D.; Tsantila, N.; Agrogiannis, G.; Papalois, A.; Karatzas, T. Changes of blood biochemistry in the rabbit animal model in atherosclerosis research; a time or stress-effect. *Lipids Health Dis.* **2011**, *10*, 139. Available online: http://www.lipidworld.com/content/10/1/139 (accessed on 15 March 2021). [CrossRef] [PubMed]
27. Kim, W.; Flamm, S.L.; Di Bisceglie, A.M.; Bodenheimer, H.C. Serum activity of alanine amino transferase (ALT) as an indicator of health and disease. *Hepatology* **2008**, *47*, 1363–1370. [CrossRef]
28. Fraser, C.M. *Merck Veterinary Manual*, 6th ed.; Ocean: Barcelona, Spain, 2007; p. 1314.
29. Shah, A.A.; Liu, Z.; Chen, Q.; Juanzi, W.; Sultana, N.; Zhong, X. Potential effect of the microbial fermented feed utilization on physicochemical traits, antioxidant enzyme and trace mineral analysis in rabbit meat. *J. Anim. Physiol. Anim. Nutr.* **2019**, *104*, 767–775. [CrossRef]
30. Yousef, M.I.; Abdallah, G.A.; Kamel, K.I. Effect of ascorbic acid and Vitamin E supplementation on semen quality and biochemical parameters of male rabbits. *Anim. Reprod. Sci.* **2003**, *76*, 99–111. [CrossRef]
31. Chauhan, M.S.; Kapila, R.; Gandhi, K.K.; Anand, S.R. Acrosome damage and enzyme leakage of goat spermatozoa during dilution, cooling and freezing. *Andrologia* **1993**, *26*, 21–26. [CrossRef]
32. Jenkins, J.R. Rabbit diagnostic testing. *J. Exot. Pet. Med.* **2008**, *17*, 4–15. [CrossRef]
33. Sivamaruthi, B.S.; Kesika, P.; Prasanth, M.I.; Chaiyasut, C. A Mini Review on Antidiabetic Properties of 833 Fermented. *Nutrients* **2018**, *10*, 1973. [CrossRef]
34. Ebringer, L.; Ferenčík, M.; Krajčovič, J. Beneficial Health Effects of Milk and Fermented Dairy. *Folia Microbiol.* **2008**, *53*, 378–394. [CrossRef]
35. Andonova, M. Role of innate defense mechanisms in acute phase response against Gram—Negative agents. *Bulg. J. Vet. Med.* **2002**, *5*, 77–92.
36. Abdel-Azeem, A.S.; Abdel-Azim, A.M.; Darwish, A.A.; Omar, E.M. Hematological and biochemical observations in four pure breeds of rabbits and their crosses under Egyptian environmental conditions. *World Rabbit Sci.* **2010**, *18*, 103–110. [CrossRef]
37. Kim, K.W.; Kang, S.S.; Woo, S.J.; Park, O.J.; Ahn, K.B.; Song, K.D.; Lee, H.J.; Yun, C.H.; Han, S.H. Lipoteichoic acid of probiotic Lactobacillus plantarum attenuates Poly I:C-induced IL-8 production in porcine intestinal epithelial cells. *Front. Microbiol.* **2017**, *8*, 1827. [CrossRef] [PubMed]
38. Sun, Y.; Dong, G.; Liao, M.; Tao, L.; LV, J. The effects of low levels of aflatoxin B1 on health, growth performance and reproductivity in male rabbits. *World Rabbit Sci.* **2018**, *26*, 123–133. [CrossRef]
39. Xu, C.; Yu, C.; Xu, L.; Miao, M.; Li, Y. High serum uric acid increases the risk for nonalcoholic fatty liver disease: A prospective observational study. *PLoS ONE* **2010**, *5*, 11578. [CrossRef] [PubMed]
40. Vizzari, F.; Massány, M.; Knížatová, N.; Corino, C.; Rossi, R.; Ondrŭska, L.; Tirpák, F.; Halo, M.; Massány, P. Effects of dietary plant polyphenols and seaweed extract mixture on male-rabbit semen: Quality traits and antioxidant markers. *Saudi J. Biol. Sci.* **2021**, *28*, 1017–1025. [CrossRef]
41. Shah, A.A.; Yuan, X.; Khan, R.U.; Shao, T. Effect of lactic acid bacteria-treated King grass silage on the performance traits and serum metabolites in New Zealand white rabbits (*Oryctlagus cuniculus*). *J. Anim. Physiol. Anim. Nutr.* **2018**, *102*, 902–908. [CrossRef] [PubMed]
42. Jabs, C.M.; Ferrell, W.J.; Robb, H.J. Plasma Changes in Endotoxin and Anaphylactic Shock (ATP, ADP and Creatine Phosphorus). *Ann. Clin. Lab. Sci.* **1979**, *9*, 121–132.

MDPI

Article

Correlated Response on Growth Traits and Their Variabilities to Selection for Ovulation Rate in Rabbits Using Genetic Trends and a Cryopreserved Control Population

Rosa Peiró [1], Celia Quirino [2], Agustín Blasco [3] and María Antonia Santacreu [3,*]

1 Centro de Conservación y Mejora de la Agrodiversidad Valenciana (COMAV), Universitat Politècnica de València (UPV), P.O. Box 22012, 46071 Valencia, Spain; ropeibar@btc.upv.es
2 Laboratório de Reprodução e Melhoramento Genético Animal, Universidade Estadual do Norte Fluminense (UENF), Campos dos Goytacazes 28013-602, Brazil; crq@uenf.br
3 Instituto de Ciencia y Tecnología Animal (ICTA), Universitat Politècnica de València (UPV), P.O. Box 22012, 46071 Valencia, Spain; ablasco@dca.upv.es
* Correspondence: msantacreu@dca.upv.es; Tel.: +34-963-879-436

Simple Summary: A successful response was obtained after selection for ovulation rate during 10 generations in rabbits. However, no correlated response in litter size was observed due to an increase in prenatal mortality. This increase could be due to the reduction in fetus weights and/or an increase in variable asynchrony among fetus weights. Therefore, the consequences of the selection procedure on weight at 28 and 63 days old (weaning and commercial time, respectively) and its variability are unknown. Using genetic trends and a cryopreserved control population for estimating correlated responses to selection, no relevant response on weight at 28 and 63 days old was observed. Similar results have been obtained for the variability of growth traits.

Abstract: The aim of this work was to estimate correlated responses in growth traits and their variabilities in an experiment of selection for ovulation rate during 10 generations in rabbits. Individual weight at 28 days old (IW28, kg) and at 63 days old (IW63, kg) was analyzed, as well as individual growth rate (IGR = IW63 − IW28, kg). The variability of each growth trait was calculated as the absolute value of the difference between the individual value and the mean value of their litter. Data were analyzed using Bayesian methodology. The estimated heritabilities of IW28, IW63 and IGR were low, whereas negligible heritabilities were obtained for growth variability traits. The common litter effect was high for all growth traits, around 30% of the phenotypic variance, whereas low maternal effect for all growth traits was obtained. Low genetic correlations between ovulation rate and growth traits were found, and also between ovulation rate and the variability of growth traits. Therefore, genetic trends methods did not show correlated responses in growth traits. A similar result was also obtained using a cryopreserved control population.

Keywords: control population; genetic parameters; growth rate; ovulation rate; slaughter weight; variability of growth traits; weaning weight

Citation: Peiró, R.; Quirino, C.; Blasco, A.; Santacreu, M.A. Correlated Response on Growth Traits and Their Variabilities to Selection for Ovulation Rate in Rabbits Using Genetic Trends and a Cryopreserved Control Population. *Animals* **2021**, *11*, 2591. https://doi.org/10.3390/ani11092591

Academic Editor: Sabine Gebhardt-Henrich

Received: 30 July 2021
Accepted: 30 August 2021
Published: 3 September 2021

1. Introduction

Intensive rabbit production is produced by a three-way cross in which males, selected for growth traits from paternal lines, are mated with crossbred females from lines selected for reproductive traits [1]. All these traits have economic importance [2] and it is important to know the genetic relationship between them. Crossbred females provide 50% of genes to terminal rabbits; therefore, maternal lines should also have an acceptable level for growth traits. On the other hand, paternal lines should also have an adequate level for reproductive traits (including litter size and ovulation rate) to ensure line maintenance and selection through time.

Selection for ovulation rate, proposed as a way to increase litter size [3], has been successful to increase the number of ova shed, but no correlated response on litter size has been observed in the unique experiment performed in rabbits [4]. However, the correlated response on growth traits is unknown. In this experiment, females belonging to the line selected for increase ovulation rate had more implanted embryos according to the increase in ovulation rate [5,6]. An increment in the number of ova shed and implanted embryos has also been observed in maternal commercial rabbit lines selected for higher litter size [7,8]. Some authors show how higher competition for space and nutrition during the fetal period, after implantation, could reduce fetal weight [9,10] and modify its variability [10]. Maternal and litter effects during gestation could have a relevant effect on young rabbits after birth in the growth period due to an increase in ovulation rate. A reduced fetal weight might lead to lower weight at birth and subsequently affect weaning and commercial weight [11].

The objective of this study was to estimate the correlated response on growth traits and their variability using genetic trends methods and also by comparing the selected line with a cryopreserved control in a rabbit line selected for ovulation rate during 10 generations. Maternal and litter effects on these traits were also evaluated.

2. Materials and Methods

2.1. Ethical Statement

All experimental procedures were approved by the Universitat Politècnica de València Research Ethics Committee, according to Council Directives 98/58/EC and 2010/63/EU.

2.2. Animals and Selection Procedure

The animals used in the experiment came from a synthetic line selected for the phenotypic value of ovulation rate in the second gestation of each female during 10 generations. The selection procedure has been described previously by Laborda and coauthors [4].

All animals were kept under constant photoperiod of 16 h light:8 h dark and controlled ventilation. The female nourished its kits during lactation period without fostering. Young rabbits for all parities were weaned at 28 days old and placed in flat-deck cages, 8 rabbits per cage, and fed ad libitum with a commercial diet (crude protein, 16.1%; crude fiber, 16.5%; ether extract, 4.4%; ash, 8.1% as-fed basis; NANTA S.A.; Valencia, Spain). The fattening period was 35 days. At 63 days old, rabbits were slaughtered or selected and placed in individual flat-deck cages and fed ad libitum with a commercial diet (crude protein, 17.5%; crude fiber, 15.5%; ether extract, 5.4%; ash, 8.1% as-fed basis; NANTA S.A.; Valencia, Spain) and they stayed there up to their reproductive age: 17–18 weeks of age.

To produce a control population, embryos from 45 donor females and 10 males from the base generation were vitrified and stored in liquid nitrogen until they were transferred. Details of the procedure are presented by Laborda and coauthors [6]. The offspring of the cryopreserved animals were used in order to avoid the effect of recovery, cryopreservation and transferring techniques on growth characteristics suggested by Cifre and coauthors [12]. The offspring of the control population was contemporary to animals from the 10th generation.

2.3. Traits

The reproductive trait analyzed during the second gestation in the selection procedure during ten generations was ovulation rate (OR), measured as the number of corpora lutea. The OR was also recorded in the last gestation of all females, that did not die or were not culled, obtaining a total of 1478 records from 856 females. A total of 20,230 records were used to analyze individual weight at 28 days old (IW28, kg) and 19,362 records were used to analyze individual weight at 63 days old (IW63, kg) and individual growth rate (IGR = IW63 − IW28, kg). An approach to study variability of growth traits was performed using deviation of growth traits, which was estimated as the absolute value of the difference between the individual value for IW28, IW63 and IGR and the mean value of their litter, DIW28, DIW63 and DIGR in kg, respectively.

All growth traits were also measured in the offspring of the control population and in contemporary animals from the 10th generation, obtaining a total of 1238 and 1142 records for IW28 and 1208 and 1123 records for IW63 and IGR, respectively.

2.4. Statistical Analyses

All analyses were performed using Bayesian methodology [13,14].

Correlated responses to selection have been estimated using two independent approaches: genetic trends and control population. Genetic trends use all data from all generations of selection but its estimations depend on the model, and therefore response to selection depends on the genetic parameters used in the model. In the control population approach, the number of data is smaller but differences between control and selected population will have only genetic bases if both populations are raised contemporarily.

2.4.1. Genetic Parameters and Genetic Trends

A repeatability model was used to analyze OR in the selection period:

$$y_{ijklm} = YS_i + PO_j + L_k + a_{ijkl} + p_{ijkl} + e_{ijklm} \tag{1}$$

where y_{ijklm} is the trait OR; YS_i is the effect of year-season of the mating day (31 levels for OR); PO_j is the effect of the parity order (four levels for OR); L_k is the effect of lactation status at mating (two levels: lactating and nonlactating does when mated); a_{ijkl} is the additive value of the animal; p_{ijkl} is the permanent environmental effect; and e_{ijklm} is the residual effect.

For individual weights and IGR, as well as for growth variability traits, the animal model used was:

$$y_{ijklmn} = YS_i + PO_j + NBA_k + d_{ijkl} + c_{ijklm} + a_{ijklmn} + e_{ijklmn} \tag{2}$$

where y_{ijklmn} is IW28, IW63 and IGR; YS_i is the fixed effect of year-season in which the animal was growing (29 levels); PO_j is the effect parity in which the animal was born (five levels); NBA_k is the effect of the number of rabbits born alive when the animal was born (17 levels); d_{ijkl} is the random dam (or female) effect between parities (851 levels); c_{ijklm} is the common litter effect (2683 levels); a_{ijklmn} is the additive value of animal; and e_{ijklmn} is the residual effect.

Univariate analysis for OR was performed to estimate the heritability of the selection trait. To account for the selection process and to estimate the heritability of each growth trait as well as the correlation between OR and growth traits, bivariate analyses including OR were performed.

Data augmentation [14,15] was performed to analyze the data in order to have the same design matrices for all traits since different models for OR and growth traits were used. Augmented data are not used for inferences but allow the simplification of computing.

After data augmentation, the model for all traits was:

$$(\boldsymbol{y}|\boldsymbol{b}, \boldsymbol{a}, \boldsymbol{p}, \boldsymbol{d}, \boldsymbol{c}, \boldsymbol{R}_0) \sim N(\boldsymbol{Xb} + \boldsymbol{Za} + \boldsymbol{Wp} + \boldsymbol{Dd} + \boldsymbol{Cc}, \boldsymbol{R}_0 \otimes \boldsymbol{I}_n) \tag{3}$$

where $\boldsymbol{y}$ is a vector of augmented data; $\boldsymbol{X}$, $\boldsymbol{Z}$, $\boldsymbol{W}$, $\boldsymbol{D}$ and $\boldsymbol{C}$ are known incidence matrices; and $\boldsymbol{R}$ is the (co)variance residual matrix. Records of different individuals were assumed to be conditionally independent, given the parameters, but a correlation between residuals of different traits of the same individual was allowed.

Hence, sorting the data by individual, the residual (co)variance matrix can be written as $\boldsymbol{R}_0 \otimes \boldsymbol{I}_n$, with $\boldsymbol{R}_0$ being the 2×2 residual (co)variance matrix between OR and the growth trait analyzed and $\boldsymbol{I}_n$ being an identity matrix of the same order as the number of individuals. Bounded uniform priors were used to represent vague previous knowledge of environmental effects, $\boldsymbol{b}$. Prior knowledge concerning the other random effects was represented by assuming that they were normally distributed, conditionally on the associated variance components. Thus, for the additive genetic effects:

$$\boldsymbol{a}|\boldsymbol{G} \sim N(\boldsymbol{0}, \boldsymbol{G}) \tag{4}$$

where $\boldsymbol{0}$ is a vector of zeroes and $\boldsymbol{G}$ is the genetic variance covariance matrix. Sorting the data by individual as before, this matrix can be written as $\boldsymbol{G_0} \otimes \boldsymbol{A}$, where $\boldsymbol{G_0}$ is the 2×2 genetic (co)variance matrix between traits analyzed and $\boldsymbol{A}$ is the known additive genetic relationship matrix between elements of the additive genetic effects vector.

The distribution of permanent environmental effects was assumed to be normal and of the form:

$$\boldsymbol{p}|\boldsymbol{P} \sim N(\boldsymbol{0}, \boldsymbol{P}) \tag{5}$$

where $\boldsymbol{0}$ is a vector of zeroes and $\boldsymbol{P}$ is the permanent effect matrix. Sorting the data by individual, this matrix can be written as $\boldsymbol{P_0} \otimes \boldsymbol{I_p}$, with $\boldsymbol{P_0}$ being the 2×2 permanent variance matrix between traits analyzed and $\boldsymbol{I_p}$ being the identity matrix with the same order of the number of levels of permanent effects.

The distribution of the random dam effects between parities was assumed to be normal and of the form:

$$\boldsymbol{d}|\boldsymbol{D} \sim N(\boldsymbol{0}, \boldsymbol{D}) \tag{6}$$

where $\boldsymbol{0}$ is a vector of zeroes and $\boldsymbol{D}$ is the random dam effects between parities matrix. Sorting the data by individual, this matrix can be written as $\boldsymbol{D_0} \otimes \boldsymbol{I_d}$, with $\boldsymbol{D_0}$ being the 2×2 permanent variance matrix between traits analyzed and $\boldsymbol{I_d}$ being the identity matrix with the same order of the number of levels of the random dam effects between parities.

Finally, the distribution of the common litter effects was assumed to be normal and of the form:

$$\boldsymbol{c}|\boldsymbol{C} \sim N(\boldsymbol{0}, \boldsymbol{C}) \tag{7}$$

where $\boldsymbol{0}$ is a vector of zeroes and $\boldsymbol{C}$ is the common litter effects matrix. Sorting the data by individual, this matrix can be written as $\boldsymbol{C_0} \otimes \boldsymbol{I_c}$, with $\boldsymbol{C_0}$ being the 2×2 common litter variance matrix between traits analyzed and $\boldsymbol{I_c}$ being the identity matrix with the same order of the number of levels of the common litter effects.

For all analyses, bounded flat priors were used for matrices $\boldsymbol{R_0}$, $\boldsymbol{G_0}$, $\boldsymbol{P_0}$, $\boldsymbol{D_0}$ and $\boldsymbol{C_0}$.

To estimate the correlations between growth traits, trivariate analyses including OR and two growth traits were performed. Hence, sorting the data by individual, the residual (co)variance matrix can be written as $\boldsymbol{R_0} \otimes \boldsymbol{I_n}$, with $\boldsymbol{R_0}$ being the 3×3 residual (co)variance matrix between OR and the growth traits analyzed and $\boldsymbol{I_n}$ being an identity matrix of the same order as the number of individuals. As previous analyses, bounded uniform priors were used for environmental effects and normally distributed priors, conditionally on the associated variance components that were used for random effects. The additive genetic effects, the permanent environmental effects, the random female effects between parities and the common litter effects were the same as described above. As in the previous model, all effects are independent among them. However, sorting the data by individual as before, the $\boldsymbol{G}$ matrix can be written $\boldsymbol{G_0} \otimes \boldsymbol{A}$, where $\boldsymbol{G_0}$ is the 3×3 genetic (co)variance matrix between traits analyzed and $\boldsymbol{A}$ is the known additive genetic relationship matrix described previously; the $\boldsymbol{P}$ matrix can be written as $\boldsymbol{P_0} \otimes \boldsymbol{I_p}$, with $\boldsymbol{P_0}$ being the 3×3 permanent variance matrix between traits analyzed and $\boldsymbol{I_p}$ being the identity matrix with the same order of the number of levels of permanent effects; the $\boldsymbol{D}$ matrix can be written as $\boldsymbol{D_0} \otimes \boldsymbol{I_d}$, with $\boldsymbol{D_0}$ being the 3×3 permanent variance matrix between traits analyzed and $\boldsymbol{I_d}$ being the identity matrix with the same order of the number of levels of the random dam effects between parities; and the $\boldsymbol{C}$ matrix can be written as $\boldsymbol{C_0} \otimes \boldsymbol{I_c}$, with $\boldsymbol{C_0}$ being the 3×3 permanent (co)variance matrix between traits analyzed and $\boldsymbol{I_c}$ being the identity matrix with the same order of the number of levels of the common litter effects.

Marginal posterior distributions of all unknowns were estimated using a Gibbs sampling procedure using the program TM [16]. Different confidence intervals were estimated:

$k_{95\%}$ is the guaranteed value of the interval [k, 1] containing the 95% of the probability, P is the probability of the estimation being higher (or lower) than 0.00, $P_{0.10}$ and $P_{0.30}$ are the probability of the estimation being higher (or lower) than 0.10 and 0.30, respectively, and P_r is the probability of relevance [13,14]. We considered a heritability to be irrelevant when it was lower than 0.10 [4–6,17]. In the case of correlation, we considered to be an irrelevant value all correlations in absolute value lower than 0.30, since the percentage of the variance explained by the other trait (r^2) is <10%. After some exploratory analyses, two chains were used, each of 1,000,000 iterations, with a burning period of 200,000 iterations. Only every 100th iteration was saved. Features of marginal posterior distributions of parameters were obtained using the package R code. Convergence was tested using the Z criterion of Geweke and Monte Carlo sampling errors were computed.

2.4.2. Selected versus Control Population

The model assumed for analyzing OR using the offspring of the control and selected population was:

$$y_{ijklmn} = Line_i + YS_j + PO_k + L_l + p_{ijklm} + e_{ijklmn} \quad (8)$$

where y_{ijklm} is the trait OR; $Line_i$ is the effect of the line (two levels: control and selected); YS_j is the effect of year-season (three levels); PO_k is the effect of parity (two levels: at second and fourth gestation); L_l is the effect of lactation state of the doe (two levels: lactating and nonlactating does when mated); p_{ijklm} is the effect of the doe (105 levels); and e_{ijklmn} is the residual of the model.

The model assumed for analyzing individual growth traits and their variabilities was:

$$y_{ijklmn} = YS_i + PO_j + NBA_k + d_{ijkl} + c_{ijklm} + e_{ijklmn} \quad (9)$$

where y_{ijklm} is IW28, IW63 and IGR; YS_i is the fixed effect of year-season in which the animal was growing (four levels); PO_j is the effect parity in which the animal was born (four levels); NBA_k is the effect of the number of rabbits born alive when the animal was born (16 levels); d_{ijkl} is the random dam effect between parities (96 levels); c_{ijklm} is the common litter effect (302 levels); and e_{ijklmn} is the residual effect.

Bounded uniform priors were used for all unknowns with the exception of the dam and common litter effects, which were considered normally distributed. Dam effect was with mean **0** and variance $\boldsymbol{I}\sigma_d^2$, where $\boldsymbol{I}$ is a unity matrix and σ_d^2 is the dam effect variance of the trait. Common effect was with mean **0** and variance $\boldsymbol{I}\sigma_c^2$, where $\boldsymbol{I}$ is a unity matrix and σ_c^2 is the common effect variance of the trait. Residuals were normally distributed with mean **0** and variance $\boldsymbol{I}\sigma_e^2$. The priors for the variances were also bounded uniform positive. Features of the marginal posterior distribution of differences between line means were estimated by using the Gibbs sampling algorithm. Similarly to previously described genetic estimations, chains of 1,000,000 samples each were used, with a burning period of 200,000. One sample out of each 100 was saved to avoid high correlations between consecutive samples. Features of marginal posterior distributions of differences between line means were obtained using the package R code. Convergence was tested using the Z criterion of Geweke and Monte Carlo sampling errors were computed.

3. Results

3.1. Genetic Parameters and Genetic Trends

Descriptive statistics for all traits are presented in Table 1. Rabbits at 28 and 63 days old weighed 0.52 and 1.76 kg, respectively. The coefficient of variation of growth traits was close to 0.20, whereas their variability, estimated as the absolute value of the difference between the individual value and the mean value of their litter, was close to 0.90.

Table 1. Raw means, coefficients of variation (CV) and number of data (N) for growth traits and their variability.

	Mean	CV	N
IW28	0.52	0.23	20,230
IW63	1.76	0.14	19,362
IGR	1.24	0.14	19,362
DIW28	0.05	0.85	20,230
DIW63	0.11	0.91	19,362
DIGR	0.09	0.99	19,362

IW28: individual weight at 28 days old (kg), IW63: individual weight at 63 days old (kg), IGR: individual growth rate (kg), DIW28: variability of individual weight at 28 days old (estimated as the absolute value of the difference between IW28 and the mean value of their litter at 28 days old; kg), DIW63: variability of individual weight at 63 days old (estimated as the absolute value of the difference between IW63 and the mean value of their litter at 63 days old; kg), DIGR: variability of individual growth rate (estimated as the absolute value of the difference between IGR and the mean value of their litter for growth rate; kg).

Phenotypic correlations between OR and the growth traits were low: 0.17, 0.15 and 0.09 for IW28, IW63 and IGR, respectively. The R-squared between OR and the other analyzed traits was lower than 5%. When the ratio between R-squared and R-squared maximum was estimated, different results were observed: 0.72 for IW28, 0.57 for IW63 and 0.29 for IGR.

Phenotypic residual correlations between OR and growth traits and also between OR and the variability of growth traits were very low (Table 2). Only correlations between weights (IW28–IW63) and between IGR and IW63 were higher than 0.60. Low correlations were also found within the variability of growth traits and also between growth traits and their variabilities. These results showed that the accuracy was low when the model included some effects to correct the data.

Table 2. Phenotypic residual correlations between ovulation rate (OR, ova) and growth traits (kg) and their variabilities (estimated as the absolute value of the difference between the individual value and the mean value of their litter; kg).

	OR	IW28	IW63	IGR	DIW28	DIW63
IW28	0.11					
IW63	0.08	0.64				
IGR	0.03	0.35	0.90			
DIW28	0.05	0.12	0.09	0.13		
DIW63	0.04	0.11	0.10	0.11	0.21	
DIGR	0.06	0.09	0.09	0.12	0.15	0.35

IW28: individual weight at 28 days old, IW63: individual weight at 63 days old, IGR: individual growth rate, DIW28: variability of individual weight at 28 days old, DIW63: variability of individual weight at 63 days old, DIGR: variability of individual growth rate.

The heritability for OR was moderate, 0.17, with a probability to be higher than 0.10 of 94% and the highest posterior density region at 95% ranging from 0.08 to 0.27 (data not shown). The heritability values of IW28, IW63 and IGR were low: 0.09, 0.12 and 0.11, respectively (Table 3). A heritability higher than 0.10 was considered relevant for growth traits and their variability. The probability of the heritability being higher than 0.10 was high for IW63 and IGR. The maternal effects of the doe over all their parities, which is calculated as the ratio of the maternal effect variance with respect to phenotypic variance (m^2), had relevance for IW28 (0.14), and it was close to zero for IW63 and IGR. However, the common litter effect, which is calculated as the ratio of the common litter effect variance with respect to phenotypic variance (c^2), explained a greater part of phenotypic variance for all growth traits, with their values being close to 0.30. Genetic variation for the variability of growth traits was negligible, lower than 0.02, similarly to the maternal effects. The common litter effects for the variability of growth traits were close to 0.10.

Table 3. Features of marginal posterior distributions of the heritability (h^2), ratio of the maternal effect variance with respect to phenotypic variance (m^2) and ratio of the common litter effect variance with respect to phenotypic variance (c^2) for growth traits and their variability.

	h^2				m^2		c^2	
Trait	**Mean**	**$HPD_{95\%}$**	**$k_{95\%}$**	**$P_{0.10}$**	**Mean**	**$k_{95\%}$**	**Mean**	**$k_{95\%}$**
IW28	0.09	0.04, 0.15	0.05	0.37	0.14	0.17	0.33	0.35
IW63	0.12	0.07, 0.19	0.08	0.79	0.04	0.06	0.29	0.31
IGR	0.11	0.06, 0.17	0.07	0.71	0.01	0.02	0.30	0.32
DIW28	0.01	0.00, 0.02	0.01	0.00	0.01	0.00	0.08	0.10
DIW63	0.00	0.00, 0.01	0.00	0.00	0.01	0.00	0.10	0.13
DIGR	0.01	0.00, 0.02	0.01	0.00	0.01	0.00	0.13	0.15

$HPD_{95\%}$: highest posterior density region at 95%, $k_{95\%}$: limit of the interval [k, 1] containing a probability of 95%, $P_{0.10}$: probability of the proportion being higher than 0.10, IW28: individual weight at 28 days old, IW63: individual weight at 63 days old, IGR: individual growth rate, DIW28: variability of individual weight at 28 days old (estimated as the absolute value of the difference between IW28 and the mean value of their litter at 28 days old), DIW63: variability of individual weight at 63 days old (estimated as the absolute value of the difference between IW63 and the mean value of their litter at 63 days old), DIGR: variability of individual growth rate (estimated as the absolute value of the difference between IGR and the mean value of their litter for growth rate).

All high posterior density regions for genetic correlations between OR and growth traits were large (Table 4). The estimate of the genetic correlation between OR and IW28 was very low (0.11), whereas between OR and IW63 and between OR and IGR it was 0.23 and 0.28, respectively. The assumed relevant value for correlation was 0.30 (in absolute value). The probability that the genetic correlation between OR and growth traits was higher than 0.30 was lower than 50%. Moderate positive genetic correlation between OR and variability of growth traits was obtained, although a low accuracy was achieved since the highest posterior density region at 95% was wide.

Table 4. Features of the estimated marginal posterior distributions of the genetic correlations between ovulation rate (OR) and growth traits and their variability.

Trait	Mean	$HPD_{95\%}$	P	$P_{0.30}$
OR-IW28	0.11	−0.26, 0.50	0.71	0.21
OR-IW63	0.23	−0.13, 0.56	0.90	0.39
OR-IGR	0.28	−0.12, 0.63	0.93	0.48
OR-DIW28	0.28	0.02, 0.59	1.00	0.56
OR-DIW63	0.62	0.40, 0.79	1.00	1.00
OR-DIGR	0.55	0.17, 0.76	1.00	0.54

$HPD_{95\%}$: highest posterior density region at 95%, $k_{95\%}$: limit of the interval [k, 1] containing a probability of 95%, P: probability of the correlation being higher than zero, $P_{0.30}$: probability of the absolute value of correlation being higher than 0.30, IW28: individual weight at 28 days old, IW63: individual weight at 63 days old, IGR: individual growth rate, DIW28: variability of individual weight at 28 days old (estimated as the absolute value of the difference between IW28 and the mean value of their litter at 28 days old), DIW63: variability of individual weight at 63 days old (estimated as the absolute value of the difference between IW63 and the mean value of their litter at 63 days old), DIGR: variability of individual growth rate (estimated as the absolute value of the difference between IGR and the mean value of their litter for growth rate).

The genetic correlation between IW28 and IW63 was positive ($P > 0 = 100\%$) and high (mean = 0.83) (Table 5). Moreover, the probability that the genetic correlation was at least 0.71 was 95%. A similar result was obtained when the genetic correlation between IW63 and IGR was analyzed; the mean was 0.95 and the probability that the genetic correlation was at least 0.92 was 95%. Moreover, genetic correlations between DIW63 and the other analyzed variability of growth traits were also higher than 0.90. However, non-relevant genetic correlations between growth traits and their variabilities were observed, since the probability of relevance is lower than 0.30.

Table 5. Features of the estimated marginal posterior distributions of the genetic (r_g) and common litter (r_c) correlations between individual weights.

	r_g					r_c	
Traits	Mean	$HPD_{95\%}$	$k_{95\%}$	P	$P_{0.30}$	Mean	$k_{95\%}$
IW28-IW63	0.83	0.70, 0.93	0.71	1.00	1.00	0.69	0.73
IW28-IGR	0.40	0.07, 0.69	0.11	0.99	0.64	0.15	0.21
IW63-IGR	0.95	0.91, 0.98	0.92	1.00	1.00	0.91	0.93
DIW28-DIW63	0.93	0.79, 0.99	0.82	1.00	1.00	0.53	0.57
DIW28-DIGR	0.62	0.33, 0.87	0.38	1.00	1.00	0.29	0.32
DIW63-DIGR	0.95	0.86, 0.99	0.87	1.00	1.00	0.80	0.86
IW28-DIW28	0.18	−0.02, 0.44	0.00	0.97	0.26	0.04	0.08
IW63-DIW63	−0.18	−0.54, 0.18	−0.50	0.58	0.29	−0.20	−0.15
IGR-DIGR	−0.21	−0.48, 0.09	−0.45	0.57	0.28	−0.23	−0.20

$HPD_{95\%}$: highest posterior density region at 95%, $k_{95\%}$: limit of the interval [k, 1] containing a probability of 95%, P: probability of the r_g being higher than zero when the mean is positive or lower than zero when it is negative, $P_{0.30}$: probability of the r_g being higher than 0.30 when the mean is positive or lower than −0.30 when the mean is negative, IW28: individual weight at 28 days old, IW63: individual weight at 63 days old, IGR: individual growth rate, DIW28: variability of individual weight at 28 days old (estimated as the absolute value of the difference between IW28 and the mean value of their litter at 28 days old), DIW63: variability of individual weight at 63 days old (estimated as the absolute value of the difference between IW63 and the mean value of their litter at 63 days old), DIGR: variability of individual growth rate (estimated as the absolute value of the difference between IGR and the mean value of their litter for growth rate).

The correlated responses to selection were estimated at the end of the selection period as the difference of the average breeding values between the end and the beginning of the period. Correlated responses on IW28, IW63 and IW2863 were low (Figure 1), being 2.3, 11.2 and 7.9 g per generation, respectively. These correlated responses corresponded to 0.4, 0.6 and 0.6% per generation, respectively. However, the correlated response on the variability of growth traits was close to zero (Figure 2).

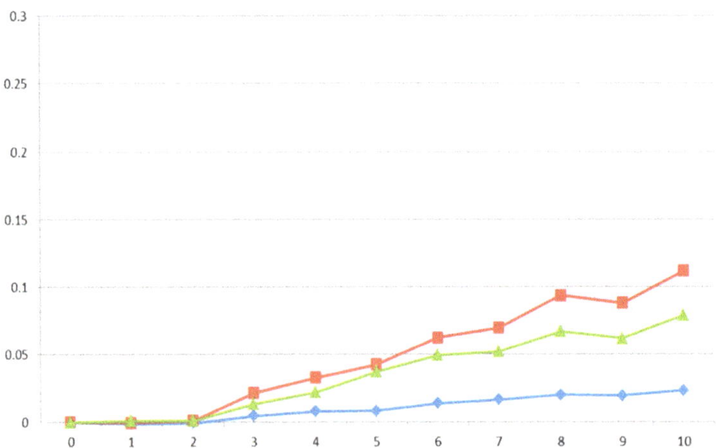

Figure 1. Genetic trends for individual weight at 28 days old (blue line; kg), individual weight at 63 days old (red line; kg) and individual growth rate (green line; kg).

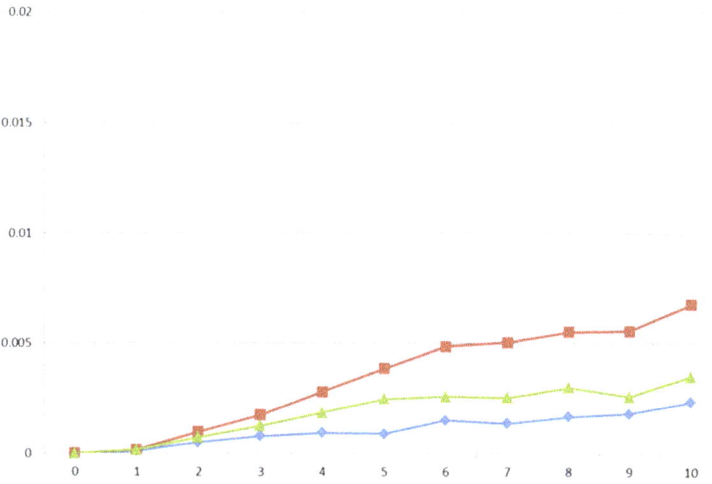

Figure 2. Genetic trends for variability of individual weight at 28 days old (estimated as the absolute value of the difference between IW28 and the mean value of their litter at 28 days old; blue line; kg), variability of individual weight at 63 days old (estimated as the absolute value of the difference between IW63 and the mean value of their litter at 63 days old; red line; kg) and variability of individual growth rate (estimated as the absolute value of the difference between IGR and the mean value of their litter for growth rate; green line; kg) of the OR line.

3.2. Control Population

Table 6 shows raw means and coefficients of variation for the growth traits measured in the control population. Rabbits at 28 and 63 days old belonging to the control population weighed 0.50 and 1.77 kg, respectively. The coefficient of variation of growth traits was smaller than the coefficient of variation of the variability of growth traits.

Table 6. Raw means, coefficients of variation (CV) and number of data (N) for growth traits and their variability in the offspring of the control population.

	Mean	CV	N
IW28	0.50	0.25	1238
IW63	1.77	0.14	1208
IGR	1.26	0.15	1208
DIW28	0.05	0.81	1238
DIW63	0.10	0.90	1208
DIGR	0.06	0.87	1208

IW28: individual weight at 28 days old (kg), IW63: individual weight at 63 days old (kg), IGR: individual growth rate (kg), DIW28: variability of individual weight at 28 days old (estimated as the absolute value of the difference between IW28 and the mean value of their litter at 28 days old; kg), DIW63: variability of individual weight at 63 days old (estimated as the absolute value of the difference between IW63 and the mean value of their litter at 63 days old; kg), DIGR: variability of individual growth rate (estimated as the absolute value of the difference between IGR and the mean value of their litter for growth rate; kg).

Features of the marginal posterior distributions of the differences between the control and selected populations for growth traits and their variabilities are presented in Table 7. A relevant response to selection (R value) was assumed when the difference between populations was at least 10% of the mean of the control population, corresponding to an increase of 1% per generation. Growth traits presented no relevant differences between the control and selected populations (0.02, 0.07 and 0.08 for IW28, IW63 and IW2863, respectively). For these traits, the probability of relevance was lower than 15%. Similarly, no correlated responses on the variability of growth traits were observed.

Table 7. Features of the estimated marginal posterior distributions of the differences between the control and selected population for growth traits and their variability.

	Control–Selected	$HPD_{95\%}$	P	R	P_R
IW28	0.02	−0.01, 0.06	0.89	0.05	0.12
IW63	0.07	0.01, 0.15	0.98	0.17	0.00
IGR	0.08	0.04, 0.13	1.00	0.13	0.02
DIW28	0.001	−0.001, 0.003	0.75	0.005	0.00
DIW63	0.001	0.000, 0.002	0.76	0.001	0.42
DIGR	0.003	0.002, 0.004	1.00	0.006	0.01

$HPD_{95\%}$: highest posterior density region at 95%, P: probability of the difference between the control and selected population being higher than zero, R = 10% of the mean of the control population (1% per generation), P_R = probability of response, which is the probability of the difference being higher than R, IW28: individual weight at 28 days old, IW63: individual weight at 63 days old, IGR: individual growth rate, DIW28: variability of individual weight at 28 days old (estimated as the absolute value of the difference between IW28 and the mean value of their litter at 28 days old), DIW63: variability of individual weight at 63 days old (estimated as the absolute value of the difference between IW63 and the mean value of their litter at 63 days old), DIGR: variability of individual growth rate (estimated as the absolute value of the difference between IGR and the mean value of their litter for growth rate).

Similarly to the estimations using genetic trends, the common litter effect, c^2, explained a greater part of the phenotypic variance for growth traits, close to 0.25, whereas low estimations were observed for the variability of growth traits. The maternal effect of the doe over all their parities, m^2, had a non-negligible value for IW28 (0.10).

4. Discussion

Direct response to selection for ovulation rate was obtained in rabbits after 10 generations of selection, 0.13 ova per generation. However, prenatal survival decreased and no correlated response on litter size at birth was observed [4]. The decrease in prenatal survival can be explained by a higher fetal competition due to an increment of implanted embryos [5]. Higher competition among fetuses for space, nutrients and blood supply in the uterus reduces birth weight [18]. This has led to the question of whether selection for ovulation rate could reduce weight at birth and subsequently weaning weight (IW28), commercial weight (IW63) and individual growth rate (IGR) [11], as well as modify their variability (DIW28, DIW63 and DIGR, respectively). On the other hand, in pigs, a line selected for ovulation rate with no response in litter size showed a high response in litter size when directly selected for litter size [19]; thus, a line selected for ovulation rate can be interesting for further research to improve litter size.

Descriptive statistics obtained in the present work are similar to those previously reported in other rabbit maternal lines [8,17,20] at the same ages. Phenotypic correlations between ovulation rate and growth traits were low, similarly to correlation between ovulation rate and the variability of growth traits. As expected, weaning and commercial weight were highly related and commercial weight and growth rate were also highly associated, in agreement with previous publications in rabbits [8,17]. Weights at birth were not available but high correlations between weight at birth and weight at weaning have been found by Argente and coauthors [18]. Finally, low phenotypic correlations were also found between growth traits and the variability of growth traits. To our knowledge, these are the first phenotypic estimations between growth traits and their variability.

4.1. Genetic Parameters

Low heritabilities, close to 0.10, were obtained for IW28 and IW63, which corresponded to weaning and commercial time, respectively. Likewise, IGR, which is the most common selection criterion in sire lines in rabbits [1], also showed a low heritability. Similar heritabilities for these three traits were obtained by Peiró and coauthors [17], who analyzed a line selected by ovulation rate and litter size, and also for IW28 by Mínguez and coauthors [20], who analyzed four maternal lines (A, V, H and LP). However, a broad range of heritabilities, ranging from 0.03 to 0.25, has been obtained for these traits in paternal rabbit lines (reviewed by Garcia and coauthor) [21]. The common litter effect is more important than the maternal effect and additive genetic effect in all growth traits, close to 0.3 for growth traits and to 0.1 for the variability of growth traits. However, low maternal effects have been obtained for all growth traits, in agreement with previous results [8,17,20,21]. Regarding the variability of growth traits, negligible estimations (P > 0.10 = 0.00) were observed for heritabilities and the maternal effect and low values were obtained for the common litter effect, around 0.10, as it is mentioned earlier. Estimations were in the range of those published previously by Peiró and coauthors [17] in a maternal line. The preweaning environment effect, common litter effect and maternal effect have a large influence on studied growth traits, although there is a reduction in the preweaning environment effect over time, as previously found by other authors [8,17,20]. This reduction should be related to kits' separation from their mothers and also for kits' distribution into different cages (it is common to separate the litter into several cages, so litters from different rabbit does can be mixed).

Genetic and phenotypic correlations were similar, and this is a generally observed phenomenon [22]. Genetic correlations between ovulation rate and growth traits and their variability were similar to the only estimation found in the literature [17]. Genetic correlation between growth traits was positive and high in agreement with [23–25]. Genetic correlation between commercial weight (IW63) and IGR was higher than between IW28 and IGR. These results were previously observed by McNitt and coauthor [23] and by Ezzeroug and coauthors [25]. Similar behavior has been observed for genetic correlations between the variability of growth traits; a higher genetic correlation was observed when DIW63 was evaluated, close to 0.90. Moderate–high genetic correlations, close to 0.60, were observed between the variability of individual weight at 28 days old and the variability of the individual growth rate. However, no relevant genetic correlations between each growth trait and its variability have been obtained. To our knowledge, there is no information about genetic correlations between the variability of growth traits.

4.2. Correlated Response

Considering all these genetic parameters, the correlated response to selection for ovulation rate during ten generations was estimated at the end of the selection period. Using genetic trends methods, no relevant correlated responses on IW28, IW63 and IW2863 were found, less than 1% of the trait per generation. These estimates agreed with previous results [17]. Similarly, no relevant correlated responses on the variability of growth traits were observed, less than 1% of the trait per generation. Using the comparison of the selected line with a cryopreserved control population method, no correlated response was observed. The use of a cryopreserved control population corroborates the results obtained with genetic trends, and neither correlated response to selection for ovulation rate during 10 generations on individual weight at 28 and 63 days old nor on their variability was observed. As the response to selection estimated by control and selected populations is the same as the one obtained using the model by genetic trends that includes maternal and common litter effects, we can infer that dam effects were not modified after selection.

5. Conclusions

The increase in ovulation rate by selection did not reduce the weight of the young rabbits at weaning (28 days) and at marketing (63 days) and did not modify their variabilities. It can be inferred that female effects, which include maternal and common litter effect, were not modified after selection.

Author Contributions: Conceptualization, A.B. and M.A.S.; formal analysis, C.Q. and R.P.; data curation, R.P. and C.Q.; writing—original draft preparation, R.P. and M.A.S.; writing—review and editing, A.B.; C.Q.; R.P. and M.A.S.; supervision, M.A.S.; project administration, A.B.; funding acquisition, A.B. and M.A.S. All authors have read and agreed to the published version of the manuscript.

Funding: This research was funded by Comisión Interministerial de Ciencia y Tecnología, grant number PID2020−115558GB−C21, and by Generalitat Valenciana research program (Prometeo 2009/125). Celia Quirino was supported by a fellowship of the Coordenação de Aperfeiçoamento de Pessoal de Nível Superior-Brasil (CAPES)-Finance Code 001.

Institutional Review Board Statement: The study was conducted according to the guidelines of the Council Directives 98/58/EC and 2010/63/EU, and approved by the Universitat Politècnica de València Research Ethics Committee.

Conflicts of Interest: The authors declare no conflict of interest.

References

1. Baselga, M. Genetic improvement of meat rabbits: Programmes and diffusion. In Proceedings of the 8th World Rabbit Congress, Puebla, Mexico, 7–10 September 2004; pp. 1–13.
2. Cartuche, L.; Pascual, M.; Gómez, E.A.; Blasco, A. Economic weights in rabbit meat production. *World Rabbit Sci.* **2014**, *22*, 165. [CrossRef]
3. Zimmerman, D.R.; Cunningham, P.J. Selection for ovulation rate in swine: Population, procedures and ovulation response. *J. Anim. Sci.* **1975**, *40*, 61–69. [CrossRef] [PubMed]
4. Laborda, P.; Mocé, M.L.; Santacreu, M.A.; Blasco, A. Selection for ovulation rate in rabbits: Genetic parameters, direct response, and correlated response on litter size. *J. Anim. Sci.* **2011**, *89*, 2981–2987. [CrossRef] [PubMed]
5. Laborda, P.; Mocé, M.L.; Blasco, A.; Santacreu, M.A. Selection for ovulation rate in rabbits: Genetic parameters and correlated responses on survival rates1. *J. Anim. Sci.* **2012**, *90*, 439–446. [CrossRef] [PubMed]
6. Laborda, P.; Santacreu, M.A.; Blasco, A.; Mocé, M.L. Selection for ovulation rate in rabbits: Direct and correlated responses estimated with a cryopreserved control population. *J. Anim. Sci.* **2012**, *90*, 3392–3397. [CrossRef] [PubMed]
7. Brun, J.M.; Bolet, G.; Ouhayoun, J. The effects of crossbreeding and selection on productive and reproductive traits in a triallel experiment between three strains of rabbits. *J. Appl. Rabbit Res.* **1992**, *15*, 181–189.
8. García, M.L.; Baselga, M. Estimation of correlated response on growth traits to selection in litter size of rabbits using cryopreserved control population and genetic trends. *Livest. Prod. Sci.* **2002**, *78*, 91–98. [CrossRef]
9. Szendrő, Z.; Cullere, M.; Atkári, T.; Zotte, A.D. The birth weight of rabbits: Influencing factors and effect on behavioural, productive and reproductive traits: A review. *Livest. Sci.* **2019**, *13*, 2457–2462. [CrossRef]
10. Argente, M.J.; Santacreu, M.A.; Climent, A.; Blasco, A. Relationships between uterine and fetal traits in rabbits selected on uterine capacity. *J. Anim. Sci.* **2003**, *81*, 1265–1273. [CrossRef] [PubMed]
11. Bolet, G.; Esparbi, J.; Falières, J. Relations entre le nombre de fœtus par corne utérine, la taille de portée à la naissance et la croissance pondérale des lapereaux. *Ann. Zootech.* **1996**, *45*, 185–200. [CrossRef]
12. Cifre, J.; Baselga, M.; Gómez, E.A.; García, M.L. Effect of embryo cryopreservation techniques on reproductive and growth traits in rabbits. *Ann. Zootech.* **1999**, *48*, 15–24. [CrossRef]
13. Blasco, A. The bayesian controversy in animal breeding. *J. Anim. Sci.* **2001**, *79*, 2023–2046. [CrossRef] [PubMed]
14. Blasco, A. *Bayesian Data Analysis for Animal Scientists*; Springer: Berlin/Heidelberg, Germany, 2017.
15. Sorensen, D.; Gianola, D. *Likelihood, Bayesian, and MCMC Methods in Quantitative Genetics*; Springer Science and Business Media: Berlin/Heidelberg, Germany, 2002. [CrossRef]
16. Legarra, A.; Varona, L.; López de Maturana, E. Program^{TM}. 2008. Retrieved on 7 January 2021. Available online: http://snp.toulouse.inra.fr/~alegarra/ (accessed on 30 August 2021).
17. Peiró, R.; Badawy, A.Y.; Blasco, A.; Santacreu, M.A. Correlated responses on growth traits after two-stage selection for ovulation rate and litter size in rabbits. *Animal* **2019**, *13*, 2457–2462. [CrossRef] [PubMed]
18. Argente, M.J.; Santacreu, M.A.; Climent, A.; Blasco, A. Phenotypic and genetic parameters of birth weight and weaning weight in rabbits born from unilaterally ovariectomized and intact does. *Livest. Prod. Sci.* **1999**, *57*, 159–167. [CrossRef]
19. Lamberson, W.; Zimmerman, D.R.; Eldridge, F.; Kopf, J. Direct responses from selection for increased litter size, decreased age and puberty or relaxed selection following selection for ovulation rate in swine. *J. Anim. Sci.* **1991**, *69*, 3129–3143. [CrossRef] [PubMed]
20. Mínguez, C.; Sanchez, J.; El Nagar, A.; Ragab, M.; Baselga, M. Growth traits of four maternal lines of rabbits founded on different criteria: Comparisons at foundation and at last periods after selection. *J. Anim. Breed. Genet.* **2016**, *133*, 303–315. [CrossRef] [PubMed]
21. Gárcia, M.L.; Argente, M.J. The genetic improvement in meat rabbits. In *Lagomorpha Characteristics*, 1st ed.; IntechOpen: London, UK, 2020. [CrossRef]
22. Chevereud, J.M. A comparasion of genetic and phenotypic correlations. *Evolution* **1988**, *42*, 958–968. [CrossRef] [PubMed]
23. McNitt, J.I.; Lukefahr, S.D. Breed and environmental effects on postweaning growth of rabbits. *J. Anim. Sci.* **1993**, *71*, 1996–2005. [CrossRef] [PubMed]

24. Gómez, E.A.; Rafel, O.; Ramon, J. Genetic relationship between growth and litter size in traits at first parity in a specialised dam line in rabbits. In Proceedings of the 6th World Congress on Genetics Applied to Livestock Production, Armidale, Australia, 11–16 January 1998; pp. 552–555.
25. Ezzeroug, R.; Belabbas, R.; Argente, M.J.; Berbar, A.; Diss, S.; Boudjella, Z.; Talaziza, D.; Boudahdir, N.; Garcia, M.L. Genetic correlations for reproductive and growth traits in rabbits. *Can. J. Anim. Sci.* **2020**, *100*, 317–322. [CrossRef]

MDPI

Article

Fatty Acid Profile of Blood Plasma at Mating and Early Gestation in Rabbit

Imane Hadjadj [1], Anna-Katharina Hankele [2], Eva Armero [3], María-José Argente [1] and María de la Luz García [1,*]

1 Centro de Investigación e Innovación Agroalimentaria y Agroambiental (CIAGRO-UMH), Miguel Hernández University, Ctra. de Beniel km 3,2., 03312 Orihuela, Spain; imane.hadjadj@alu.umh.es (I.H.); mj.argente@umh.es (M.-J.A.)
2 ETH Zurich, Animal Physiology, Institute of Agricultural Sciences, Universitaetstr. 2, 8092 Zurich, Switzerland; anna-katharina.hankele@usys.ethz.ch
3 Department of Agronomic Engineering, Technical University of Cartagena, Paseo Alfonso XIII 48, 30203 Cartagena, Spain; eva.armero@upct.es
* Correspondence: mariluz.garcia@umh.es; Tel.: +34-96-6749-709

Simple Summary: Fatty acids can be used as an energy substrate by oocytes and embryos. Ovulation rate and normal preimplantation embryos are limiting factors to increased litter size in rabbits. Knowledge of the fatty acid profile in blood plasma at mating and in early gestation and its relationship with the ovulation rate and early embryonic development can help improve doe productivity. In our study, palmitic, linoleic, oleic and stearic acids show the highest concentrations. Moreover, monounsaturated fatty acids are correlated with ovulation rate and normal embryos. The more SFA, the greater the embryonic development. This study could be useful for designing enriched feeds in animal production and for embryological studies, as the rabbit is an experimental model.

Abstract: The aim of this study was to analyse the fatty acid (FA) profile of blood plasma at mating and 72 hpm by gas chromatography. Moreover, the correlation between FA and ovulation rate, normal embryos and compacted morulae was estimated. Palmitic, linoleic, oleic and stearic were the highest FA concentrations at mating and 72 hpm. Most long chain saturated and PUFA were higher at 72 hpm than at mating, while MUFA were higher at mating. SFA, MUFA and PUFA were high and positively correlated. Correlation was 0.643 between MUFA at mating and ovulation rate, and 0.781 between MUFA and normal embryos, respectively. Compacted morulae were slightly correlated with SFA at mating (0.465). In conclusion, the FA profile of plasma varies depending on the reproductive cycle of the rabbit female, adapting to energetic requirements at mating and early gestation. Moreover, positive correlations are found between fatty acids and ovulation rate and embryo development and quality.

Keywords: embryo; MUFA; ovulation; PUFA; SFA

Citation: Hadjadj, I.; Hankele, A.-K.; Armero, E.; Argente, M.-J.; de la Luz García, M. Fatty Acid Profile of Blood Plasma at Mating and Early Gestation in Rabbit. *Animals* **2021**, *11*, 3200. https://doi.org/10.3390/ani11113200

Academic Editor: Johnny Roughan

Received: 14 October 2021
Accepted: 8 November 2021
Published: 9 November 2021

Publisher's Note: MDPI stays neutral with regard to jurisdictional claims in published maps and institutional affiliations.

1. Introduction

From ovulation to implantation, oocytes and embryos are supported by maternal secretions from the oviduct and uterus (cow [1], sheep [2], murine [3], horse [4], sow [5] and rabbit [6]). The oviductal and uterine fluids mainly originate from blood filtrates [7] and secretions of endometrial luminal and glandular epithelial cells [8]. Thus, a complex milieu containing proteins, amino acids, carbohydrates and fatty acids (FA) is constituted to maintain embryo viability.

Fatty acids represent compact reserves of stored energy for oocytes and embryos [9]. Additionally, FA play a crucial role in modifying the physical properties and functions of biological membranes, and have potential effects on oocyte growth and maturation and embryo development and transport through reproductive traits [10–12]. The ability for the exogenous uptake of FA by the embryo has been demonstrated in several studies [13–15].

Specifically, rabbit embryos can already absorb FA at zygote stage [13], and it is reported that exogenously supplied fatty acids are beneficial for growth and development in cultivated oocytes and embryos [10]. Moreover, FA enriched diets consumed around the fertilization period and the first days of pregnancy and conditions in the oviductal and uterine environment [14] can affect embryo quality [15]. Rabbit is a livestock species reared for the production of hair, skin or meat and also as an experimental reference for other species, such as pigs or humans [16]. A detailed understanding of how ovulation rate and early embryo survival and development are affected by the FA profile could improve the productivity of rabbit females and further basic knowledge as an experimental model. Therefore, the aim of this work was to study the fatty acid profile at mating and in early gestation and its relationship with ovulation rate and embryo quality and development in rabbits.

2. Materials and Methods

2.1. Experiment Animals and Design

A total of 15 non-lactating multiparous rabbit females were used. Does were 9–10 months of age. Females belonged to two lines selected divergently by litter size variability [17]. Females were kept on the experimental farm at the Universidad Miguel Hernández de Elche (Spain). They were reared in individual cages and fed ad libitum with a commercial diet (crude protein, 17.5%; crude fibre, 15.5%; ether extract, 5.4%; ash, 8.1%) during their reproductive life.

2.2. Blood Sampling

Females that had finished their fourth lactation were weighed and mated. Following the blood sampling procedure described in [18], two blood samples of 3 mL were drawn from the central artery of each doe's ear at mating and 72 h post-mating (hpm). The blood sample was collected into a tube with tripotassium ethylenediaminetetraacetic acid (K3-EDTA). All samples were immediately centrifuged at 4000× g rpm for 15 min, and plasma was stored at −80 °C until required for lipid analyses.

2.3. Reproductive Traits

At 72 hpm, females were euthanized by intravenous administration of sodium thiopental in a dose of 50 mg/kg of body weight (Thiobarbital, B. Braun Medical S.A., Barcelona, Spain). The entire reproductive tract was immediately removed in order to measure reproductive traits.

Total ovulation rate (OR) was estimated as the number of corpora lutea. Total embryos were recovered by perfusion of oviducts and uterine horns with 10 mL of Dulbecco's phosphate-buffered saline containing 0.2% of BSA. Embryos were classified as normal embryos (NE) when they presented homogeneous cellular mass and intact embryo coat, using a binocular stereoscopy microscope (Leica Mz 9.5; ×600). Embryos were classified as early morulae or compacted morulae. Compacted morulae (CM) were expressed as percentage of NE.

2.4. Fatty Acid Analyses

All samples were analysed in duplicate. A 200 μL sample of plasma was taken in a screw cap glass tube. One milliliter of 0.5 M NaOH-methanol was added to each sample and the sample was boiled at 90 °C for 15 min. The sample was cooled in an ice bath to room temperature, 2 mL of BF3-methanol was added, and the sample was boiled at 90 °C for 20 s. The sample was again cooled in an ice bath to room temperature, 1 mL of isooctane was added, and the mixture was shaken. In the measurement of recovery ratio of the internal standard, 1 mL iso-octane containing 0.05 mg 13:0 ME and 0.01 mg 19:0 ME was added. Five milliliters of saturated NaCl solution was then added, and the mixture was centrifuged at 4000× g rpm for 10 min at 4 °C. When the iso-octane layer separated from the aqueous lower phase, the iso-octane layer was transferred to a glass vial. After iso-octane was evaporated under a stream of nitrogen gas, 250 μL of hexane was added, and the sample was injected into the GC system. The fatty acids were measured using a gas chromatograph (GC-17A, Shimadzu, Kyoto, Japan) coupled with a flame ionization detector (FID) equipped with a capillary column (CP Sil 88 100 m × 0.25 mm i.d., 0.20 μm film thickness; Agilent technologies, Madrid, Spain). The carrier gas was helium (flow 1.2 mL/min) with a split injection of 1:1. The temperature profiles were as follows: initial temperature, 45 °C for 4 min; heating first rate, 13 °C/min until 175 °C (27 min isolation) and a second rate, 4 °C until 215 °C (30 min isolation); injector temperature, 250 °C; and detector temperature, 260 °C. The fatty acids were identified by comparing the retention times with those of Fame Standard Mix (CRM47885, Sigma Aldrich, St. Louis, MA, USA).

2.5. Statistical Analyses

The model for analyzing FA profile included the effects of moment (two levels: at mating and at 72 hpm), line, random effect of female and weight of female as covariate.

All analyses were performed using Bayesian methodology [19]. Bounded uniform priors were used for all effects with the exception of the female effect, which was considered normally distributed, with mean 0 and variance $I\sigma_f^2$, where I is a unity matrix, and σ_f^2 is the variance of the female effect. Female and residual effects were considered to be independent. Residuals were a priori normally distributed, with mean 0 and variance $I\sigma_e^2$. The priors for the variances were also bounded uniform. Features of the marginal posterior distributions for all unknowns were estimated using Gibbs sampling. The Rabbit program developed by the Institute for Animal Science and Technology (Valencia, Spain) was used for all procedures. We used a chain of 60,000 samples, with a burn-in period of 10,000. Only one out of every 10 samples was saved for inferences. Convergence was tested using Geweke's Z criterion and Monte Carlo sampling errors were computed using time-series procedures.

Residual correlations between FA at mating and at 72 hpm after correction by line and weight of female were estimated. A principal component analysis was carried out. These analyses were performed using the SAS statistical package.

3. Results

Table 1 shows the features of the marginal posterior distributions of the difference between FA profile measured at mating and at 72 hpm. A total of 31 different FA were identified in blood plasma, and three FA were not detected (octanoic, cis-10 pentadecanoic and adrenic acid). The highest FA concentrations at mating and at 72 hpm were palmitic (425.01 ng/mL), linoleic (408.00 ng/mL), cis-9 oleic (385.30 ng/mL) and stearic (238.08 ng/mL), while palmitoleic (29.31 ng/mL), heptadecanoic (21.26 ng/mL), α-linolenic (13.77 ng/mL), trans-9 elaidic (13.41 ng/mL), myristic (12.91 ng/mL) and pentadecanoic (12.28 ng/mL) were in lower concentration.

Short and medium chain FA were shown in very low concentrations and similar concentrations were found at mating and at 72 hpm. Only lauric was higher at mating than at 72 hpm (D = + 0.39 ng/mL; $p > 0.80$).

Most long chain fatty acids were higher at 72 hpm than at mating. Myristic, pentadecanoic, arachidic, heneicosanoic, behenic and tricosylic increased their concentration in this period. Conversely, heptadecanoic and lignoceric were higher at mating than at 72 hpm. Similar concentrations were found for palmitic, stearic and SFA between both moments.

Monounsaturated FA (MUFA) were higher at mating than 72 hpm (D = + 55.75 ng/mL; $p > 0.80$). The increases ranged from 8% to 54% for myristoleic, trans-9 elaidic, cis-9 oleic, palmitoleic and cis-10 heptadecanoic.

Polyunsaturated FA (PUFA) showed a relevant lower concentration at mating than at 72 hpm (D = −30.67 ng/mL, $p > 0.90$). Specifically, these PUFA were linolelaidic (D = −1.26 ng/mL), linoleaic (D = −21.70 ng/mL), α-Linolenic (D = −4.61 ng/mL), cis-11,14 eicosadienoic (D = −1.41 ng/mL) and arachidonic (D = −1.09 ng/mL). Only γ-Linolenic was higher at mating, while similar concentrations were found for cis-11,14,17 eicosatrienoic and cis-4,7,10,13,16,19 docohexaenoic.

Table 1. Means of the estimated marginal posterior distribution of the differences between mating and at 72 hpm for fatty acid profile (D).

Pathway	Biochemical Name	Means (ng/mL)	D (ng/mL)
Short chain fatty acids	Butyric (C4:0)	0.42	+0.08
	Hexanoic (C6:0)	0.02	+0.00
Medium chain fatty acids	Octanoic (C8:0)	ND	
	Decanoic (C10:0)	0.30	+0.06
	Undecanoic (C11:0)	0.45	+0.09
	Lauric (C12:0)	1.84	+0.39
Long chain fatty acids	Myristic (C14:0)	12.91	−5.57
	Pentadecanoic (C15:0)	12.28	−1.04
	Palmitic (C16:0)	425.01	+4.80
	Heptadecanoic (C17:0)	21.26	+2.14
	Stearic (C18:0)	238.08	−5.19
	Arachidic (C20:0)	3.50	−0.76
	Heneicosanoic (C21:0)	2.12	−0.98
	Behenic (C22:0)	5.08	−2.05
	Tricosylic (C23:0)	8.09	−5.55
	Lignoceric (C24:0)	0.85	+0.61
	ΣSFA	733.96	−14.61
Monounsaturated fatty acids	Myristoleic (C14:1c9)	2.63	+0.80
	Cis-10 pentadecenoic (C15:1c10)	ND	
	Trans-9 elaidic (C18:1t9)	13.41	+5.75
	Cis-9 oleic (C18:1c9)	385.30	+32.13
	Palmitoleic (C16:1c9)	29.31	+8.75
	Cis-10 heptadecenoic (C17:1c10)	5.91	+2.22
	Cis-11 eicosenoic (C20:1c11)	4.04	+0.41
	Erucic (C22:1c13)	0.25	+0.07
	Nervonic (C24:1c15)	2.17	+0.42
	ΣMUFA	445.07	+55.75
Polyunsaturated fatty acids	Linolelaidic (C18:2t9t12)	3.14	−1.26
	Linoleic (C18:2c9c12)	408.00	−21.70
	γ-Linolenic (C18:3c6c9c12)	2.23	+0.81
	α-Linolenic (C18:3c9c12c15)	13.77	−4.61
	Cis-11,14 eicosadienoic (C20:2)	6.65	−1.41
	Cis-11,14,17 eicosatrienoic (C20:3)	4.14	+0.84
	Arachidonic (C20:4c5c8c11c14)	1.61	−1.09
	Cis-4,7,10,13,16,19 docosahexaenoic (C22:6c4c7c10c13c16c19)	0.77	+0.02
	Adrenic (C22:4c7c10c13c16)	ND	
	ΣPUFA	440.76	−30.67

ND: not detected; SFA: saturated fatty acids; MUFA: monounsaturated fatty acids; PUFA: polyunsaturated fatty acids; Red: probability of the difference between mating and 72 hpm being > 0 when D > 0 or being < 0 when D < 0 is higher than 0.9; Green: probability of the difference between mating and 72 hpm being > 0 when D > 0 or being < 0 when D < 0 ranges from 0.80 to 0.90.

The correlations between FA at mating and at 72 hpm and reproductive traits are presented in Table 2. Moreover, a principal component analysis was performed (Figure 1). The first three principal components explained 83% of total variation (53%, 18% and 12%, respectively). The predominant variables defining the first principal component were SFA,

MUFA and PUFA both at mating and at 72 hpm, except for MUFA measured at mating. They were far from the origin and close to the axis. Thus, high and positive correlations were found between them, ranging from 0.492 to 0.899. These correlations were significant, except for MUFA at mating and PUFA at 72 hpm (Table 2).

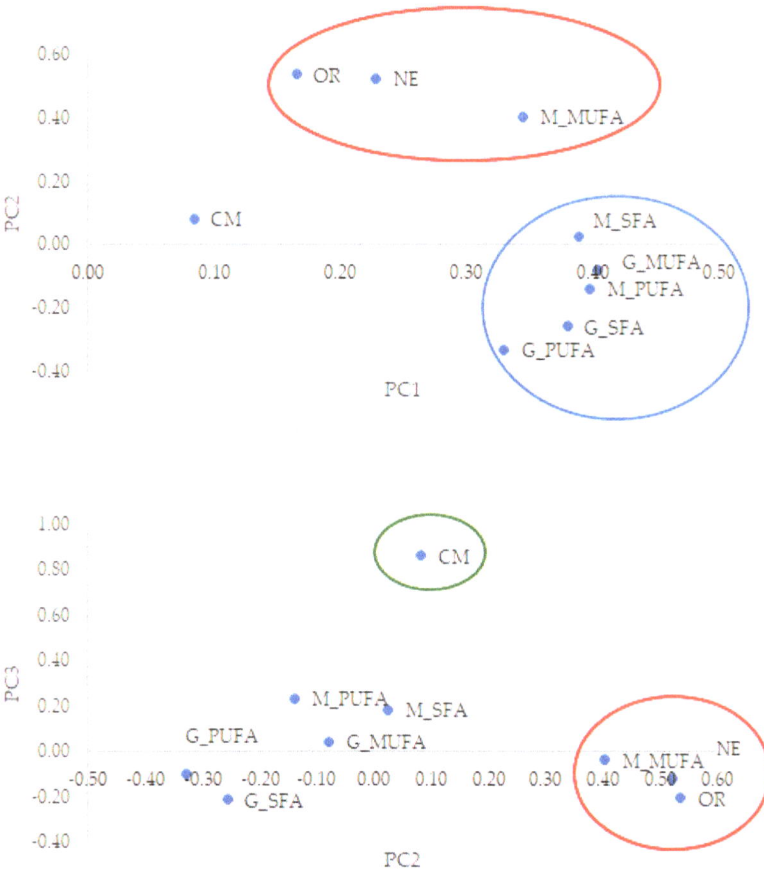

Figure 1. Projection of the traits in the plane defined by the 1st and 2nd principal component (PC) and the 2nd and 3rd PC. Variables of the 1st, 2nd and 3rd PC are surrounded by a blue, red and green circle, respectively. M_SFA: saturated fatty acid (FA) at mating; M_MUFA: monounsaturated FA at mating; M_PUFA: polyunsaturated FA at mating; G_SFA: saturated FA at 72 hpm; G_MUFA: monounsaturated FA at 72 hpm; G_PUFA: polyunsaturated FA at 72 hpm; OR: ovulation rate; NE: normal embryos; CM: compacted morulae.

Table 2. Correlations for saturated fatty acid (SFA), monounsaturated fatty acid (MUFA), polyunsaturated fatty acid measured at mating and at 72 h post-mating, ovulation rate (OR), normal embryos (NE) and compacted morulae (CM).

		Mating		72 hpc					
		MUFA	**PUFA**	**SFA**	**MUFA**	**PUFA**	**OR**	**NE**	**MC**
Mating	SFA	0.675 *	0.899 *	0.813 *	0.808 *	0.543 *	0.267	0.268	0.465 +
	MUFA		0.605 *	0.492 +	0.721 *	0.342	0.643 *	0.781 *	0.169
	PUFA			0.807 *	0.861 *	0.740 *	0.123	0.341	0.328
72 hpc	SFA				0.826 *	0.773 *	0.092	0.267	−0.123
	MUFA					0.727 *	0.275	0.334	0.183
	PUFA						0.108	0.092	0.050
	OR							0.630 *	0.027
	NE								0.047

* Significant at level 0.05; + significant at level 0.10.

Ovulation rate, normal embryos and MUFA at mating were located near the second principal component and close to each other (Figure 1). Thus, correlation between ovulation rate and normal embryos was 0.630. Correlation was 0.643 between MUFA at mating and ovulation rate, and 0.781 between MUFA and normal embryos, respectively ($p < 0.05$).

Compacted morulae are the predominant variable defining the third principal component. Compacted morulae were slightly correlated with SFA at mating (0.465), and uncorrelated with the other FA.

4. Discussion

The results of this study provide a detailed FA profile at mating, when ovulation takes place and in early gestation. Palmitic, linoleic, oleic and stearic acid are the highest FA concentrations at mating and 72 hpm. It has been reported that oleic, palmitic and linoleic are capable of supporting growth of one-cell rabbit embryos to viable morulae [10], as these FA may serve as a storage pool of metabolic precursors presented in oviductal and uterine fluids and embryos [20].

As expected, arachidonic acid was found in lower concentration at 72 hpm. Arachidonic acid is a crucial precursor of prostaglandins [8], which are found in low concentration in the first days of pregnancy due to their luteolytic action [21,22].

The absence or low concentration of short and medium FA, docohexaenoic and adrenic acid agrees with the findings of other studies [23].

The FA profile is different at mating than at 72 hpm. MUFA concentration is higher at mating, while PUFA is in higher concentration at 72 hpm. MUFA are mainly used by follicle components as primary energy sources (see review in [24]), whereas PUFA are nutritionally essential for embryo development [20]. High quality oocytes exhibit high levels of oleic acid (in cows, [25]). The PUFA concentrations, especially linoleic acid, are high not only in plasma but also in oviductal fluid and embryos [20]. PUFA in general, and linolenic acid in particular, support essential development processes in mammalian embryos [26]. Linoleic acid acts as a precursor for eicosanoids and regulates the processes of endocytosis or exocytosis, ion channel modulation, DNA polymerase inhibition and gene expression [27]. Moreover, linoleic acid stimulates protein kinase C, which is necessary for cell differentiation and growth [28,29]. In mammals, the absence of enzymes to introduce double bonds at carbon atoms beyond C-9 in FA chains determines linoleic acid as essential FA because they are not able to synthesize it. Thus, it must be included in the diet [30]. In this sense, rabbit females can increase deposition of PUFA in the periovarian adipose tissue after a long-term dietary supplementation with fish oil. This deposition could favour the PUFA accessibility to their ovarian structures as corpora lutea, whose activity, measured by the progesterone production, is increased during embryo preimplantation period [31].

To the best of our knowledge, this is the first time that the correlation between FA profile at mating and 72 hpm and ovulation rate and normal embryo and development have been studied. Ovulation rate and number of normal embryos are positively correlated with MUFA at mating. Moreover, the correlation between the percentage of compact morulae and SFA is slightly positive. Overlap of different embryo developmental stages is commonly observed [32] and could be related to the duration of ovulation or the oviductal and uterine fluid compositions [33]. Early morulae and compacted morulae can be found at 72 hpm. Palmitic acid, as the main SFA, was identified in abundant concentrations in more developed embryos [27] and is reported to be essential for FA elongation and desaturation in embryo development [34]. Moreover, incubation with palmitic acid showed that embryos can oxidize palmitic acid even at the single-cell stage, with subsequent increases, particularly from four-cell embryos onwards [20].

The results obtained in fatty acid composition could add useful knowledge of ovulation and early embryo metabolism. These data may also be useful practically in helping to develop non-destructive tests of oocyte and embryo quality and in improving culture media, cryopreservation and the success of IVF treatment.

5. Conclusions

The fatty acid profile of plasma varies depending on the reproductive cycle of the rabbit female, adapting to energy requirements at mating and in early gestation. In addition, correlations are found between fatty acids and ovulation rate and embryonic development and quality.

Author Contributions: Conceptualization, M.d.l.L.G., E.A. and M.-J.A.; methodology, M.d.l.L.G. and M.-J.A.; formal analysis, I.H., M.d.l.L.G.; data curation, I.H., M.d.l.L.G.; writing—original draft preparation, M.d.l.L.G.; writing—review and editing, I.H., A.-K.H., E.A., M.-J.A.; funding acquisition, M.d.l.L.G. All authors have read and agreed to the published version of the manuscript.

Funding: This research was funded by Valencia Regional Government, grant number AICO/2019/169. María de la Luz García was supported by postdoctoral grant, number BEST/2019, from Valencia Regional Government.

Institutional Review Board Statement: All experimental procedures were approved by the Miguel Hernández University of Elche Research Ethics Committee, according to Council Directives 98/58/EC and 2010/63/EU (reference number 2019/VSC/PEA/0017).

Informed Consent Statement: Not applicable.

Data Availability Statement: Data are available upon request to the corresponding author.

Acknowledgments: We thank our technician, Raquel Muelas, for the technical support.

Conflicts of Interest: The authors declare no conflict of interest.

References

1. Groebner, A.E.; Rubio-Aliaga, I.; Schulke, K.; Reichenbach, H.D.; Daniel, H.; Wolf, E.; Meyer, H.H.D.; Ulbrich, S.E. Increase of essential amino acids in the bovine uterine lumen during preimplantation development. *Reproduction* **2011**, *141*, 685–695. [CrossRef] [PubMed]
2. Gao, H.; Wu, G.; Spencer, T.E.; Johnson, G.A.; Li, X.; Bazer, F.W. Select nutrients in the ovine uterine lumen. I. Amino acids, glucose, and ions in uterine lumenal flushings of cyclic and pregnant ewes. *Biol. Reprod.* **2009**, *80*, 86–93. [CrossRef] [PubMed]
3. Harris, S.E.; Gopichandran, N.; Picton, H.M.; Leese, H.J.; Orsi, N. Nutrient concentrations in murine follicular fluid and the female reproductive tract. *Theriogenology* **2005**, *64*, 992–1006. [CrossRef] [PubMed]
4. Drews, B.; Milojevic, V.; Giller, K.; Ulbrich, S. Fatty acid profile of blood plasma and oviduct and uterine fluid during early and late luteal phase in the horse. *Theriogenology* **2018**, *114*, 258–265. [CrossRef] [PubMed]
5. Iritani, A.; Sato, E.; Nishikawa, Y. Secretion rates and chemical composition of oviduct and uterine fluids in sows. *J. Anim. Sci.* **1974**, *39*, 582–588. [CrossRef] [PubMed]
6. Beier, H.M. Oviductal and uterine fluids. *J. Reprod. Fertil.* **1974**, *37*, 221–237. [CrossRef]
7. Oliphant, G.; Reynolds, A.B.; Smith, P.F.; Ross, P.R.; Marta, J.S. Immunocytochemical localization and determination of hormone-induced synthesis of the sulfated oviductal glycoproteins. *Biol. Reprod.* **1984**, *31*, 165–174. [CrossRef]
8. Aguilar, J.; Reyley, M. The uterine tubal fluid: Secretion, composition and biological effects. *Anim. Reprod.* **2005**, *2*, 91–105.
9. Sturmey, R.; Reis, A.; Leese, H.; McEvoy, T. Role of fatty acids in energy provision during oocyte maturation and early embryo development. *Reprod. Domest. Anim.* **2009**, *44*, 50–58. [CrossRef]

10. Kane, M. Fatty acids as energy sources for culture of one-cell rabbit ova to viable morulae. *Biol. Reprod.* **1979**, *20*, 323–332. [CrossRef]
11. Stubbs, C.D.; Smith, A.D. The modification of mammalian membrane polyunsaturated fatty acid composition in relation to membrane fluidity and function. *Biochim. Biophys. Acta BBA Rev. Biomembr.* **1984**, *779*, 89–137. [CrossRef]
12. Warzych, E.; Lipinska, P. Energy metabolism of follicular environment during oocyte growth and maturation. *J. Reprod. Dev.* **2020**, *66*, 1–7. [CrossRef] [PubMed]
13. Waterman, R.A.; Wall, R.J. Lipid interactions with in vitro development of mammalian zygotes. *Gamete Res.* **1988**, *21*, 243–254. [CrossRef] [PubMed]
14. Ashworth, C.J.; Toma, L.M.; Hunter, M.G. Nutritional effects on oocyte and embryo development in mammals: Implications for reproductive efficiency and environmental sustainability. *Philos. Trans. R. Soc. B Biol. Sci.* **2009**, *364*, 3351–3361. [CrossRef]
15. Rodríguez, M.; García-García, R.; Arias-Álvarez, M.; Millán, P.; Febrel, N.; Formoso-Rafferty, N.; López-Tello, J.; Lorenzo, P.; Rebollar, P. Improvements in the conception rate, milk composition and embryo quality of rabbit does after dietary enrichment with n-3 polyunsaturated fatty acids. *Animal* **2018**, *12*, 2080–2088. [CrossRef]
16. García, M.-L.; Argente, M.-J. The genetic improvement in meat rabbits. *Lagomorpha Charact. Work. Title* **2020**, *5*, 1–18. [CrossRef]
17. Argente, M.J.; García, M.L.; Zbyňovská, K.; Petruška, P.; Capcarova, M.; Blasco, A. Correlated response to selection for litter size environmental variability in rabbits' resilience. *Animal* **2019**, *13*, 2348–2355. [CrossRef]
18. Beloumi, D.; Blasco, A.; Muelas, R.; Santacreu, M.A.; García, M.D.L.L.; Argente, M.-J. Inflammatory correlated response in two lines of rabbit selected divergently for litter size environmental variability. *Animals* **2020**, *10*, 1540. [CrossRef] [PubMed]
19. Blasco, A. *Bayesian Data Analysis for Animal Scientists*; Springer: Cham, Switzerland, 2017; ISBN 978-3-319-54273-7.
20. Khandoker, M.A.M.Y.; Tsujii, H.; Karasawa, D. A kinetics study of fatty acid composition of embryos, oviductal and uterine fluids in the rabbit. *Asian-Australas. J. Anim. Sci.* **1998**, *11*, 60–64. [CrossRef]
21. Kehl, S.J.; Carlson, J.C. Assessment of the luteolytic potency of various prostaglandins in the pseudopregnant rabbit. *Reproduction* **1981**, *62*, 117–122. [CrossRef]
22. García, M.L. Embryo manipulation techniques in the rabbit. In *New Insights into Theriogenology*, 1st ed.; Payan-Carreira, R., Ed.; IntechOpen: London, UK, 2018; pp. 113–133. [CrossRef]
23. Schindler, M.; Pendzialek, S.M.; Grybel, K.; Seeling, T.; Santos, A.N. Metabolic profiling in blastocoel fluid and blood plasma of diabetic rabbits. *Int. J. Mol. Sci.* **2020**, *21*, 919. [CrossRef] [PubMed]
24. Zarezadeh, R.; Mehdizadeh, A.; Leroy, J.L.; Nouri, M.; Fayezi, S.; Darabi, M. Action mechanisms of n-3 polyunsaturated fatty acids on the oocyte maturation and developmental competence: Potential advantages and disadvantages. *J. Cell. Physiol.* **2019**, *234*, 1016–1029. [CrossRef] [PubMed]
25. Kim, J.Y.; Kinoshita, M.; Ohnishi, M.; Fukui, Y. Lipid and fatty acid analysis of fresh and frozen-thawed immature and in vitro matured bovine oocytes. *Reproduction* **2001**, *122*, 131–138. [CrossRef] [PubMed]
26. Khandoker, M.A.M.Y.; Tsujii, H.; Karasawa, D. Fatty acid composition of blood serum, oocytes, embryos and reproductive tract fluids of rat and comparison with BSA. *Anim. Sci. Technol.* **1997**, *68*, 1070–1074. [CrossRef]
27. Tsujii, H.; Matsuoka, Y.; Obata, R.; Hossain, M.S.; Takagi, Y. Fatty acid composition of lipids in day 7—13 blastocysts, serum and uterine fluid of rabbits. *Reprod. Med. Biol.* **2009**, *8*, 107–112. [CrossRef]
28. Murakami, K.; Chan, S.Y.; Routtenberg, A. Protein kinase C activation by cis-fatty acid in the absence of Ca2+ and phospholipids. *J. Biol. Chem.* **1986**, *261*, 15424–15429. [CrossRef]
29. Nishizuka, Y. The molecular heterogeneity of protein kinase C and its implications for cellular regulation. *Nature* **1988**, *334*, 661–665. [CrossRef]
30. Rodríguez, M.; Rebollar, P.G.; Mattioli, S.; Castellini, C. n-3 PUFA sources (precursor/products): A review of current knowledge on rabbit. *Animals* **2019**, *9*, 806. [CrossRef]
31. Rebollar, P.; García-García, R.; Arias-Álvarez, M.; Millán, P.; Rey, A.; Rodríguez, M.; Formoso-Rafferty, N.; de la Riva, S.; Masdeu, M.; Lorenzo, P. Reproductive long-term effects, endocrine response and fatty acid profile of rabbit does fed diets supplemented with n-3 fatty acids. *Anim. Reprod. Sci.* **2014**, *146*, 202–209. [CrossRef]
32. García, M.; Blasco, A.; Argente, M. Embryologic changes in rabbit lines selected for litter size variability. *Theriogenology* **2016**, *86*, 1247–1250. [CrossRef]
33. Hunter, M.; Robinson, R.; Mann, G.; Webb, R. Endocrine and paracrine control of follicular development and ovulation rate in farm species. *Anim. Reprod. Sci.* **2004**, *82*, 461–477. [CrossRef] [PubMed]
34. Jeffcoat, R. The biosynthesis of unsaturated fatty acids and its control in mammalian liver. *Essays Biochem.* **1979**, *15*, 1–36. [PubMed]

Article

Cornelian Cherry Pulp Has Beneficial Impact on Dyslipidemia and Reduced Bone Quality in Zucker Diabetic Fatty Rats

Radoslav Omelka [1,*], Jana Blahova [1], Veronika Kovacova [2], Martina Babikova [1], Vladimira Mondockova [1], Anna Kalafova [3], Marcela Capcarova [3] and Monika Martiniakova [2,*]

1 Department of Botany and Genetics, Faculty of Natural Sciences, Constantine the Philosopher University in Nitra, 949 74 Nitra, Slovakia; jana.blahova@ukf.sk (J.B.); martina.babikova@ukf.sk (M.B.); vmondockova@ukf.sk (V.M.)

2 Department of Zoology and Anthropology, Faculty of Natural Sciences, Constantine the Philosopher University in Nitra, 949 74 Nitra, Slovakia; vkovacova@ukf.sk

3 Department of Animal Physiology, Faculty of Biotechnology and Food Sciences, Slovak University of Agriculture in Nitra, 949 76 Nitra, Slovakia; anna.kalafova@uniag.sk (A.K.); marcela.capcarova@uniag.sk (M.C.)

* Correspondence: romelka@ukf.sk (R.O.); mmartiniakova@ukf.sk (M.M.); Tel.: +421-376-408-737 (R.O.)

Received: 6 November 2020; Accepted: 16 December 2020; Published: 19 December 2020

Simple Summary: Fruits of Cornelian cherry (*Cornus mas* L.) are often used as an antidiabetic supplement mainly due to their hypoglycemic properties. A very important aspect and secondary complication of type 2 diabetes mellitus (T2DM) represents diabetic bone disease. In our study, the impacts of *Cornus mas* pulp on lipid profile and bone quality parameters were evaluated in Zucker diabetic fatty (ZDF) rats as a well-matched T2DM animal model. We demonstrated, for the first time, that Cornelian cherry pulp could be used as a potential therapeutic agent to alleviate T2DM-reduced bone quality and impaired bone health. Moreover, the hypolipidemic effect of this fruit was also confirmed in our study.

Abstract: Cornelian cherry (*Cornus mas* L.) is a medicinal plant with a range of biological features. It is often used as a nutritional supplement in the treatment of diabetes mellitus. Our study was aimed to first investigate the effects of Cornelian cherry pulp on bone quality parameters in Zucker diabetic fatty (ZDF) rats. Moreover, lipid-lowering properties of this fruit were also evaluated. Adult rats (n = 28) were assigned into four groups of seven individuals each: L group (non-diabetic lean rats), C group (diabetic obese rats), and E1 and E2 groups (diabetic obese rats receiving 500 and 1000 mg/kg body weight of Cornelian cherry pulp, respectively, for 10 weeks). Significantly lower levels of triglyceride, total cholesterol and alkaline phosphatase activity were determined in the E2 group versus the C group. A higher dose of *Cornus mas* also had a beneficial impact on femoral weight, cortical bone thickness, relative volume of trabecular bone and trabecular thickness. We observed elevated density of Haversian systems and accelerated periosteal bone apposition in both treated groups (E1 and E2). Our results clearly demonstrate that Cornelian cherry pulp has a favorable effect on lipid disorder and impaired bone quality consistent with type 2 diabetes mellitus in a suitable animal model.

Keywords: Cornelian cherry; diet; diabetes mellitus; bone quality; biochemistry; microcomputed tomography; histomorphometry; Zucker diabetic fatty (ZDF) rats

---

1. Introduction

Type 2 diabetes mellitus (T2DM) is a chronic endocrine disorder with increasing prevalence worldwide. It is characterized by insulin deficiency or ineffective insulin production by the pancreas, resulting in persistent hyperglycemia [1,2]. T2DM negatively affects glucose transport into the liver, muscle cells and fat cells [3]. These conditions also adversely influence the skeletal system via diabetic bone disease. Harmful impacts are associated with bone strength, bone remodeling and stem cell differentiation and lead to varied bone mineral density (BMD) and changed bone structure [4]. Hyperglycemia can suppress BMD through altered osteoblast gene expression, reduced osteoblast function and number, higher oxidative stress and adipogenesis, downregulation of vitamin D receptors and increased production of glycation end-products (AGEs) [5]. AGEs inhibit bone remodeling and indirectly attenuate osteoblast activity and apoptosis of osteocytes.

In general, bone remodeling is a major mechanism for maintaining a healthy skeleton in adults, and bone remodeling dysfunction may contribute to bone loss and/or impaired bone quality [6,7]. Bone turnover markers, which reflect the bone resorption and formation processes, are usually reduced in individuals with T2DM (mainly osteocalcin as bone formation marker and C-terminal telopeptide of type I collagen as bone resorption marker) [8]; however, not all studies yielded consistent findings. Some reports found no differences in the bone turnover markers or even their increased levels in T2DM [9]. The differences between various studies may be explained by changes in metabolic status, diabetes duration and treatment at the time of the measurement [10]. Despite conflicting findings, the evidence seems to point towards a suppression of bone remodeling in diabetes and also in the pre-diabetic state [8].

The association between T2DM and BMD is also controversial. Most studies showed decreased or unchanged BMD; however, some others revealed increased BMD [11–13]. Another diabetic complication contributing to detrimental bone effects is microangiopathy. Impaired vascularity and T2DM-enhanced inflammation can cause improper distribution of nutrients, oxygen, hormones or growth factors to the bone cells. It can lead to impaired bone healing [14,15], loss of bone strength and increased bone fragility [16].

Zucker diabetic fatty (ZDF) rats are considered to be an appropriate animal model of type 2 diabetes mellitus, with the spontaneous mutation on chromosome 5, which encodes leptin receptors. This mutation causes obesity, glucose intolerance, hyperglycemia, hyperlipidemia and hyperinsulinemia [17,18]. ZDF rats show lower trabecular bone volume and trabecular number, higher trabecular separation and cortical porosity, improper vascularization [19], and decreased osteoblast differentiation and mineral capacity [15].

Active substances extracted from plants can be useful to prevent or support the therapy of various pathological conditions, including diabetes mellitus [20]. Generally, plant products have high nutritional value and possess beneficial impacts for health. Cornelian cherry (*Cornus mas* L.) is a flowering plant that belongs to Cornaceae family, with occurrence in Southern and Central Europe, as well as Western Asia. Cornelian cherry fruit is used for food, syrup and jam production, and in the fermented form, it is used to make alcoholic beverage [21,22]. It is rich in proteins, vitamins, antioxidants, mineral matters [23–25], polyphenols (e.g., anthocyanins, flavonoids, catechins and tannins) and other essential supplements [22]. All of these compounds have antioxidant, antimicrobial, hypolipidemic, hyperinsulinemic [26], anti-inflammatory and anticancer properties and also offer positive support to the nervous and cardiovascular systems [25].

The main aim of this in vivo study was to first analyze the effect of Cornelian cherry pulp on bone quality parameters in ZDF rats, since the impact of this fruit as a potential therapeutic agent to alleviate T2DM-reduced bone quality has not been investigated yet. Moreover, lipid-lowering properties of *Cornus mas* were also evaluated.

2. Materials and Methods

2.1. Sample Preparation

The experiment was authorized under the number 2288/16-221 by the Ethical Committee and the State Veterinary and Food Administration of the Slovak Republic. Cornelian cherries were supplied by the Institute of Biodiversity Conservation and Biosafety of the Slovak University of Agriculture in Nitra (Slovakia). The fresh ripe fruits were washed, isolated from the stones, crushed and stored at −20 °C. Aliquots of fruits, homogenized in distilled water, were used in the experiment.

2.2. Animals

Adult ZDF rats (n = 28) were obtained from the Institute of Experimental Pharmacology and Toxicology, Slovak Academy of Sciences. The experiment was performed at the Slovak University of Agriculture in Nitra. The rats were housed in pairs, in cages of 1500 U Eurostandard Type IV S (Tecniplast, Buguggiate, Italy), with 12 h light/12 h dark cycle, at a room temperature of 22 ± 2 °C, with free access to food and drinking water. The complete rat feed mixture KKZ-P/M (register number 6147, Dobra Voda, Slovakia; composition of the mixture is given in the study by Capcarova et al. [27]) was used for feeding. The animals were divided into 4 groups: L group (n = 7) consisted of non-diabetic lean rats and served as a negative control; C group (n = 7) included diabetic obese rats which were considered as a positive control; and E1 (n = 7) and E2 (n = 7) groups consisted of diabetic obese rats receiving 500 and 1000 mg/kg body weight (bw) of Cornelian cherry pulp, respectively, for 10 weeks. Groups E1 and E2 received the exact dose of *Cornus mas* directly into the stomach, every day, using sterile oral rodent gavage (Instech, Plymouth, MA, USA), whereas groups L and C received distilled water by the same way. The doses of *Cornus mas* were selected based on the study by Capcarova et al. [27]. The animals were sacrificed by intraperitoneal injection of xylazine/zoletil cocktail and all samples were collected after deep anesthesia.

2.3. Biochemistry

Blood glucose levels were determined with a FreeStyle Optium Neo Glucose and Ketone monitoring system (Abbott Diabetes Care Ltd., Maidenhead, UK), using test strips (FreeStyle, Abbott Diabetes Care Ltd., Maidenhead, UK). The levels of total cholesterol, LDL cholesterol, HDL cholesterol, triglyceride and alkaline phosphatase (ALP) were measured by a Biolis 24i Premium analyzer (Tokyo Boeki MediSys Inc., Tokyo, Japan), with commercially available kits (Randox Laboratories Ltd., Crumlin, UK).

2.4. Macroscopic Measurements

Prior to 3D (microcomputed tomography) and 2D (histomorphometry) imaging, the femoral bones (n = 56) were weighed, and their lengths were measured. Furthermore, the total body weight of all animals was also determined.

2.5. Microcomputed Tomography

Microcomputed tomography (microCT, µCT 50, Scanco Medical, Brüttisellen, Switzerland) was used to designate selected quantitative 3D parameters of cortical bone (relative bone volume, relative bone volume with marrow cavity, BMD, cortical bone thickness and bone surface) and trabecular bone sections (relative bone volume, trabecular number, trabecular thickness, BMD and bone surface). The measurement parameters and scanning regions of interest were the same as those in the study by Omelka et al. [28].

2.6. Histomorphometry

The femurs were cut at the diaphysis, and the sections were fixed, dehydrated, degreased and embedded according to the method defined by Martiniaková et al. [29]. A sawing microtome (Leitz 1600,

Leica, Wetzlar, Germany) was used to prepare thin sections, as previously described [30]. The qualitative 2D characteristics were recorded according to Enlow and Brown [31] and de Ricqlés et al. [32]. Measured parameters included area of primary osteons' vascular canals, area of Haversian canals and area of Haversian systems. They were determined by Motic Images Plus 2.0 ML software (Motic China Group Co., Ltd., Nanjing, China).

2.7. Data Analysis

Statistical analysis was conducted by using SPSS Statistics 26.0 software (IBM, New York, NY, USA). The data were expressed as mean ± standard deviation. Significant differences in parameters investigated were detected by ANOVA, with post hoc (Games-Howell and/or Tukey's) tests. A p value less than 0.05 was considered to be statistically significant.

3. Results

3.1. Biochemistry

Significantly lower levels of triglyceride were determined in both the E1 and E2 groups versus the diabetic control one (C group). In addition, decreased total cholesterol level and ALP activity were recorded in the E2 group, in comparison with the C group. Nevertheless, the blood glucose level did not change significantly in the treated groups against the C group. Significant changes in all parameters investigated were observed between lean rats (L group) and those from the C group, as well as between treated groups (E1 and E2) versus L group. The results are summarized in Figure 1A–F.

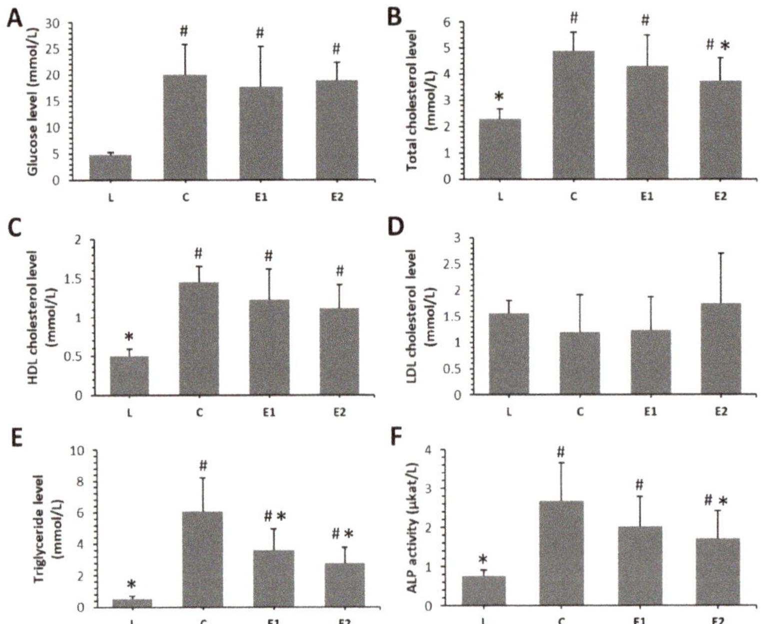

Figure 1. Biochemical parameters examined (**A–F**) in Zucker diabetic fatty (ZDF) rats from the L group (non-diabetic lean rats), C group (diabetic obese rats), and E1 and E2 groups (diabetic obese rats receiving 500 and 1000 mg/kg body weight of Cornelian cherry pulp, respectively, for 10 weeks). HDL cholesterol—high-density lipoprotein cholesterol; LDL cholesterol—low-density lipoprotein cholesterol; ALP—alkaline phosphatase. * Significant differences versus C group ($p < 0.05$). # Significant differences versus L group ($p < 0.05$).

3.2. Macroscopic Measurements

Our results showed non-significant impact of *Cornus mas* pulp on the total body weight of ZDF rats. On the contrary, significantly decreased values for femoral weight and length were observed in C and E1 groups, when compared to the L group. No significant changes in femoral weight were determined between the L and E2 groups, suggesting a positive impact of a higher dose of Cornelian cherry on this parameter. Significantly lower values for femoral length were recorded in the treated groups versus the lean control one. The results are shown in Figure 2A–C.

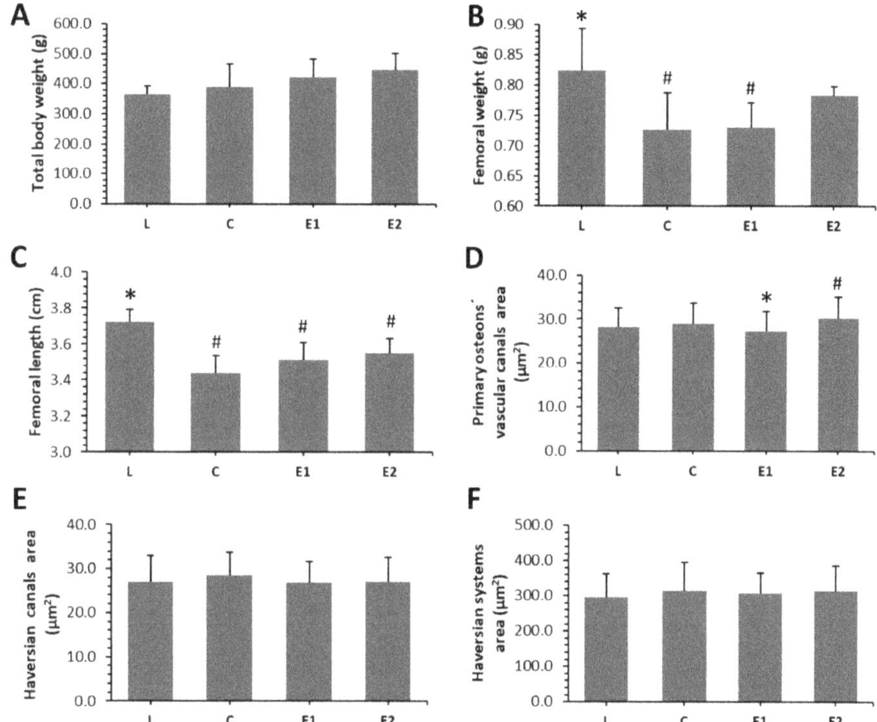

Figure 2. Macroscopical (**A**–**C**) and histomorphometrical (**D**–**F**) parameters examined in ZDF rats from L, C, E1 and E2 groups * Significant differences versus C group ($p < 0.05$). # Significant differences versus L group ($p < 0.05$).

3.3. Microcomputed Tomography

Microcomputed tomography revealed that treatment with Cornelian cherry had an insignificant effect on the relative volume of cortical bone, BMD and cortical bone surface. On the other hand, significantly decreased cortical bone thickness was determined in the C and E1 groups, as compared to the L group. No significant changes between the L and E2 groups point to a beneficial impact of a higher dose of *Cornus mas* on cortical bone thickness. The results are summarized in Figure 3A–E.

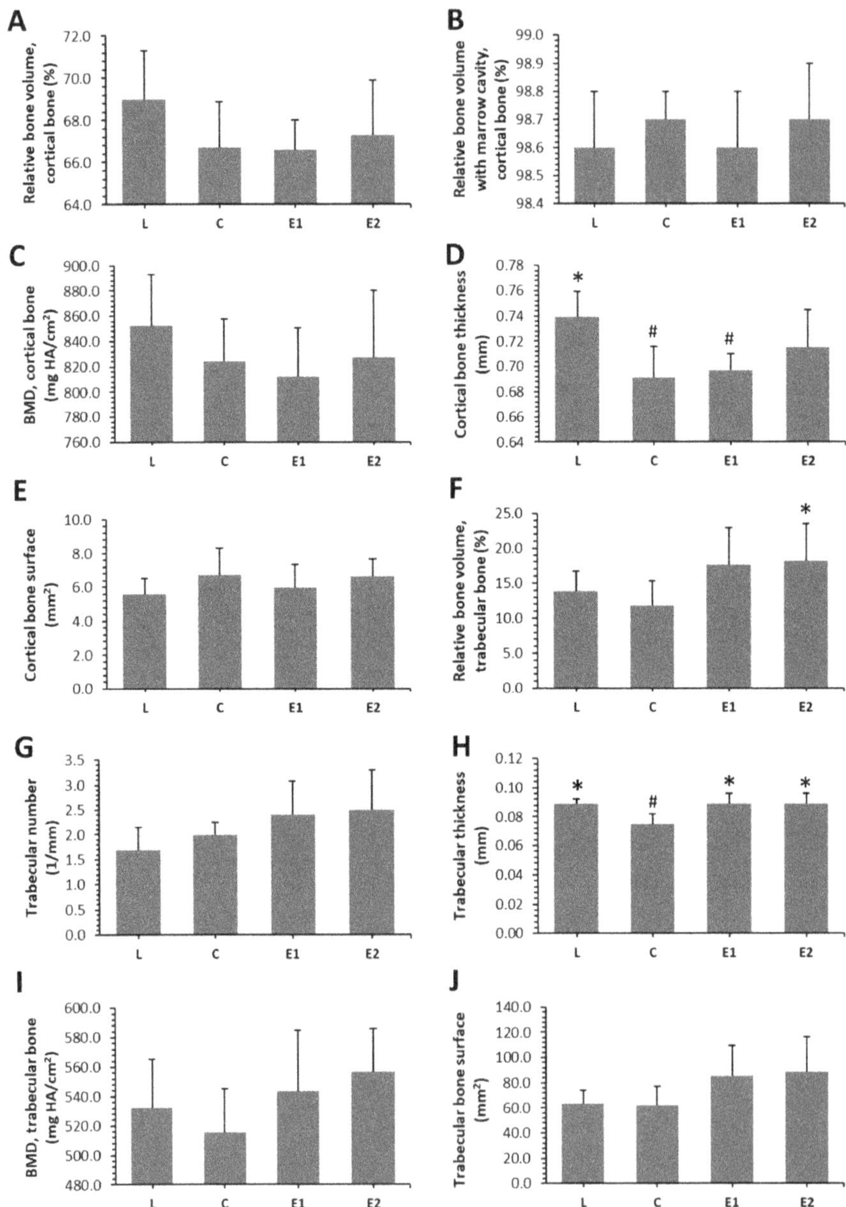

Figure 3. Parameters examined by microcomputed tomography in cortical (**A**–**E**) and trabecular (**F**–**J**) bone regions in ZDF rats from L, C, E1 and E2 groups. BMD—bone mineral density; HA—hydroxyapatite. * Significant differences versus C group ($p < 0.05$). # Significant differences versus L group ($p < 0.05$).

Representative reconstructed 3D images of cortical bone are illustrated in Figure 4A–D. A significantly increased relative volume of trabecular bone was observed in the E2 group versus the C group. Significantly higher trabecular thickness was determined in both treated groups, in comparison with the C group. No significant impact of Cornelian cherry administration on trabecular number, trabecular bone surface and BMD was recorded. The data are summarized in Figure 3F–J. Representative reconstructed 3D images of trabecular bone are illustrated in Figure 4E–H.

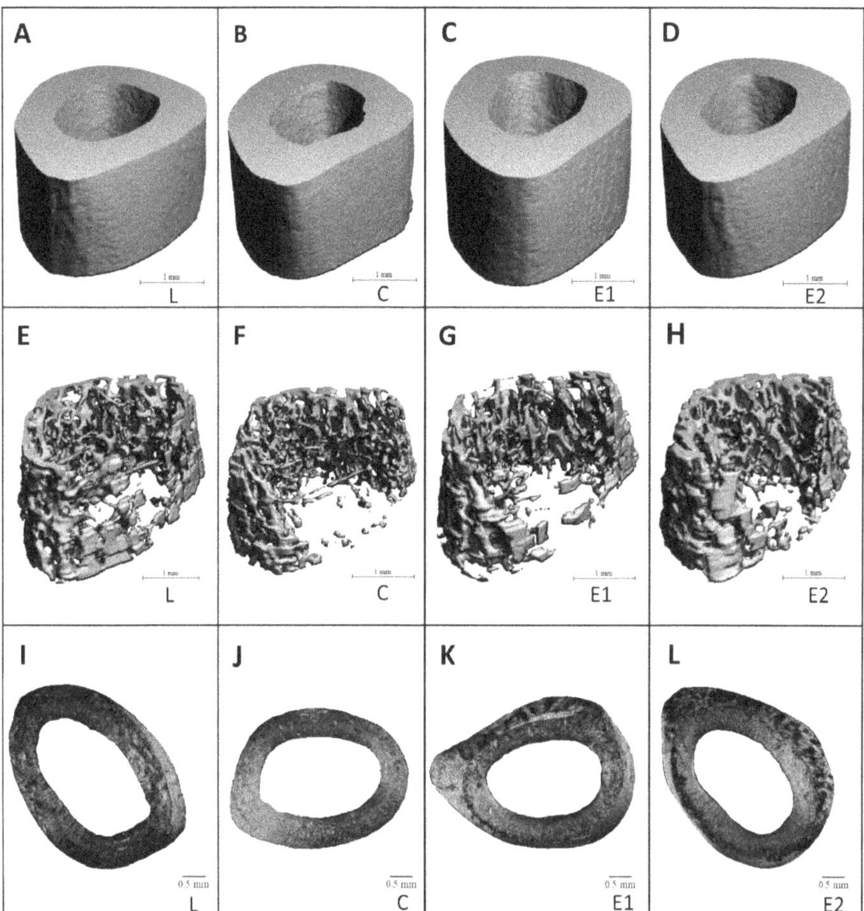

Figure 4. Representative 3D images of cortical (**A–D**) and trabecular bone regions (**E–H**), and representative 2D images of cortical bone (**I–L**) in ZDF rats from L, C, E1 and E2 groups.

3.4. Histomorphometry

Endosteal and periosteal surfaces of the compact bone were composed of non-vascular bone tissue in all groups. Primary vascular radial bone tissue was observed near the endosteum and in the middle of substantia compacta. However, a higher density of Haversian systems was recorded in the E1 and E2 groups versus the C group. In addition, intensive periosteal bone apposition was determined in both treated groups (Figure 4I–L). In total, 796 primary osteons' vascular canals, 337 Haversian canals and 337 Haversian systems were measured. We determined a significantly decreased area of primary osteons' vascular canals in the E1 versus the C and E2 groups. Interestingly, the area of these canals was higher in the E2 group when compared to the L group. On the contrary, the area of Haversian canals and Haversian systems was not affected by Cornelian cherry treatment. The data are summarized in Figure 2D–F.

4. Discussion

A previous study [27] demonstrated that administration of *Cornus mas* at the same doses which were used in our research caused significant reduction of water intake in ZDF rats versus the diabetic control ones. However, feed intake in diabetic control group did not differ significantly when compared to both treated groups. Moreover, ZDF rats from the E1 and E2 groups had significantly decreased blood glucose level versus the C group only in the pre-diabetic state (fifth to seventh week). At the end of the experiment (after 10 weeks), blood glucose did not reach significant values among the C, E1 and E2 groups, thus corresponding to our findings. We determined insignificantly lower blood glucose level in both treated groups versus the C group. Capcarova et al. [27] also observed that the rats from the E1 and E2 groups exhibit similar values of insulin than those from the diabetic control group. Therefore, other biochemical parameters that would reflect lipid-lowering properties of Cornelian cherry in diabetes were evaluated in our study. Enhanced levels of triglyceride, total cholesterol, HDL cholesterol, LDL cholesterol and ALP activity were recorded in the diabetic control versus the lean control group, which is consistent with the findings of Pang et al. [33]. Our results revealed that a higher dose of *Cornus mas* reduces triglyceride, total cholesterol levels and ALP activity. It can be related to its hypotriglyceridemic and hypocholesterolemic properties, as well as its protective effect against liver and bone disease. Soltani et al. [24] also identified a decreased value of triglyceride in T2DM patients receiving the fruit extract of *Cornus mas*. In alloxan-induced diabetic rats, hydroalcoholic fruits of Cornelian cherry also lowered triglyceride level [34]. In the study by Asgary et al. [21], reduced values of triglyceride and ALP activity were observed in diabetic rats after *Cornus mas* administration, as well. In general, ALP is considered as a good sensitive marker to access bone formation. However, elevated ALP in T2DM can indicate not only bone failure but also liver disease, or both. For that reason, it would be necessary to investigate the levels of other bone formation and bone resorption markers in the blood of ZDF rats to obtain more accurate results.

Cornelian cherry treatment insignificantly increased the total body weight of ZDF rats in our study. One of diabetes symptoms is a decrease in body weight caused by insulin resistance and impairment in whole-body glucose uptake. Dayar et al. [35] also observed non-significant impact of wild type Cornelian cherries administration (5 g/kg/day for six weeks) on the body weight in obese Zucker rats. Our study seems to be the first evaluating the effect of Cornelian cherry pulp on femoral weight and length. No significant changes in femoral weight were determined between L and E2 groups which could be consistent with a positive impact of higher dose of this fruit on bone weight. Generally, one of the most important components of Cornelian cherry fruit are anthocyanins. Van der Heijden et al. [36] did not find significant changes in the total body weight, weight of liver and adipose tissue in obese mice receiving high-energy diet with 36% extract of anthocyanins. Neither Shimizu et al. [37] revealed a significant impact of anthocyanins-rich bilberry extract (500 mg/kg bw/day for eight weeks) on the total body weight and weight of uterus in ovariectomized (OVX) rats. According to Damiano et al. [38], the body weight was significantly increased in ZDF rats compared with controls, whereas red orange and lemon extract rich in anthocyanins in ZDF rats significantly restored these values.

We demonstrated, for the first time, that higher dose of *Cornus mas* has also favorable effect on femoral weight, cortical bone thickness, relative volume of trabecular bone and trabecular thickness. In addition, elevated trabecular thickness was also observed in E1 versus C group. Thus, the increase in trabecular volume was mostly not due to an increase in their number but to their thickness.

Higher density of Haversian systems and accelerated periosteal bone apposition were determined in our treated groups. Through Haversian intracortical remodeling, fatigue microdamages are repaired (targeted remodeling), but it is also present at locations where microdamage do not occur (non-targeted remodeling) [39]. According to Bell et al. [40], enhanced density of smaller Haversian systems decreases microdamage propagation. In general, larger Haversian systems are more vulnerable to microfractures, and therefore undertake targeted remodeling [39,41]. In agreement with our results, Cornelian cherry pulp has strengthened the bone through both mechanisms Haversian remodeling and periosteal bone modeling. However, additional analyses related to mechanical properties of the cortical bone (e.g., maximum displacement, fracture load, stiffness and energy absorption) should be performed in order to confirm these results.

We determined lower area of primary osteons' vascular canals in E1 versus C, E2 groups. However, the area of these canals was higher in E2 group against lean control one. It is known that polyphenolic compounds in Cornelian cherry support angiogenesis, function of endothelium, and enhance the proliferation and migration of cells in the blood vessel [42,43]. We have found that the effect of *Cornus mas* on blood vessels in vascular canals of primary osteons depends on the dose used. On the contrary, Cornelian cherry treatment did not affect the size of Haversian canals and Haversian systems. The vascular structures in Haversian canals show typical capillary properties and are often paired. In general, they are fenestrated, lined with partial layer of endothelial cells and bounded by a continual thick basal membrane that restricts the transport of ions across the capillary [44]. It is known that Haversian system is the classic result of the process of bone remodeling. Bone remodeling is closely associated with vascular remodeling that means blood vessels in Haversian canals are able to modify its structure unlike those present in primary osteons.

The most important causes of bone damage and reduced bone strength, resulting from T2DM-related chronic hyperglycemia, are oxidative stress, production of reactive oxygen species (ROS) and AGEs, inflammation [45]. According to our results, Cornelian cherry pulp had favorable effect on impaired bone quality consistent with T2DM. Given the documented chemical composition and pharmacological activities of *Cornus mas*, it is likely that the mechanism of its action interferes with all aforementioned processes. The main anthocyanin compounds of *Cornus mas* include cyanidin 3-O-galactoside and pelargonidin 3-O-galactoside. Quercetin 3-O-glucuronide is the major flavonoid constituent, followed by kaempferol 3-O-galactoside. Among flavanols, catechin is predominant in fruits [46]. Generally, anthocyanins decrease the blood glucose level and peripheral insulin resistance in animal models of diabetes [47]. We suppose that improvement of blood glucose levels in both E1, E2 groups only in pre-diabetic state could have beneficial effects on bone deterioration in ZDF rats at the end of the experiment, as it is documented by our results from other biochemical parameters, microcomputed tomography and histomorphometry. The anthocyanins and other antioxidant components also inhibit oxidative stress and stress-sensitive signaling pathways. They have been proposed to attenuate ROS, especially by the ability to act as reducing agents in the electron-transfer reaction pathway [48]. It has been documented that the number of hydroxyl groups present on the glycosylated B-ring structure of anthocyanins is associated with their scavenging ability, and may influence their antioxidant capacity. According to Dzydzan et al. [47], the administration of Cornelian cherry extracts also decreased AGEs level in blood plasma which is consistent with the fact that polyphenols are able to inhibit AGEs formation. Moreover, polyphenols exhibit anti-inflammatory activity through peroxisome proliferator-activated receptor alpha (PPARα) expression in the liver and are associated with a decrease in triglyceride and pro-inflammatory cytokines levels [46]. However, recent data suggest that polyphenols can exert their beneficial effects by a compendium of other mechanisms, such the activation of transcription factors involved in antioxidant responsive capacity, metal chelating, and their capacity to bind to several proteins and thus impacting cellular homeostasis [49]. In addition, polyphenols have been shown to protect bone health through the modulation of osteoblastogenesis, osteoclastogenesis and osteoimmunological action [50].

5. Conclusions

We demonstrated, for the first time, that Cornelian cherry pulp could be used as a food supplement to alleviate T2DM-reduced bone quality and impaired bone health. Administration of this fruit positively influenced femoral weight, cortical bone thickness, relative volume of trabecular bone, trabecular thickness, Haversian remodeling and periosteal bone modeling. Moreover, hypolipidemic effect of *Cornus mas* was also confirmed in our study, using ZDF rats as an appropriate animal model.

Author Contributions: Conceptualization, M.C. and M.M.; methodology, R.O. and A.K.; software, J.B., V.K. and R.O.; validation, R.O., M.C. and M.M.; formal analysis, J.B., M.B. and V.M.; investigation, R.O., J.B., V.K., A.K., M.C. and M.M.; resources, R.O. and M.M.; data curation, M.B. and V.M.; writing—original draft preparation, R.O., J.B. and M.M.; writing—review and editing, R.O., J.B., V.K., M.B., V.M., A.K., M.C. and M.M.; visualization, R.O., J.B., M.B. and M.M.; supervision, R.O., M.C. and M.M.; project administration, R.O. and M.M.; funding acquisition, R.O. and M.M. All authors have read and agreed to the published version of the manuscript.

Funding: This research was funded by the Ministry of Education, Science, Research and Sport of the Slovak Republic, grant numbers VEGA 1/0505/18 and VEGA 1/0444/20.

Conflicts of Interest: The authors declare no conflict of interest. The funders had no role in the design of the study; in the collection, analyses, or interpretation of data; in the writing of the manuscript; or in the decision to publish the results.

References

1. Olokoba, A.B.; Obateru, O.A.; Olokoba, L.B. Type 2 Diabetes Mellitus: A Review of Current Trends. *Oman Med. J.* **2012**, *27*, 269–273. [CrossRef] [PubMed]
2. Wu, Y.; Ding, Y.; Tanaka, Y.; Zhang, W. Risk Factors Contributing to Type 2 Diabetes and Recent Advances in the Treatment and Prevention. *Int. J. Med. Sci.* **2014**, *11*, 1185–1200. [CrossRef] [PubMed]
3. Deshmukh, C.D.; Jain, A. Diabetes Mellitus: A Review. *Int. J. Pure Appl. Biosci.* **2015**, *3*, 224–230.
4. Sundararaghavan, V.; Mazur, M.M.; Evans, B.; Liu, J.; Ebraheim, N.A. Diabetes and bone health: Latest evidence and clinical implications. *Ther. Adv. Musculoskelet. Dis.* **2017**, *9*, 67–74. [CrossRef] [PubMed]
5. Ghodsi, M.; Larijani, B.; Keshtkar, A.A.; Nasli-Esfahani, E.; Alatab, S.; Mohajeri-Tehrani, M.R. Mechanisms involved in altered bone metabolism in diabetes: A narrative review. *J. Diabetes Metab. Disord.* **2016**, *15*. [CrossRef] [PubMed]
6. Gaddini, G.W.; Turner, R.T.; Grant, K.A.; Iwaniec, U.T. Alcohol: A Simple Nutrient with Complex Actions on Bone in the Adult Skeleton. *Alcohol. Clin. Exp. Res.* **2016**, *40*, 657–671. [CrossRef] [PubMed]
7. Martiniakova, M.; Babikova, M.; Omelka, R. Pharmacological agents and natural compounds: Available treatments for osteoporosis. *J. Physiol. Pharmacol.* **2020**, *71*. [CrossRef]
8. Costantini, S.; Conte, C. Bone health in diabetes and prediabetes. *World J. Diabetes* **2019**, *10*, 421–445. [CrossRef]
9. Hamilton, E.J.; Rakic, V.; Davis, W.A.; Paul Chubb, S.A.; Kamber, N.; Prince, R.L.; Davis, T.M.E. A five-year prospective study of bone mineral density in men and women with diabetes: The Fremantle Diabetes Study. *Acta Diabetol.* **2012**, *49*, 153–158. [CrossRef]
10. Starup-Linde, J.; Vestergaard, P. Biochemical bone turnover markers in diabetes mellitus—A systematic review. *Bone* **2016**, *82*, 69–78. [CrossRef]
11. Sharifi, F.; Ahmadi Moghaddam, N.; Mousavi-Nasab, N. The Relationship Between Type 2 Diabetes mellitus And Bone Density In Postmenopausal Women. *Int. J. Endocrinol. Metab. IJEM* **2006**, *4*, 117–122.
12. Dumic-Cule, I.; Ivanac, G.; Lucijanic, T.; Katicic, D. Type 2 diabetes and osteoporosis: Current knowledge. *Endocr. Oncol. Metab.* **2018**, *4*, 23–29. [CrossRef]
13. Thakur, A.K.; Dash, S. Estimation of bone mineral density among type 2 diabetes mellitus patients in western Odisha. *Int. J. Res. Med. Sci.* **2018**, *6*, 459–464. [CrossRef]
14. Filipowska, J.; Tomaszewski, K.A.; Niedźwiedzki, Ł.; Walocha, J.A.; Niedźwiedzki, T. The role of vasculature in bone development, regeneration and proper systemic functioning. *Angiogenesis* **2017**, *20*, 291–302. [CrossRef]
15. Marin, C.; Luyten, F.P.; Van der Schueren, B.; Kerckhofs, G.; Vandamme, K. The Impact of Type 2 Diabetes on Bone Fracture Healing. *Front. Endocrinol.* **2018**, *9*. [CrossRef]
16. Linda Kao, W.H.; Kammerer, C.M.; Schneider, J.L.; Bauer, R.L.; Mitchell, B.D. Type 2 diabetes is associated with increased bone mineral density in Mexican-American women. *Arch. Med. Res.* **2003**, *34*, 399–406. [CrossRef]
17. Janle, E.M.; Kissinger, P.T.; Pesek, J.F. Short Interval Monitoring of Glucose in Zucker Diabetic Fatty (ZDF) Rats. *Curr. Sep.* **1995**, *14*, 58–63.
18. Capcarova, M.; Kalafova, A.; Schwarzova, M.; Soltesova Prnova, M.; Svik, K.; Schneidgenova, M.; Slovak, L.; Bovdisova, I.; Toman, R.; Lory, V.; et al. The high-energy diet affecting development of diabetes symptoms in Zucker diabetic fatty rats. *Biologia* **2018**, *73*, 659–671. [CrossRef]

19. Zeitoun, D.; Caliaperoumal, G.; Bensidhoum, M.; Constans, J.M.; Anagnostou, F.; Bousson, V. Microcomputed tomography of the femur of diabetic rats: Alterations of trabecular and cortical bone microarchitecture and vasculature-a feasibility study. *Eur. Radiol. Exp.* **2019**, *3*, 17. [CrossRef]
20. Santini, A.; Tenore, G.C.; Novellino, E. Nutraceuticals: A paradigm of proactive medicine. *Eur. J. Pharm. Sci.* **2017**, *96*, 53–61. [CrossRef]
21. Asgary, S.; Rafieian-Kopaei, M.; Shamsi, F.; Najafi, S.; Sahebkar, A. Biochemical and histopathological study of the anti-hyperglycemic and anti-hyperlipidemic effects of cornelian cherry (*Cornus mas* L.) in alloxan-induced diabetic rats. *J. Complement. Integr. Med.* **2014**, *11*, 63–69. [CrossRef]
22. Tiptiri-Kourpeti, A.; Fitsiou, E.; Spyridopoulou, K.; Vasileiadis, S.; Iliopoulos, C.; Galanis, A.; Vekiari, S.; Pappa, A.; Chlichlia, K. Evaluation of Antioxidant and Antiproliferative Properties of Cornus mas L. Fruit Juice. *Antioxidants* **2019**, *8*, 377. [CrossRef] [PubMed]
23. Narimani-Rad, M.; Lotfi, A.; Mesgari Abbasi, M.; Abdollahi, B. The Effect of Cornelian Cherry (*Cornus mas* L.) Extract on Serum Ghrelin and Corticosterone Levels in Rat Model. *J. Pharm. Biomed. Sci.* **2013**, *3*, 7–9.
24. Soltani, R.; Gorji, A.; Asgary, S.; Sarrafzadegan, N.; Siavash, M. Evaluation of the Effects of *Cornus mas* L. Fruit Extract on Glycemic Control and Insulin Level in Type 2 Diabetic Adult Patients: A Randomized Double-Blind Placebo-Controlled Clinical Trial. *Evid. Based Complement. Altern. Med.* **2015**, *2015*, 1–5. [CrossRef] [PubMed]
25. Czerwińska, M.E.; Melzig, M.F. Cornus mas and Cornus Officinalis—Analogies and Differences of Two Medicinal Plants Traditionally Used. *Front. Pharmacol.* **2018**, *9*, 894. [CrossRef]
26. Lotfi, A.; Aghdam Shahryar, H.; Rasoolian, H. Effects of cornelian cherry (*Cornus mas* L.) fruit on plasma lipids, cortisol, T3 and T4 levels in hamsters. *J. Anim. Plant Sci.* **2014**, *24*, 459–462.
27. Capcarova, M.; Kalafova, A.; Schwarzova, M.; Schneidgenova, M.; Švík, K.; Prnova, M.; Slovák, L.; Kovacik, A.; Lóry, V.; Zórad, S.; et al. Cornelian cherry fruit improves glycaemia and manifestations of diabetes in obese Zucker diabetic fatty rats. *Res. Vet. Sci.* **2019**, *126*. [CrossRef]
28. Omelka, R.; Martiniakova, M.; Svik, K.; Slovak, L.; Payer, J.; Oppenbergerova, I.; Kovacova, V.; Babikova, M.; Soltesova-Prnova, M. The effects of eggshell calcium (Biomin H®) and its combinations with alfacalcidol (1α-hydroxyvitamin D3) and menaquinone-7 (vitamin K2) on ovariectomy-induced bone loss in a rat model of osteoporosis. *J. Anim. Physiol. Anim. Nutr.* **2020**. [CrossRef]
29. Martiniaková, M.; Boboňová, I.; Omelka, R.; Grosskopf, B.; Stawarz, R.; Toman, R. Structural changes in femoral bone tissue of rats after subchronic peroral exposure to selenium. *Acta Vet. Scand.* **2013**, *55*, 8. [CrossRef]
30. Martiniaková, M.; Chovancová, H.; Omelka, R.; Grosskopf, B.; Toman, R. Effects of a single intraperitoneal administration of cadmium on femoral bone structure in male rats. *Acta Vet. Scand.* **2011**, *53*, 49. [CrossRef]
31. Enlow, D.H.; Brown, S.O. A comparative histological study of fossil and recent bone tissues. Part I. *Tex. J. Sci.* **1956**, *8*, 405–412.
32. de Ricqles, A.J.; Meunier, F.J.; Castanet, J.; Francillon-Vieillot, H. Comparative microstructure of bone. In *Bone Matrix and Bone Specific Products*; CRC Press, Inc.: Boca Raton, FL, USA, 1991; Volume 3, pp. 1–78.
33. Pang, Y.; Hu, J.; Liu, G.; Lu, S. Comparative medical characteristics of ZDF-T2DM rats during the course of development to late stage disease. *Anim. Models Exp. Med.* **2018**, *1*, 203–211. [CrossRef] [PubMed]
34. Mirbadalzadeh, R.; Shirdel, Z. Antihyperglycemic and Antihyperlipidemic effects of *Cornus mas* extract in diabetic rats compared with glibenclamide. *Elixir Hormo Signal* **2012**, *47*, 8969–8972.
35. Dayar, E.; Cebova, M.; Lietava, J.; Panghyova, E.; Pechanova, O. Beneficial Effects of Cornelian Cherries on Lipid Profile and NO/ROS Balance in Obese Zucker Rats: Comparison with CoQ10. *Molecules* **2020**, *25*, 1922. [CrossRef] [PubMed]
36. van der Heijden, R.A.; Morrison, M.C.; Sheedfar, F.; Mulder, P.; Schreurs, M.; Hommelberg, P.P.H.; Hofker, M.H.; Schalkwijk, C.; Kleemann, R.; Tietge, U.J.F.; et al. Effects of Anthocyanin and Flavanol Compounds on Lipid Metabolism and Adipose Tissue Associated Systemic Inflammation in Diet-Induced Obesity. Available online: https://www.hindawi.com/journals/mi/2016/2042107/ (accessed on 5 November 2020).
37. Shimizu, S.; Matsushita, H.; Morii, Y.; Ohyama, Y.; Morita, N.; Tachibana, R.; Watanabe, K.; Wakatsuki, A. Effect of anthocyanin-rich bilberry extract on bone metabolism in ovariectomized rats. *Biomed. Rep.* **2018**, *8*, 198–204. [CrossRef]

38. Damiano, S.; Lombari, P.; Salvi, E.; Papale, M.; Giordano, A.; Amenta, M.; Ballistreri, G.; Fabroni, S.; Rapisarda, P.; Capasso, G.; et al. A red orange and lemon by-products extract rich in anthocyanins inhibits the progression of diabetic nephropathy. *J. Cell. Physiol.* **2019**, *234*, 23268–23278. [CrossRef]
39. Burr, D.B. Targeted and nontargeted remodeling. *Bone* **2002**, *30*, 2–4. [CrossRef]
40. Bell, K.L.; Loveridge, N.; Reeve, J.; Thomas, C.D.; Feik, S.A.; Clement, J.G. Super-osteons (remodeling clusters) in the cortex of the femoral shaft: Influence of age and gender. *Anat. Rec.* **2001**, *264*, 378–386. [CrossRef]
41. Bruce Martin, R. Fatigue Damage, Remodeling, and the Minimization of Skeletal Weight. *J. Theor. Biol.* **2003**, *220*, 271–276. [CrossRef]
42. Stoclet, J.-C.; Chataigneau, T.; Ndiaye, M.; Oak, M.-H.; El Bedoui, J.; Chataigneau, M.; Schini-Kerth, V.B. Vascular protection by dietary polyphenols. *Eur. J. Pharmacol.* **2004**, *500*, 299–313. [CrossRef]
43. Rafieian-Kopaei, M.; Asgary, S.; Adelnia, A.; Setorki, M.; Khazaei, M.; Kazemi, S.; Shamsi, F. The effects of cornelian cherry on atherosclerosis and atherogenic factors in hypercholesterolemic rabbits. *J. Med. Plants Res.* **2011**, *5*, 7.
44. Gasser, J.; Kneissel, M. Bone Physiology and Biology. In *Bone Toxicology*; Springer: Cham, Switzerland, 2017; pp. 27–94, ISBN 978-3-319-56190-5.
45. Conte, C.; Bouillon, R.; Napoli, N. Chapter 40—Diabetes and bone. In *Principles of Bone Biology*, 4th ed.; Bilezikian, J.P., Martin, T.J., Clemens, T.L., Rosen, C.J., Eds.; Academic Press: Cambridge, MA, USA, 2020; pp. 941–969, ISBN 978-0-12-814841-9.
46. Dinda, B.; Kyriakopoulos, A.M.; Dinda, S.; Zoumpourlis, V.; Thomaidis, N.S.; Velegraki, A.; Markopoulos, C.; Dinda, M. *Cornus mas* L. (cornelian cherry), an important European and Asian traditional food and medicine: Ethnomedicine, phytochemistry and pharmacology for its commercial utilization in drug industry. *J. Ethnopharmacol.* **2016**, *193*, 670–690. [CrossRef] [PubMed]
47. Dzydzan, O.; Bila, I.; Kucharska, A.Z.; Brodyak, I.; Sybirna, N. Antidiabetic effects of extracts of red and yellow fruits of cornelian cherries (*Cornus mas* L.) on rats with streptozotocin-induced diabetes mellitus. *Food Funct.* **2019**, *10*, 6459–6472. [CrossRef] [PubMed]
48. Speer, H.; D'Cunha, N.M.; Alexopoulos, N.I.; McKune, A.J.; Naumovski, N. Anthocyanins and Human Health—A Focus on Oxidative Stress, Inflammation and Disease. *Antioxidants* **2020**, *9*, 366. [CrossRef]
49. González, I.; Morales, M.A.; Rojas, A. Polyphenols and AGEs/RAGE axis. Trends and challenges. *Food Res. Int.* **2020**, *129*, 108843. [CrossRef]
50. Đudarić, L.; Fužinac-Smojver, A.; Muhvić, D.; Giacometti, J. The role of polyphenols on bone metabolism in osteoporosis. *Food Res. Int.* **2015**, *77*, 290–298. [CrossRef]

Publisher's Note: MDPI stays neutral with regard to jurisdictional claims in published maps and institutional affiliations.

Article

Antagonistic Impact of Acrylamide and Ethanol on Biochemical and Morphological Parameters Consistent with Bone Health in Mice

Monika Martiniakova [1,*], Anna Sarocka [1], Veronika Kovacova [1], Edyta Kapusta [2], Zofia Goc [2], Agnieszka Gren [2], Grzegorz Formicki [2] and Radoslav Omelka [1,*]

[1] Faculty of Natural Sciences, Constantine the Philosopher University in Nitra, 949 74 Nitra, Slovakia; sarocka.anna@gmail.com (A.S.); vkovacova@ukf.sk (V.K.)

[2] Faculty of Exact and Natural Sciences, Pedagogical University of Cracow, 30 084 Cracow, Poland; edyta.kapusta@up.krakow.pl (E.K.); zofia.goc@up.krakow.pl (Z.G.); agnieszka.gren@up.krakow.pl (A.G.); grzegorz.formicki@up.krakow.pl (G.F.)

* Correspondence: mmartiniakova@ukf.sk (M.M.); romelka@ukf.sk (R.O.); Tel.: +421-376-408-718 (M.M.)

Received: 22 September 2020; Accepted: 7 October 2020; Published: 9 October 2020

Simple Summary: Alcohol consumption, the drinking of beverages containing ethanol, represents a growing problem worldwide. Alcohol intake is often combined with an improper diet based on highly processed starch products that are rich in acrylamide. Both acrylamide and alcohol have a harmful impact on bone health. We previously demonstrated that adverse effects of ethanol on cortical bone structure were partly reduced by a relatively high dose of acrylamide in mice after one remodelling cycle. The present research was designated to reveal whether the antagonistic impact of both aforementioned toxins can also be achieved using a lower dose of acrylamide. According to our results, individual administrations of acrylamide and ethanol had adverse impacts on biochemical and morphological parameters consistent with bone health in mice. However, the most detrimental effects of ethanol were again alleviated by acrylamide at the dose used in this study.

Abstract: The aim of present study was to verify antagonistic effect of acrylamide (AA) and ethanol (Et) on bone quality parameters. Adult mice (n = 20) were segregated into four groups following 2 weeks administration of toxins: group E1, which received AA (20 mg/kg body weight daily); group E2, which received 15% Et (1.7 g 100% Et/kg body weight daily); group E12, which received simultaneously both toxins; and a control group. An insignificant impact of individual applications of AA, Et or their simultaneous supplementation on the total body weight of mice and the length and weight of their femoral bones was identified. In group E1, higher levels of alanine aminotransferase (ALT), aspartate aminotransferase (AST), triglyceride (TG), a decreased level of glutathione (GSH) and elevated endocortical bone remodelling were determined. A significantly lower relative volume of cortical bone, bone mineral density (BMD), elevated endocortical bone remodelling and cortical porosity, higher levels of ALT, AST, lower values for total proteins (TP), GSH, alkaline phosphatase (ALP), calcium, and phosphorus were recorded in group E2. In the mice from group E12, the highest endocortical bone remodelling, decreased values for BMD, TP, GSH and ALP and increased levels of ALT and AST were found. Our findings confirmed the antagonistic impact of AA and Et at doses used in this study on biochemical and morphological parameters consistent with bone health in an animal model.

Keywords: acrylamide; alcohol; diet; bone health; biochemical analysis; morphological analysis; microcomputed tomography; mice

1. Introduction

Alcohol beverages containing ethanol (Et) are widely consumed throughout the world. Acrylamide (AA) belongs to the most common toxins in the human diet. It is formed during frying, deep frying and baking foods rich in carbohydrates, and especially in amino acid asparagine [1]. Simultaneous consumption of Et and AA-rich food is widespread among humans. Detrimental effects of AA and Et on various organ systems including the skeleton have been described in several studies [2–6]. Their toxicity contributed to differences in selected biochemical and morphological parameters of various organs. AA is present in all forms of foods prepared at high temperatures, including potato chips, fried potatoes, coffee, cornflakes and bread [1]. It is known that AA is carcinogenic to animals and might pose a risk to human health. According to Benziane et al. [7], peroral exposure to AA (5 mg or 10 mg for 2 months) induces kidney damage, hepatocellular insufficiency, and chronic liver disease, resulting in primary immunodeficiency and activation of the immune system in rats. Excessive alcohol consumption may also cause several pathological conditions, such as liver failure, brain damage, and various form of cancer [8].

With respect to the bone, acute peroral administration of AA (1 mg/kg body weight (bw) in a 24 h and 48 h period) affected the microstructure of cortical and trabecular bone tissues of mice. The cortical bone was more resorbed because of a higher number of resorption lacunae. In the trabecular bone, increased values for relative bone volume and trabecular number were determined [3]. Subacute exposure to AA (40 mg/kg bw for 2 weeks) had adverse effects only on cortical bone microstructure. Acrylamidated mice were shown to have increased levels of alanine aminotransferase (ALT), aspartate aminotransferase (AST), calcium (Ca), and decreased glutathione (GSH) [6]. Subchronic exposure to Et (1.7 g 100%/kg bw for 8 weeks) negatively influenced both the cortical and trabecular bone tissue structures of mice. In the cortical bone, increased porosity and decreased values were established for relative bone volume and bone mineral density (BMD). In the trabecular bone, lower relative bone volume, trabecular number, trabecular thickness, and bone surface were determined [4]. In the study by Broulik et al. [2], femoral bones of alcohol-fed rats (7.6 g of 95% Et/kg bw daily for 3 months) were characterised by a reduction in BMD, mechanical strength, cortical bone thickness, as well as in Ca and phosphate content. The levels of alkaline phosphatase (ALP) and AST did not differ between alcohol-fed rats and control ones. According to Bartlett et al. [9], chronic alcohol consumption induced changes in calcium regulating hormones, mineral homeostasis and mechanical loading and was consistent with an accumulation of reactive oxygen species (ROS) [10]. In addition, osteoblasts of rats injected intragastrically with liquid diets containing Et (12 g Et/kg daily for 4 weeks) had upregulated expression of nicotinamide adenine dinucleotide phosphate (NADPH) oxidase [11].

We previously demonstrated that harmful effects of Et on cortical bone structure were partly reduced by a relatively high dose of AA [6]. The present research was designated to reveal whether antagonistic impact of these toxins can also be achieved using a lower dose of AA after one remodelling cycle.

2. Materials and Methods

2.1. Animals

The First Local Ethic Committee on Experiments on Animals in Cracow approved all the relevant procedures (resolution number 175/2012). Twenty clinically healthy 12-week-old Swiss mice (males) were used in our experiment. Mice were housed in individual flat-deck wire cages under a 12 h light/dark cycle, a temperature of 20–24 °C and humidity of 55% ± 10% with free access to a standard diet (Agropol, Motycz, Poland) and water on an ad libitum basis. Animals were segregated into 4 groups following 2 weeks of administration of toxins: group E1 received AA (daily dose of 20 mg/kg bw); group E2 received 15% Et (daily dose of 1.7 g 100% Et/kg bw); group E12 was simultaneously supplemented by both toxins (20 mg AA/kg bw + 15% Et) per day; and a control (C) group without AA and/or Et administration per os.

The AA and Et dosages were selected based on studies performed by other authors [6,12–14]. Both toxins were dissolved in physiological saline and administered by syringe at established doses perorally to mice. Group C of animals received only physiological saline solution.

2.2. Biochemical Analysis

A day after the last toxin application, mice were placed in a state of deep anaesthesia for sacrifice by administration of Vetbutal (35 mg/kg bw; Biowet, Poland), and fasting whole blood samples were taken from the carotid artery. Then, whole blood was processed, and plasma ALP, ALT, AST, total protein (TP), triglyceride (TG), Ca and phosphorus (P) were measured using commercially available kits (Stamar, Dąbrowa Gornicza, Poland). Blood GSH levels were determined by Ellman's method [15].

2.3. Macroscopical Analysis

Bone specimens (both femoral bones) were sampled during necropsy. All femoral bones (n = 40) were weighed with a precision of 0.01 g on analytical scales, and their lengths were measured using a sliding instrument. In addition, total body weight of mice from all groups was also determined.

2.4. Micro-CT Analysis

Microcomputed tomography (micro-CT) was used to conduct quantitative 3D analysis of cortical and trabecular bone tissues. Within the cortical bone, regions of interest (ROIs), beginning at 5.2 mm from the growth plate at the distal end and extending 1.5 mm in the femoral midshaft, were analysed. The ROIs of trabecular bone started at 1.2 mm from the growth plate at the distal end and continued for 1.5 mm. High-resolution scans were obtained with a voxel size of 6.8 μm. The scanning conditions included 70 kV, 200 μA, 300 ms, 0.5 mm, and an aluminium filter. The micro-CT scans were obtained by μCT 50 (Scanco Medical). The standard analysis programme by Scanco was used to assess specific bone parameters as follows: relative bone volume with and without marrow cavity (%), bone mineral density (BMD; mg HA/ccm), bone surface (mm^2), cortical bone thickness (mm), trabecular number (1/mm), and trabecular thickness (mm).

2.5. Histomorphological Analysis

Histomorphological 2D analysis of cortical bone tissue was performed according to the previously published method [16]. In brief, femoral bones were embedded in epoxy resin, and thin sections were cut with a sawing microtome [17]. Internationally accepted classification systems of Enlow and Brown [18] and Ricqlés et al. [19] were applied to evaluate the qualitative 2D characteristics. The quantitative 2D parameters were calculated using Motic Images Plus 2.0 ML (Motic China Group Co., Ltd.) software. The area (μm^2) of primary osteons' vascular canals, Haversian canals and secondary osteons was measured in all views of thin sections in order to minimise statistical differences in the individual.

2.6. Statistics

Statistical analysis was performed using IBM SPSS Statistics 26.0 software (IBM, New York, NY, USA). The measured values were expressed as mean ± standard deviation. The differences in investigated parameters among all groups were calculated using analysis of variance (ANOVA) with Games–Howell's and/or Tukey's post hoc tests. The compared groups served as an independent variable; the measured parameters were dependent variables. The continuous data had a normal distribution. All P-values were considered significant if less than 0.05.

3. Results

3.1. Biochemical Analysis

Enhanced levels of ALT, AST, TG and decreased GSH were recorded in group E1 when compared to group C. In mice exposed to Et, higher values for ALT and AST and lower levels of TP, GSH, ALP, Ca and P were determined versus control mice. In group E12, decreased values for TP, GSH and ALP and higher levels of ALT and AST were found as compared to group C (Figure 1a–h).

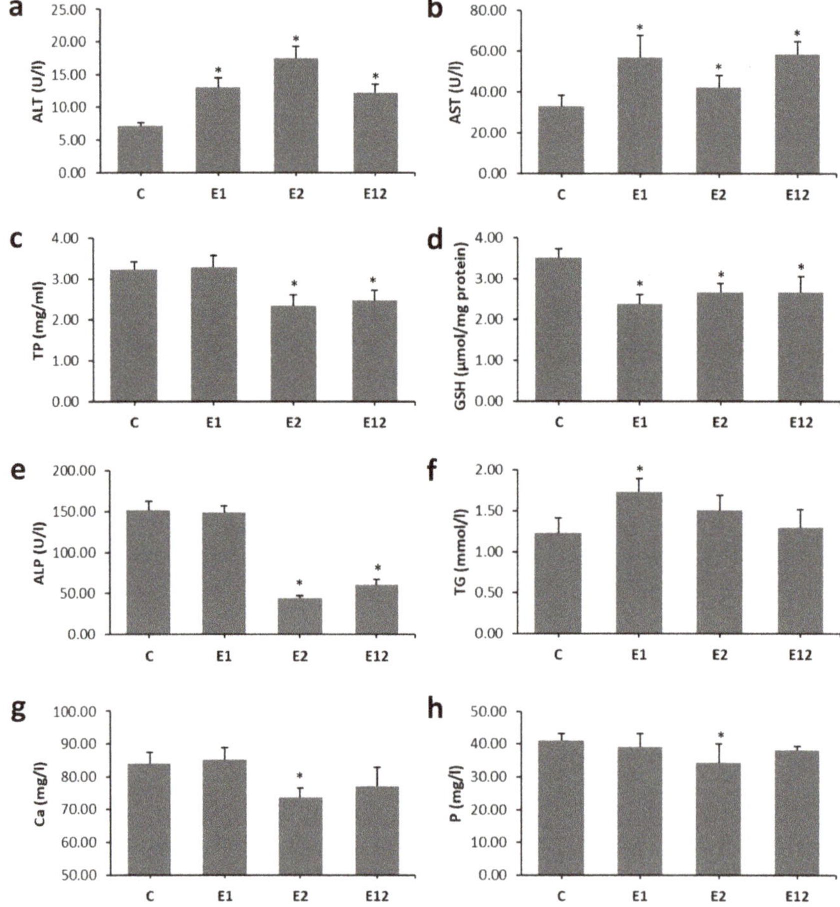

Figure 1. Biochemical parameters consistent with the bone quality in mice from groups C, E1, E2, and E12. (**a**)—alanine aminotransferase, (**b**)—aspartate aminotransferase, (**c**)—total proteins, (**d**)—glutathione, (**e**)—alkaline phosphatase, (**f**)—triglyceride, (**g**)—calcium, (**h**)—phosphorus. * Significant differences in relation to control ($P < 0.05$).

3.2. Macroscopical Analysis

We did not find a significant effect of individual applications of AA, Et or their simultaneous administration on the total body weight of mice or the length and weight of their femoral bones. The results are summarised in Figure 2a–c.

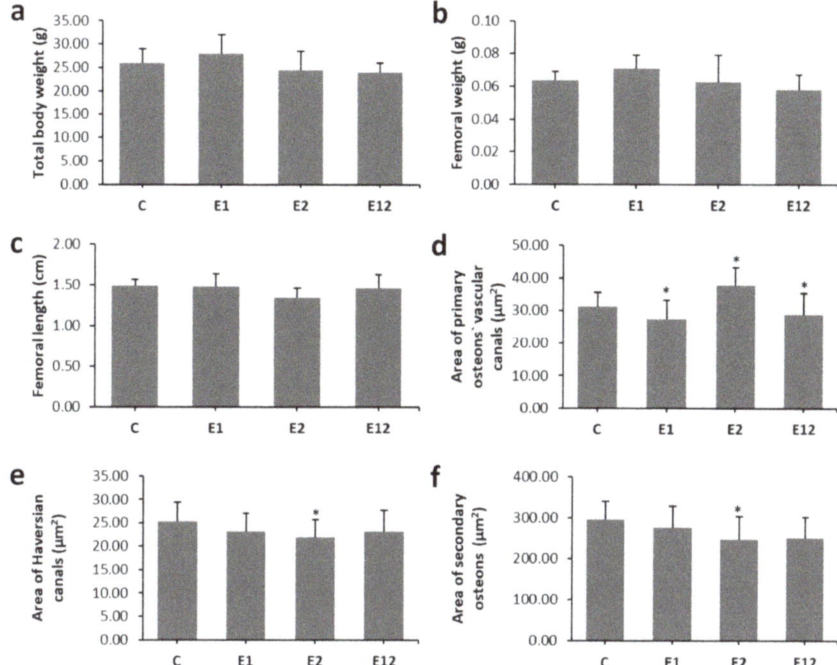

Figure 2. Macroscopical analysis of bones (**a**–**c**) and quantitative 2D analysis of cortical bone tissue (**d**–**f**) in mice from groups C, E1, E2, and E12. * Significant differences in relation to control ($P < 0.05$).

3.3. Micro-CT Analysis

Micro-CT analysis of the cortical bone showed significantly reduced relative bone volume with and without marrow cavity and BMD in mice from group E2. Lower values for relative bone volume with marrow cavity and BMD were also observed in group E12 versus group C. On the contrary, all measured 3D parameters of the cortical bone were not significantly different between groups E1 and C (Figure 3a–e). Trabecular bone microarchitecture did not change significantly among all groups. The results are documented in Figure 3f–j. Representative 3D images of the cortical and trabecular bone tissues are shown in Figure 4a–d,e–h, respectively.

3.4. Histomorphological Analysis

In mice from group C, non-vascular bone tissue formed both surfaces (endosteal and periosteal) of femoral bones. Several secondary osteons were observed in the lateral parts near the endosteum and in the middle parts of the cortical bone. Only medial parts consisted of non-vascular bone tissue. Mice from group E1 had enhanced endocortical remodelling. About 73% more secondary osteons were recorded near endosteal surfaces when compared to group C. Higher density of secondary osteons (about 46%) was also detected near the endosteum in mice from group E2. However, we identified many resorption lacunae consistent with an increased cortical porosity in Et-fed mice. The highest density of secondary osteons of about 140% was determined near endosteal surfaces in group E12. We also recorded several resorption lacune near the endosteum; however, the number of lacune was lower in comparison with group E2 (Figure 4i–l). In total, 757 primary osteons' vascular canals, 99 Haversian canals and 99 secondary osteons were measured. The area of primary osteon's vascular canals was significantly decreased in groups E1 and E12 versus group C. On the other hand, mice exposed to Et showed a higher area of primary osteon's vascular canals. Nevertheless, the area of Haversian canals and secondary osteons was lower in these mice (Figure 2d–f).

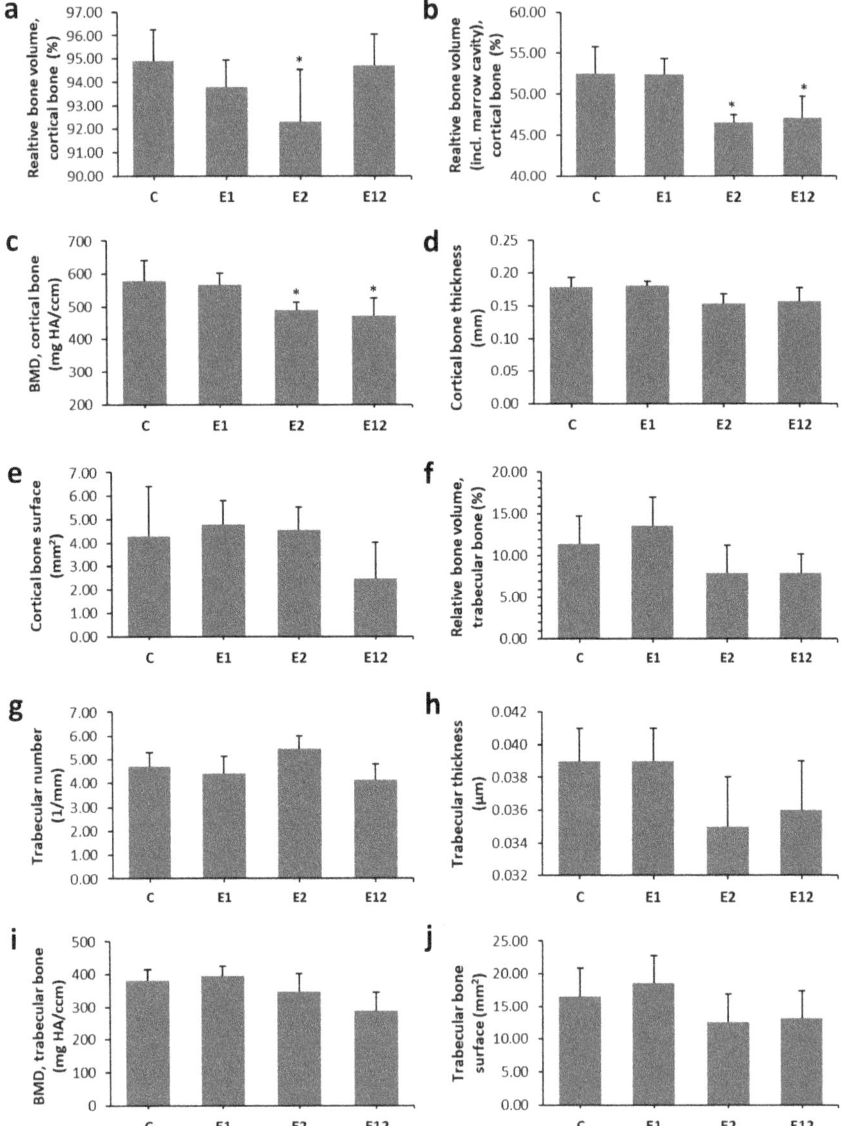

Figure 3. Micro-CT analysis of cortical (**a**–**e**) and trabecular bone tissues (**f**–**j**) in mice from groups C, E1, E2, and E12. * Significant differences in relation to control ($P < 0.05$).

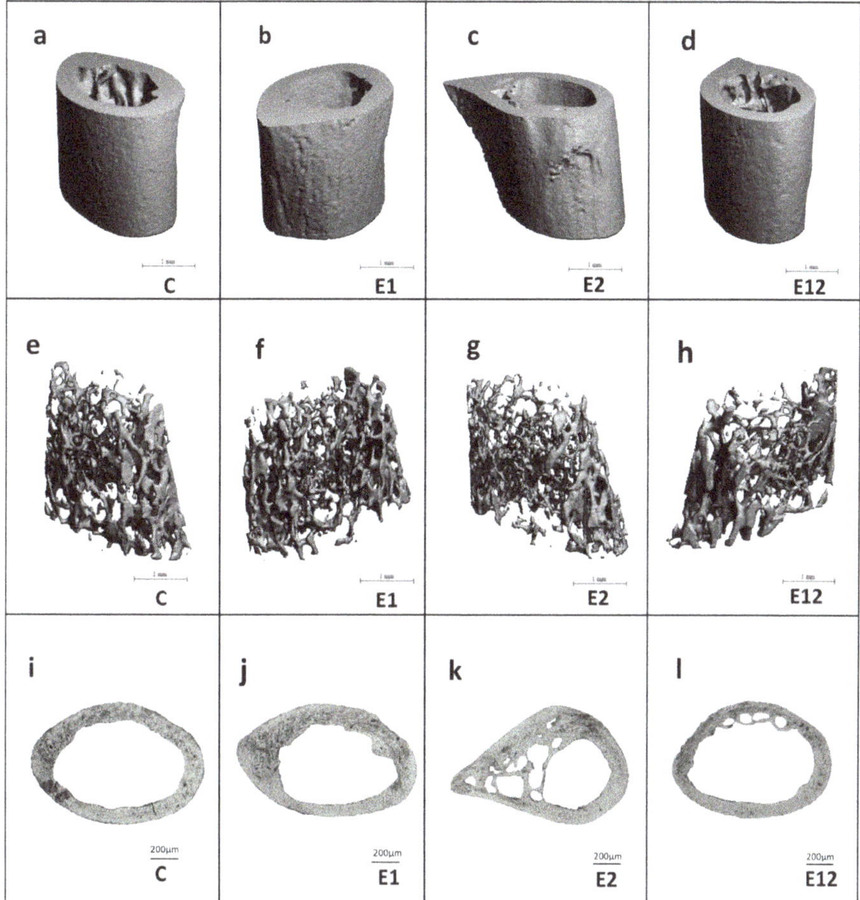

Figure 4. Representative 3D images of cortical (**a**–**d**) and trabecular bone tissues (**e**–**h**) in mice from groups C, E1, E2, and E12. Representative 2D images of cortical bone tissue (**i**–**l**) in mice from groups C, E1, E2, E12.

4. Discussion

Evidence accumulated in recent years has shown that physiological serum levels of liver enzymes (within and just above the normal range) are consistent with an increased risk of incident metabolic diseases. It is also well known that chronic liver disease is associated with profound adverse effects on bone health and homeostasis [20]. Excessive AA and Et consumption can lead to both liver and bone damage.

Our results from biochemical analysis indicate liver disease in all experimental groups of mice exposed to AA and/or Et. The liver failure caused by aforementioned toxins was also revealed in other research and was consistent with enhanced levels of ALT and AST in acrylamidated rats (10 mg AA/kg bw for 21 days) [21] and Et-fed mice (5–6% Et for 10 days to 12 weeks) [22]. Because the liver produces different molecules, it was hypothesised that damage of the liver function will result in osteoporosis by influencing the development of bone-active liver molecules [23]. Even a presence of decreased GSH in all experimental groups could lead to hepatocyte disruption. In groups E2 and E12, the hepatotoxic effect was increased, which could be related to a reduced level of ALP. In the study by Broulik et al. [2], a lower value of ALP was also recorded in rats exposed to Et (7.6 g 95% Et/kg bw for 12 weeks). Harmful hepatocyte changes may also be consistent with lower serum levels of Ca and P in group E2. It is known that acute liver failure is associated with Ca and P imbalance [24,25]. A decreased level of TP in Et-fed mice might also indicate liver disorder. In humans, lower hepatic synthesis of proteins can be a consequence of alcoholism-associated malnutrition [26]. A higher value of TG was found in group E1. Ghorbel et al. [27] also revealed an increased TG level in rats fed with AA. According to Mahmood et al. [28], low doses of AA triggered substantial increases in TG levels and contributed to increased synthesis of plasma lipoproteins and lipid mobilisation from the liver.

According to our results, the total body weight of mice and the length and weight of their femoral bones were not influenced by AA and/or Et supplementation. Similarly, AA caused only intermittent differences in the body weight of B6C3F1 mice receiving AA in the drinking water ad libitum for 2 years (doses of 0.0875, 0.175, 0.35, and 0.70 mM AA) [29]. Furthermore, available epidemiological evidence does not support a completely consistent association between body weight and regular intake of Et. Nevertheless, when Et is consumed with food, Et intake may constitute a risk factor for increased body fat due to passive overconsumption of energy in the form of fat, as well as a decrease in total fat oxidation in the presence of Et [30]. Broulik et al. [2] determined lower final body weight in Et-fed rats versus control ones; however, the difference did not reach statistical significance.

In our study, significantly decreased BMD and relative volume of cortical bone with and without marrow cavity were recorded in Et-fed mice. Similar findings were demonstrated in rats exposed to Et (3, 6, 13, and 35% Et for 4 months [31]; 36% Et for 42 days) and Et-fed mice (10–36% Et for 78 days) [32,33]. In general, decreased rates of bone formation followed by low bone mass and a lower BMD are also seen in alcoholics [26]. Significantly decreased BMD and relative volume of cortical bone with marrow cavity have also been identified in mice in group E12. Interestingly, trabecular bone microarchitecture did not vary significantly among all groups. It is known that the trabecular and cortical bones have different bone remodelling levels. Cortical bone has a large volume of matrix and a limited surface area; hence, signals deep within the matrix may not locate a surface as lightly to initiate remodelling, allowing for the microdamage accumulation mostly in less remodelled interstitial bone [34]. Generally, trabecular bone is remodelled more vigorously. However, the surface-to-volume ratio is much greater [35] and the length of one remodelling cycle in the trabecular bone is longer than in the cortical bone.

Mice from groups E1, E2 and E12 demonstrated enhanced intracortical bone remodelling. In general, bone remodelling is the principal mechanism for maintaining a healthy skeleton in adults, and dysfunction in bone remodelling may contribute to bone loss and/or lower bone quality [36]. According to Piemontese et al. [37], intracortical remodelling is associated with increased cortical porosity and formation of secondary osteons that exhibit histologic hallmarks of remodelling activity in mice. Increased intracortical remodelling was also recorded in our previous researches in mice subacutely exposed to AA [6] and in Et-fed mice after 8 weeks [4].

Our findings showed vasoconstriction of primary osteon's vascular canals in groups E1 and E12. Blood vessels located in vascular canals provide nutrients for the bone [38] and, in response to continuous functional changes, can modify its structure (vascular remodelling) [39]. Generally, AA reduces the high-density lipoprotein (HDL) [40]. Low HDL is related to arterial and vessel narrowing or blockage [41]. On the other hand, Et has a significant influence on the cardiovascular system, such as peripheral vasodilation [42], which can be consistent with increased sizes of primary osteon's vascular canals in group E2. On the contrary, the area of Haversian canals and secondary osteons was significantly lower in Et-fed mice. Secondary osteons, as products of bone remodelling, are composed of hydroxyapatite crystals and collagen fibres around the central Haversian canal. According to Romero et al. [5], Et consumption reduces the density of mature collagen fibres. Therefore, lower sizes of secondary osteons could be connected with decreased bone toughness [43].

5. Conclusions

This study clearly demonstrates that individual administrations of AA (20 mg/kg bw) and Et (15%) had detrimental effects on biochemical and morphological parameters consistent with bone quality in mice. However, the most adverse impacts of Et were alleviated by AA. Although the dose of AA used in this study had fewer negative effects on cortical bone structure compared to the previously examined relatively high dose (40 mg/kg bw), the antagonistic impact of AA and Et on the bone health parameters was again confirmed. We can conclude that the harmful effects of Et on bone quality can be effectively reduced by AA even at lower dose. Simultaneous consumption of Et and AA-rich food is therefore less bad for bone health than individual Et intake.

Author Contributions: Conceptualisation, M.M. and G.F.; methodology, Z.G. and R.O.; software, Z.G. and R.O.; validation, M.M., R.O., A.G. and G.F.; formal analysis, E.K., A.S. and V.K.; investigation, M.M., A.S., E.K., Z.G. and R.O.; resources, M.M. and G.F.; data curation, R.O. and A.G.; writing—original draft preparation, M.M.; writing—review and editing, M.M., A.S., V.K., E.K., Z.G., A.G., G.F. and R.O.; visualisation, A.S., V.K. and R.O.; supervision, M.M., G.F. and R.O.; project administration, M.M. and R.O.; funding acquisition, M.M. and R.O. All authors have read and agreed to the published version of the manuscript.

Funding: This research was funded by the grants of the Ministry of Education, Science, Research and Sport of the Slovak Republic (VEGA 1/0505/18, VEGA 1/0444/20 and KEGA 041UKF-4/2020).

Conflicts of Interest: The authors declare no conflict of interest.

References

1. Semla, M.; Goc, Z.; Martiniaková, M.; Omelka, R.; Formicki, G. Acrylamide: A common food toxin related to physiological functions and health. *Physiol. Res.* **2017**, *66*, 205–217. [CrossRef]
2. Broulík, P.D.; Vondrová, J.; Růzicka, P.; Sedlácek, R.; Zíma, T. The effect of chronic alcohol administration on bone mineral content and bone strength in male rats. *Physiol. Res.* **2010**, *59*, 599–604.
3. Sarocka, A.; Babosova, R.; Kovacova, V.; Omelka, R.; Semla, M.; Kapusta, E.; Goc, Z.; Formicki, G.; Martiniakova, M. Acrylamide-induced changes in femoral bone microstructure of mice. *Physiol. Res.* **2017**, *66*, 1067–1071. [CrossRef] [PubMed]
4. Martiniakova, M.; Sarocka, A.; Babosova, R.; Grosskopf, B.; Kapusta, E.; Goc, Z.; Formicki, G.; Omelka, R. Changes in the microstructure of compact and trabecular bone tissues of mice subchronically exposed to alcohol. *J. Biol Res. Thessal.* **2018**, *25*, 8. [CrossRef]
5. Renato Romero, J.; Krause Neto, W.; Sabbag da Silva, A.; Luiz Dos Santos, E.; Aurélio Added, M.; Pianca, E.; Florencio Gama, E.; Rodrigues de Souza, R. Chronic cachaça consumption affects the structure of tibial bone by decreasing bone density and density of mature collagen fibers in middle-aged Wistar rats. *Aging Male* **2018**, 1–6. [CrossRef]
6. Sarocka, A.; Kovacova, V.; Omelka, R.; Grosskopf, B.; Kapusta, E.; Goc, Z.; Formicki, G.; Martiniakova, M. Single and simultaneous effects of acrylamide and ethanol on bone microstructure of mice after one remodeling cycle. *BMC Pharm. Toxicol.* **2019**, *20*, 38. [CrossRef]
7. Belhadj Benziane, A.; Dilmi Bouras, A.; Mezaini, A.; Belhadri, A.; Benali, M. Effect of oral exposure to acrylamide on biochemical and hematologic parameters in Wistar rats. *Drug Chem. Toxicol.* **2019**, *42*, 157–166. [CrossRef]
8. Goc, Z.; Kapusta, E.; Formicki, G.; Martiniaková, M.; Omelka, R. Effect of taurine on ethanol-induced oxidative stress in mouse liver and kidney. *Chin. J. Physiol.* **2019**, *62*, 148–156. [CrossRef]
9. Bartlett, P.J.; Antony, A.N.; Agarwal, A.; Hilly, M.; Prince, V.L.; Combettes, L.; Hoek, J.B.; Gaspers, L.D. Chronic alcohol feeding potentiates hormone-induced calcium signalling in hepatocytes. *J. Physiol.* **2017**, *595*, 3143–3164. [CrossRef]
10. Park, B.-J.; Lee, Y.-J.; Lee, H.-R. Chronic liver inflammation: Clinical implications beyond alcoholic liver disease. *World J. Gastroenterol.* **2014**, *20*, 2168–2175. [CrossRef] [PubMed]
11. Chen, J.-R.; Lazarenko, O.P.; Shankar, K.; Blackburn, M.L.; Lumpkin, C.K.; Badger, T.M.; Ronis, M.J.J. Inhibition of NADPH oxidases prevents chronic ethanol-induced bone loss in female rats. *J. Pharm. Exp.* **2011**, *336*, 734–742. [CrossRef] [PubMed]

12. Burek, J.D.; Albee, R.R.; Beyer, J.E.; Bell, T.J.; Carreon, R.M.; Morden, D.C.; Wade, C.E.; Hermann, E.A.; Gorzinski, S.J. Subchronic toxicity of acrylamide administered to rats in the drinking water followed by up to 144 days of recovery. *J. Env. Pathol. Toxicol.* **1980**, *4*, 157–182.
13. Bull, R.J.; Robinson, M.; Stober, J.A. Carcinogenic activity of acrylamide in the skin and lung of Swiss-ICR mice. *Cancer Lett.* **1984**, *24*, 209–212. [CrossRef]
14. Doerge, D.R.; Gamboa da Costa, G.; McDaniel, L.P.; Churchwell, M.I.; Twaddle, N.C.; Beland, F.A. DNA adducts derived from administration of acrylamide and glycidamide to mice and rats. *Mutat. Res. Genet. Toxicol. Environ. Mutagen.* **2005**, *580*, 131–141. [CrossRef] [PubMed]
15. Ellman, G.L. Tissue sulfhydryl groups. *Arch. Biochem. Biophys.* **1959**, *82*, 70–77. [CrossRef]
16. Martiniaková, M.; Boboňová, I.; Omelka, R.; Grosskopf, B.; Stawarz, R.; Toman, R. Structural changes in femoral bone tissue of rats after subchronic peroral exposure to selenium. *Acta Vet. Scand.* **2013**, *55*, 8. [CrossRef]
17. Martiniaková, M.; Chovancová, H.; Omelka, R.; Grosskopf, B.; Toman, R. Effects of a single intraperitoneal administration of cadmium on femoral bone structure in male rats. *Acta Vet. Scand* **2011**, *53*, 49. [CrossRef]
18. Enlow, D.H.; Brown, S.O. A comparative histological study of fossil and recent bone tissues. Part I. *Tex. J. Sci.* **1956**, *8*, 405–412.
19. De Ricqles, A. Comparative microstructure of bone. In *Bone Matrix and Bone Specific Products*; CRC Press Inc.: Boca Raton, FL, USA, 1991; pp. 1–78.
20. Breitling, L.P. Liver enzymes and bone mineral density in the general population. *J. Clin. Endocrinol. Metab.* **2015**, *100*, 3832–3840. [CrossRef]
21. Allam, A.A.; El-Ghareeb, A.W.; Abdul-Hamid, M.; Bakery, A.E.; Gad, M.; Sabri, M. Effect of prenatal and perinatal acrylamide on the biochemical and morphological changes in liver of developing albino rat. *Arch. Toxicol.* **2010**, *84*, 129–141. [CrossRef]
22. Gao, B.; Xu, M.-J.; Bertola, A.; Wang, H.; Zhou, Z.; Liangpunsakul, S. Animal models of alcoholic liver disease: Pathogenesis and clinical relevance. *Gene Expr.* **2017**, *17*, 173–186. [CrossRef] [PubMed]
23. Nakchbandi, I.A. Osteoporosis and fractures in liver disease: Relevance, pathogenesis and therapeutic implications. *World J. Gastroenterol.* **2014**, *20*, 9427–9438. [CrossRef] [PubMed]
24. Dawson, D.J.; Babbs, C.; Warnes, T.W.; Neary, R.H. Hypophosphataemia in acute liver failure. *Br. Med. J. Clin. Res. Ed.* **1987**, *295*, 1312–1313. [CrossRef] [PubMed]
25. Camilli, J.A.; da Cunha, M.R.; Bertran, C.A.; Kawachi, E.Y. Subperiosteal hydroxyapatite implants in rats submitted to ethanol ingestion. *Arch. Oral Biol.* **2004**, *49*, 747–753. [CrossRef]
26. González-Reimers, E.; Quintero-Platt, G.; Rodríguez-Rodríguez, E.; Martínez-Riera, A.; Alvisa-Negrín, J.; Santolaria-Fernández, F. Bone changes in alcoholic liver disease. *World J. Hepatol.* **2015**, *7*, 1258–1264. [CrossRef]
27. Ghorbel, I.; Elwej, A.; Chaabene, M.; Boudawara, O.; marrakchi, R.; Jamoussi, K.; Boudawara, T.S.; Zeghal, N. Effects of acrylamide graded doses on metallothioneins I and II induction and DNA fragmentation: Bochemical and histomorphological changes in the liver of adult rats. *Toxicol. Ind. Health* **2017**. [CrossRef]
28. Mahmood, S.A.F.; Amin, K.; Rahman, H.S.; Othman, H.H. The pathophysiological effects of acrylamide in albino wister rats. *Int. J. Med. Res. Health Sci.* **2016**, *5*, 42–48.
29. TR-575: Technical Report Pathology Tables and Curves. Available online: https://ntp.niehs.nih.gov/data/tables/tr/500s/tr575/index.html (accessed on 8 September 2020).
30. Sayon-Orea, C.; Martinez-Gonzalez, M.A.; Bes-Rastrollo, M. Alcohol consumption and body weight: A systematic review. *Nutr. Rev.* **2011**, *69*, 419–431. [CrossRef]
31. Turner, R.T.; Kidder, L.S.; Kennedy, A.; Evans, G.L.; Sibonga, J.D. Moderate alcohol consumption suppresses bone turnover in adult female rats. *J. Bone Miner. Res.* **2001**, *16*, 589–594. [CrossRef]
32. Mercer, K.E.; Wynne, R.A.; Lazarenko, O.P.; Lumpkin, C.K.; Hogue, W.R.; Suva, L.J.; Chen, J.-R.; Mason, A.Z.; Badger, T.M.; Ronis, M.J.J. Vitamin D Supplementation protects against bone loss associated with chronic alcohol administration in female mice. *J. Pharm. Exp.* **2012**, *343*, 401–412. [CrossRef]
33. Trevisiol, C.H.; Turner, R.T.; Pfaff, J.E.; Hunter, J.C.; Menagh, P.J.; Hardin, K.; Ho, E.; Iwaniec, U.T. Impaired osteoinduction in a rat model for chronic alcohol abuse. *Bone* **2007**, *41*, 175–180. [CrossRef] [PubMed]
34. Li, J.; Bao, Q.; Chen, S.; Liu, H.; Feng, J.; Qin, H.; Li, A.; Liu, D.; Shen, Y.; Zhao, Y.; et al. Different bone remodeling levels of trabecular and cortical bone in response to changes in Wnt/β-catenin signaling in mice. *J. Orthop. Res.* **2017**, *35*, 812–819. [CrossRef] [PubMed]

35. Eriksen, E.F. Cellular mechanisms of bone remodeling. *Rev. Endocr. Metab. Disord.* **2010**, *11*, 219–227. [CrossRef] [PubMed]
36. Gaddini, G.W.; Turner, R.T.; Grant, K.A.; Iwaniec, U.T. Alcohol: A simple nutrient with complex actions on bone in the adult skeleton. *Alcohol. Clin. Exp. Res.* **2016**, *40*, 657–671. [CrossRef] [PubMed]
37. Piemontese, M.; Almeida, M.; Robling, A.G.; Kim, H.-N.; Xiong, J.; Thostenson, J.D.; Weinstein, R.S.; Manolagas, S.C.; O'Brien, C.A.; Jilka, R.L. Old age causes de novo intracortical bone remodeling and porosity in mice. *JCI Insight* **2017**, *2*. [CrossRef]
38. Greenlee, D.M.; Dunnell, R.C. Identification of fragmentary bone from the Pacific. *J. Archaeol. Sci.* **2010**, *37*, 957–970. [CrossRef]
39. Pries, A.R.; Reglin, B.; Secomb, T.W. Remodeling of blood vessels: Responses of diameter and wall thickness to hemodynamic and metabolic stimuli. *Hypertension* **2005**, *46*, 725–731. [CrossRef]
40. Kuchay, M.S.; Mishra, S.K.; Farooqui, K.J.; Bansal, B.; Wasir, J.S.; Mithal, A. Hypercalcemia of advanced chronic liver disease: A forgotten clinical entity! *Clin. Cases Min. Bone Metab.* **2016**, *13*, 15–18. [CrossRef]
41. Miller, M.; Seidler, A.; Kwiterovich, P.O.; Pearson, T.A. Long-term predictors of subsequent cardiovascular events with coronary artery disease and "desirable" levels of plasma total cholesterol. *Circulation* **1992**, *86*, 1165–1170. [CrossRef]
42. Palaparthy, R.; Saini, B.K.; Gulati, A. Modulation of diaspirin crosslinked hemoglobin induced systemic and regional hemodynamic response by ethanol in normal rats. *Life Sci.* **2001**, *68*, 1383–1394. [CrossRef]
43. Felder, A.A.; Phillips, C.; Cornish, H.; Cooke, M.; Hutchinson, J.R.; Doube, M. Secondary osteons scale allometrically in mammalian humerus and femur. *R. Soc. Open Sci.* **2017**, *4*, 170431. [CrossRef] [PubMed]

MDPI

Article

Dose-Dependent Impact of Bee Pollen Supplementation on Macroscopic and Microscopic Structure of Femoral Bone in Rats

Monika Martiniakova [1,*], Ivana Bobonova [2], Robert Toman [3], Branislav Galik [4], Maria Bauerova [2] and Radoslav Omelka [2,*]

[1] Department of Zoology and Anthropology, Faculty of Natural Sciences, Constantine the Philosopher University in Nitra, 949 74 Nitra, Slovakia
[2] Department of Botany and Genetics, Faculty of Natural Sciences, Constantine the Philosopher University in Nitra, 949 74 Nitra, Slovakia; ivana.bobonova@gmail.com (I.B.); mbauerova@ukf.sk (M.B.)
[3] Department of Veterinary Sciences, Faculty of Agrobiology and Food Resources, Slovak University of Agriculture in Nitra, 949 76 Nitra, Slovakia; robert.toman@uniag.sk
[4] Department of Animal Nutrition, Faculty of Agrobiology and Food Resources, Slovak University of Agriculture in Nitra, 949 76 Nitra, Slovakia; branislav.galik@uniag.sk
* Correspondence: mmartiniakova@ukf.sk (M.M.); romelka@ukf.sk (R.O.); Tel.: +421-37-6408-737 (M.M.)

Citation: Martiniakova, M.; Bobonova, I.; Toman, R.; Galik, B.; Bauerova, M.; Omelka, R. Dose-Dependent Impact of Bee Pollen Supplementation on Macroscopic and Microscopic Structure of Femoral Bone in Rats. *Animals* **2021**, *11*, 1265. https://doi.org/10.3390/ani11051265

Academic Editors: María-Luz García and María-José Argente

Received: 4 March 2021
Accepted: 26 April 2021
Published: 28 April 2021

Simple Summary: Bee pollen is considered an interesting feed supplement with beneficial health impacts. It contains many basic nutritional compounds that improve growth performance, development and immune response of animals. However, its effect on bone structure has been studied to a limited extent and the results published so far are ambiguous. Therefore, the impact of bee pollen supplementation on selected bone characteristics of rats was investigated in our study. We determined a dose-dependent effect of bee pollen administration on macroscopic and microscopic structure of femoral bone. Several negative effects of bee pollen supplementation at the level of 0.75% on bone features have been demonstrated, while the level of 0.5% did not influence these properties in rats.

Abstract: Bee pollen has been successfully used as a feed additive with beneficial impacts on productive, reproductive, and immune conditions of animals. However, its effect on bone structure and bone health remains controversial. Therefore, the purpose of our study was to examine the impact of bee pollen supplementation on macroscopic and microscopic structure of a femoral bone using rats as suitable animal models. Male rats (1 month-old) were assigned into three groups: control (C group) that was fed a standard diet without bee pollen and two bee pollen supplemented groups (P1 and P2 groups) that received an experimental diet including 0.5% and 0.75% of bee pollen, respectively, for 3 months. A number of unfavorable effects of 0.75% bee pollen administration on bone weight, cortical bone thickness, calcium content, alkaline phosphatase activity, sizes of primary osteons' vascular canals, Haversian canals and secondary osteons in the cortical bone have been recorded, whereas these bone parameters were significantly decreased in the P2 group versus the C group. On the contrary, the concentration of 0.5% did not affect any of bone features mentioned above. In conclusion, the impact of bee pollen supplementation on femoral bone structure of rats depends on the dose used.

Keywords: bee pollen; nutrition; bone structure; bone health; rat

1. Introduction

Bee pollen is a conglomerate of flower pollen collected by the bees and mixed with both honeybee salivary enzymes and nectar [1]. It is considered a well-known feed supplement. This natural product is rich in proteins, reducing sugars, lipids, essential amino acids, fatty acids, minerals, vitamins, phytosterols, and flavonoids [2,3]. Aforementioned components are responsible for its high nutritional value [4]. Due to the fact that other bioactive compounds are also represented (e.g., carotenoids, bioactive peptides, organic acids), having beneficial health effects, it is a useful tool for therapeutic approach as well. Favorable health impacts of bee pollen are the result of the presence of polyphenol compounds with antioxidant activity, phytosterols and polyunsaturated fatty acids with antitumor properties, as well as flavonoid glycosides with increased immunological activity [2,5]. It is also regarded as a promising anti-microbial agent [3]. Bee pollen supplementation enhanced growth performance, carcass quality, and immunity of farm animals [2]. However, its effect on bone macroscopic and microscopic structure remains controversial. Yamaguchi et al. [6] revealed a higher calcium (Ca) concentration in the femoral bone of rats treated with bee pollen extract (5 and 10 mg/100 g body weight). No effect of dietary bee pollen inclusion (0.5% and 1.0% of feed) on Ca content and bone mineralization has been observed in broiler chickens [7,8]. Similarly, the concentration of Ca was not affected by bee pollen supplementation at the level of 0.2% in rats [9]. On the contrary, Tomaszewska et al. [1] reported a decreased bone mineralization, bone length, bone weight, and mean relative wall thickness in Japanese quails receiving a bee pollen dietary supplemented diet (1.0%). Bobonova et al. [10] also indicated a reduced bone weight in female rats fed with a bee pollen supplemented diet at the level of 0.75%.

Due to ambiguous results published so far, the purpose of this study was to examine the impact of bee pollen supplementation at two doses (0.5% and 0.75% of feed) on selected macroscopic and microscopic properties of femoral bone using male rats as suitable animal model.

2. Materials and Methods

One-month-old male Wistar rats (n = 15) were bred at the Slovak University of Agriculture in Nitra. They were individually housed in plastic cages (Techniplast, Italy) under standardized conditions with 12:12 h light-dark regime, 55 ± 10% of humidity and temperature of 22 ± 2 °C. Complete granular diet for rats (Bonagro, Czech Republic) and water were supplied on *ad libitum* basis. The rats were assigned into three groups of five animals each: control (C group) was fed standard complete diet without bee pollen and two bee pollen groups (P1 and P2 groups) received experimental diet including 0.5% and 0.75% of bee pollen, respectively. Dried rape (*Brassica napus L.*) bee pollen was obtained from local beekeepers from western Slovakia. Nutrient composition of bee pollen (in g/kg of dry matter) was as follows: organic matter 429.2, crude proteins 190.9, crude fat 29.5, crude fiber 13.4, calcium 1.8, and phosphorus 5.3. The experiment lasted for 3 months, identical to our previous studies [9,10]. The doses used were selected according to the studies by Yamaguchi et al. [6] and Bobonova et al. [10].

After euthanasia of rats, blood samples were collected and plasma alanine aminotransferase (ALT), aspartate aminotransferase (AST) and alkaline phosphatase (ALP) were assayed by commercially available kits (Stamar, Poland). Subsequently, both femoral bones (n = 30) were dissected, weighed, and their lengths were measured. The total body weight of rats from all groups was also recorded.

Thin sections for histological analysis of cortical bone tissue were prepared using special microtome (Leica, Germany) according to Martiniakova et al. [9]. They were evaluated by reputable classification systems [11,12]. Measured parameters included area of primary osteons' vascular canals (n = 398), Haversian canals (n = 323), and secondary osteons (n = 323). Motic Images Plus 2.0 ML software (Motic China Group Co., Nanjing, China) was used to determine them. Cortical bone thickness was also evaluated. Twenty random areas were selected per thin section, and average thickness was calculated for each bone.

Atomic absorption spectrophotometry (Perkin Elmer 4100 ZL, Waltham, MA, USA) was used to determine the concentrations of Ca, magnesium (Mg), iron (Fe) and zinc (Zn) in the femoral bone. Briefly, bone samples were dried, digested in nitric acid, and diluted with distilled water [13].

Statistical analysis was conducted using SPSS Statistics 26.0 software. The data were expressed as mean ± standard deviation. Differences in all parameters examined were detected by ANOVA with Tukey's post-hoc test. Statistical significance was assessed at $p < 0.05$.

3. Results

Total body weight (bw) of rats and femoral lengths were not affected by bee pollen supplementation, although an insignificant increase in body weight can be observed in both P1 and P2 groups. Feed intake in C group did not differ in comparison with P1 and P2 groups. On the other hand, significantly decreased femoral weight and cortical bone thickness were recorded in the P2 group versus the C group. Significant changes in the latter parameters mentioned above were also revealed between P1 and P2 groups (Figure 1A–D).

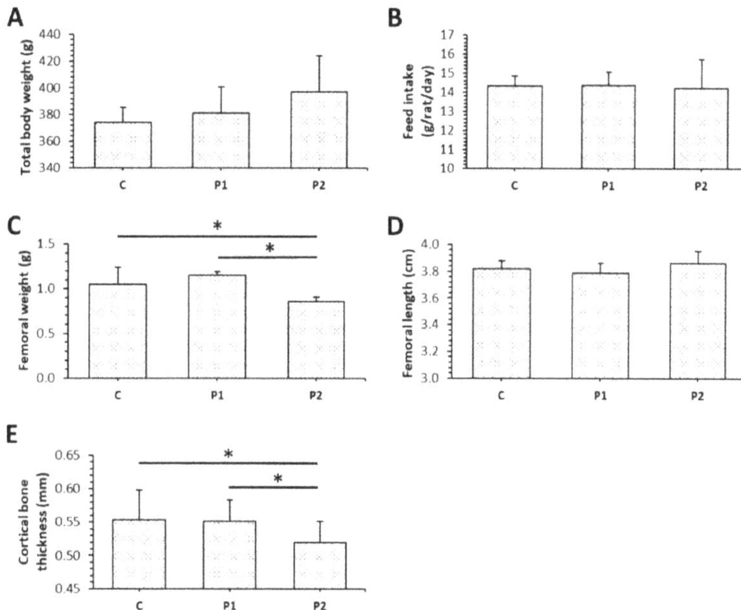

Figure 1. Total body weight (**A**), feed intake (**B**), femoral weight (**C**) and length (**D**), cortical bone thickness (**E**) in rats from C, P1, and P2 groups. * Significant differences ($p < 0.05$).

The levels of ALT did not vary between bee pollen supplemented groups and the control group. On the contrary, significantly lower values for AST were demonstrated in both P1 and P2 groups. In addition, ALP activity was significantly reduced in the P2 group when compared to the control (Figure 2A–C).

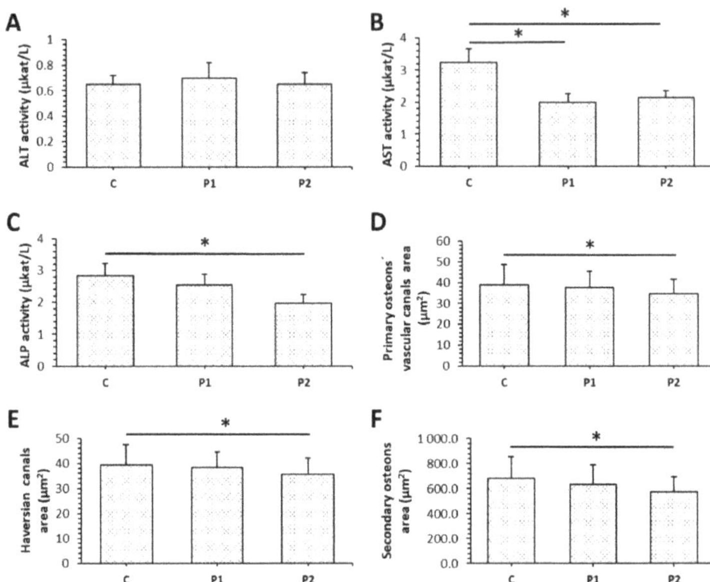

Figure 2. Levels of ALT (**A**), AST (**B**), ALP (**C**), and sizes of primary osteons' vascular canals (**D**), Haversian canals (**E**), secondary osteons (**F**) in rats from C, P1, and P2 groups. * Significant differences ($p < 0.05$).

Both surfaces (periosteal, endosteal) of the cortical bone were composed of non-vascular bone tissue in all groups studied. Primary vascular radial bone tissue was observed near the *endosteum* and in the middle part of *substantia compacta*. In the central part of the cortical bone, several secondary osteons have been identified in C, P1, and P2 groups. However, more primary and secondary osteons were recorded in *pars cranialis* (near periosteal border) in both bee pollen treated groups (Figure 3A–C).

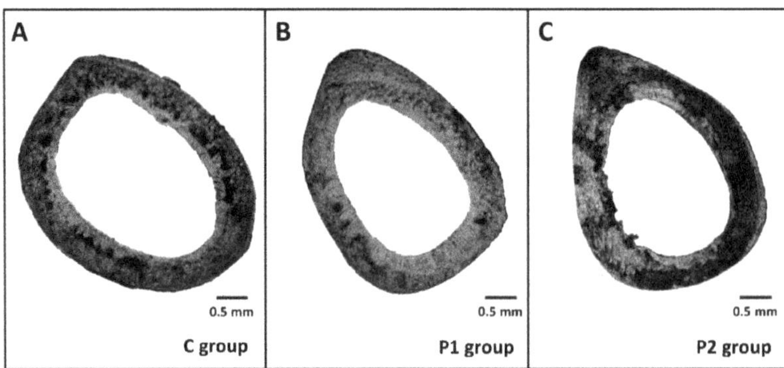

Figure 3. Cortical bone microscopic structure of rats from C (**A**), P1 (**B**) and P2 (**C**) groups.

Our results from histomorphological analysis of the cortical bone are summarized in Figure 2D–F. In the P2 group, a significantly decreased area of primary osteons' vascular canals, Haversian canals, and secondary osteons was determined in comparison with the control. Lower doses of bee pollen did not influence sizes of aforementioned structures.

The concentration of Ca in the femoral bone was significantly reduced in the P2 group against the C group (Figure 4A). Treatment with bee pollen did not affect the content of other mineral elements examined (Figure 4B–D).

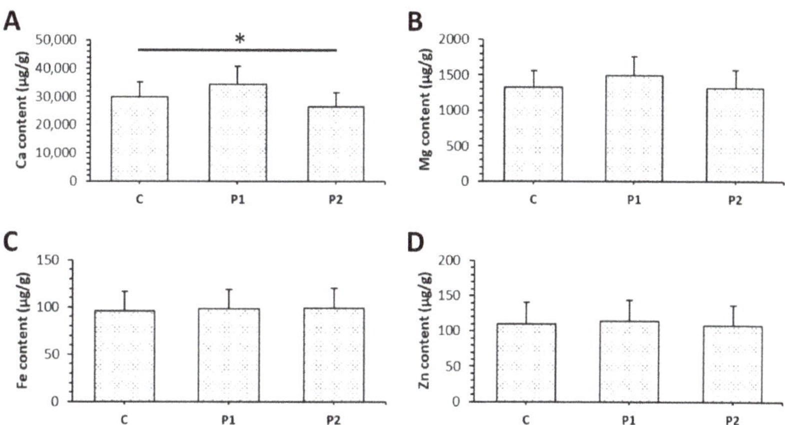

Figure 4. The content of Ca (**A**), Mg (**B**), Fe (**C**) and Zn (**D**) in femoral bones of rats from C, P1, and P2 groups. * Significant differences ($p < 0.05$).

4. Discussion

Bee pollen contains a lot of basic nutritional compounds important for productive, reproductive, and immune functions of animals. It has been used as a potential growth stimulator in many studies. According to Haro et al. [14], significantly increased total body weight was recorded in male rats receiving 1.0% of bee pollen for 10 days. The enhancement of final body weight was also observed in rabbits supplemented with 200 mg/kg bw of bee pollen for 5 weeks [15], those receiving 350 mg/kg bw of the pollen for 4 weeks [16], and also in broiler chickens treated with bee pollen at the level of 0.6% for 6 weeks [17]. On the other hand, no significant difference in total body weight was demonstrated between rats administered with bee pollen and their control in our study. Similarly, supplementation with bee pollen at the level of 0.2% (lasting 90 days) did not significantly influence body weight of rats [9]. Our findings suggest no differences in feed intake among all groups studied. However, Farag and El-Rayes [17] stated significantly decreased feed consumption in broiler chickens receiving bee pollen diets (0.2, 0.4 and 0.6% for 6 weeks) when compared to the control group. The dietary inclusion of bee pollen (200 mg/kg bw for 5 weeks) significantly reduced feed intake in rabbits [15]. It seems that longer administration with bee pollen leads to the elimination of significant changes in total body weight and feed intake between supplemented animals and their controls. Surprisingly, decreased femoral weight and cortical bone thickness were recorded in the P2 group. These results are partially consistent with those reported by Tomaszewska et al. [1], who noticed that 1.0% bee pollen administration (for 42 days) significantly reduced bone weight and mean relative bone thickness in Japanese quails, despite the lack of changes in their final body weight. Lower femoral weight was also determined in female rats receiving 0.75% of bee pollen for 90 days [10].

Dietary intake of bee pollen did not influence ALT activity in our study. However, significantly decreased values for AST were observed in both P1 and P2 groups. According to Attia et al. [15], levels of liver enzymes (AST, ALT) declined significantly in rabbits supplemented with bee pollen at a dose of 200 mg/kg bw. At the same time, Elnany and Elkholy [18] reported a reduction in AST, ALT values in rabbits treated with bee pollen (200, 300 and 400 mg/kg bw for 8 weeks). Treatment with bee pollen at the level of 0.6%

(for 6 weeks) also decreased levels of both aforementioned enzymes in broiler chickens [17], thus indicating an ability of bee pollen to decline liver damage. Moreover, we determined a significantly reduced ALP activity in rats receiving 0.75% of bee pollen. The activity of ALP found in serum is a composite of isoenzymes, especially detected in liver and bone (more than 80%). Taking into account the bone, lower levels of ALP could be associated with Zn and Mg deficiency [19] and/or depressed osteoblast activity leading to a reduced bone formation [20]. Since no significant differences in Zn and Mg contents have been found between P2 and C groups, the decreased ALP activity indicated an inhibited bone formation rate in the P2 group. That evidence is also supported by significantly lower Ca concentration and reduced cortical bone thickness in those rats. In general, decreased cortical bone thickness is associated with increased stresses that could cause bone resorption [21], and simultaneously reduced Ca content in the femoral bone only confirms this fact. Our findings differ from those reported by Yamaguchi et al. [6], who mentioned a higher Ca concentration and increased ALP activity in the femoral bone of rats administered with bee pollen extract obtained from *Cistus ladaniferus* (5 and 10 mg/100 g bw). Yamaguchi et al. [22] also detected a stimulatory effect of bee pollen *Cistus ladaniferus* extract on bone formation and its inhibitory impact on bone resorption in vitro using osteoblast cells. However, Tomaszewska et al. [1] also stated a reduced mineralization in the tibia of Japanese quails supplemented with 1.0% of multifloral bee pollen. In addition, these bones were the weakest and have the greatest susceptibility to plastic deformation. Unlike previous studies, Kleczek et al. [7] and Oliveira et al. [8] reported no effect of bee pollen treatment (0.5% and 1.0% of feed) on Ca content and bone mineralization in broiler chickens. Rape bee pollen inclusion of 0.2% did not influence the concentration of Ca in the femoral bone of rats as well [9].

Higher number of primary and secondary osteons in *pars cranialis* of the cortical bone in both P1 and P2 groups might suggest an enhanced bone remodeling there. It is generally accepted that cranial and lateral sides contain fewer secondary osteons when compared to caudal and medial sides due to different tension and compression strains [23]. For this reason, we believe that enhanced density of secondary osteons in the cranial side could be associated with bee pollen treatment. On the other hand, significantly reduced sizes of primary osteons' vascular canals, Haversian canals, and secondary osteons were demonstrated in rats after 0.75% bee pollen supplementation. This evidence might be consistent with a vasoconstriction of blood vessels in both canals and impaired blood flow. Generally, blood vessels supplying bones organize both bone remodeling and bone regeneration by delivering oxygen, nutrients, growth factors, or hormones to bone cells [24]. Decreased sizes of secondary osteons in rats from the P2 group could indicate a lower bone toughness. Significantly smaller osteons have been namely associated with decreased bone strength [25]. Interestingly, the lower dose (0.5%) of bee pollen did not affect sizes of aforementioned structures in the cortical bone of rats, similar to the treatment with bee pollen at the level of 0.2% in our previous study [9].

In summary, our results partially support the findings of Tomaszewska et al. [1], who also detected possible unfavorable impacts of bee pollen treatment on cortical bone properties using Japanese quails as an animal model. In terms of our study limitations, the number of rats used in the experiment was limited and only males were used. Therefore, it would be necessary to perform further research involving higher number of rats of both sexes. Additional analyses with different doses of bee pollen and longer duration of supplementation would also be valuable.

5. Conclusions

Our study revealed a dose-dependent impact of bee pollen supplementation on the femoral bone structure of rats. Several negative effects of bee pollen administration at the level of 0.75% on bone weight, cortical bone thickness, Ca content, ALP activity, sizes of primary osteons' vascular canals, Haversian canals and secondary osteons have been determined, while the concentration of 0.5% had no effect. The results encourage caution in dosing and nutritional recommendations regarding bee pollen supplementation for preventive use. However, additional experiments are needed to verify the results obtained.

Author Contributions: Conceptualization, M.M. and R.T.; methodology, M.M. and R.O.; software, I.B., M.B. and R.O.; validation, M.M., R.T. and R.O.; formal analysis, I.B. and M.B.; investigation, M.M., I.B., R.T., B.G. and R.O.; resources, M.M. and R.O.; data curation, R.O.; writing—original draft preparation, M.M.; writing—review and editing, M.M. and R.O.; visualization, M.M., I.B. and R.O.; supervision, M.M. and R.O.; project administration, R.O. All authors have read and agreed to the published version of the manuscript.

Funding: This research was funded by Ministry of Education, Science, Research and Sport of the Slovak Republic, grant number VEGA 1/0505/18.

Institutional Review Board Statement: The study was conducted according to the guidelines of the Declaration of Helsinki, and approved by the Ethics Committee and the State Veterinary and Food Administration of the Slovak Republic (protocol code 2072/08-221/3).

Conflicts of Interest: The authors declare no conflict of interest.

References

1. Tomaszewska, E.; Knaga, S.; Dobrowolski, P.; Lamorski, K.; Jabłoński, M.; Tomczyk-Warunek, A.; Jard Kadhim, M.; Hułas-Stasiak, M.; Borsuk, G.; Muszyński, S. The Effect of Bee Pollen on Bone Biomechanical Strength and Trabecular Bone Histomorphometry in Tibia of Young Japanese Quail (Coturnix Japonica). *PLoS ONE* **2020**, *15*, e0230240. [CrossRef]
2. Abdelnour, S.A.; Abd El-Hack, M.E.; Alagawany, M.; Farag, M.R.; Elnesr, S.S. Beneficial Impacts of Bee Pollen in Animal Production, Reproduction and Health. *J. Anim. Physiol. Anim. Nutr.* **2019**, *103*, 477–484. [CrossRef]
3. Didaras, N.A.; Karatasou, K.; Dimitriou, T.G.; Amoutzias, G.D.; Mossialos, D. Antimicrobial Activity of Bee-Collected Pollen and Beebread: State of the Art and Future Perspectives. *Antibiotics* **2020**, *9*, 811. [CrossRef] [PubMed]
4. Yucel, B.; Topal, E.; Kosoglu, M. Bee Products as Functional Food. In *Superfood and Functional Food—An Overview of Their Processing and Utilization*; IntechOpen: London, UK, 2017. [CrossRef]
5. Mărgăoan, R.; Stranț, M.; Varadi, A.; Topal, E.; Yücel, B.; Cornea-Cipcigan, M.; Campos, M.G.; Vodnar, D.C. Bee Collected Pollen and Bee Bread: Bioactive Constituents and Health Benefits. *Antioxidants* **2019**, *8*, 568. [CrossRef] [PubMed]
6. Yamaguchi, M.; Hamamoto, R.; Uchiyama, S.; Ishiyama, K.; Hashimoto, K. Anabolic Effects of Bee Pollen Cistus Ladaniferus Extract on Bone Components in the Femoral-Diaphyseal and -Metaphyseal Tissues of Rats in Vitro and in Vivo. *J. Health Sci.* **2006**, *52*, 43–49. [CrossRef]
7. Kleczek, K.; Majewska, K.; Makowski, W.; Michalik, D. The Effect of Diet Supplementation with Propolis and Bee Pollen on the Physicochemical Properties and Strength of Tibial Bones in Broiler Chickens. *Arch. Anim. Breed.* **2012**, *55*, 97–103. [CrossRef]
8. De Oliveira, M.; da Silva, D.; Loch, F.C.; Martins, P.C.; Dias, D.M.B.; Simon, G.A. Effect of Bee Pollen on the Immunity and Tibia Characteristics in Broilers. *Braz. J. Poult. Sci.* **2013**, *15*, 323–327. [CrossRef]
9. Martiniaková, M.; Boboňová, I.; Omelka, R.; Ďúranová, H.; Babosová, R.; Stawarz, R.; Toman, R. Low Administration of Bee Pollen in the Diet Affects Bone Microstructure in Male Rats. *Potravinarstvo* **2014**, *8*, 277–283. [CrossRef]
10. Boboňová, I.; Martiniaková, M.; Chovancová, H.; Omelka, R.; Toman, R.; Hájková, Z.; Stawarz, R. The Effect of Bee Pollen on Macroscopic Structure of Femora in Adult Female Rats after an Experimental Addition in Diet. *Anim. Welf. Ethol. Hous. Syst.* **2013**, *9*, 409–413.
11. De Ricqles, A.J.; Meunier, F.J.; Castanet, J.; Francillon-Vieillot, H. Comparative microstructure of bone. In *Bone Matrix and Bone Specific Products*; CRC Press, Inc.: Boca Raton, FL, USA, 1991; pp. 1–78.
12. Enlow, D.H.; Brown, S.O. A comparative histological study of fossil and recent bone tissues. Part I. *Tex. J. Sci.* **1956**, *8*, 405–412.
13. Martiniakova, M.; Omelka, R.; Stawarz, R.; Formicki, G. Accumulation of Lead, Cadmium, Nickel, Iron, Copper, and Zinc in Bones of Small Mammals from Polluted Areas in Slovakia. *Pol. J. Environ. Stud.* **2012**, *21*, 153–158.
14. Haro, A.; López-Aliaga, I.; Lisbona, F.; Barrionuevo, M.; Alférez, M.J.; Campos, M.S. Beneficial Effect of Pollen and/or Propolis on the Metabolism of Iron, Calcium, Phosphorus, and Magnesium in Rats with Nutritional Ferropenic Anemia. *J. Agric. Food Chem.* **2000**, *48*, 5715–5722. [CrossRef] [PubMed]
15. Attia, Y.A.; Bovera, F.; El-Tahawy, W.S.; El-Hanoun, A.M.; Al-Harthi, M.A.; Habiba, H.I. Productive and reproductive performance of rabbits does as affected by bee pollen and/or propolis, inulin and/or mannan-oligosaccharides. In Proceedings of the World Rabbit Science, Editorial Universitat Politècnica de València, Valencia, Spain, 23 December 2015; Volume 23, pp. 273–282.
16. Abdel-Hamid, T.M.; El-Tarabany, M.S. Effect of Bee Pollen on Growth Performance, Carcass Traits, Blood Parameters, and the Levels of Metabolic Hormones in New Zealand White and Rex Rabbits. *Trop. Anim. Health Prod.* **2019**, *51*, 2421–2429. [CrossRef] [PubMed]
17. Farag, S.; El-Rayes, T. Effect of Bee-Pollen Supplementation on Performance, Carcass Traits and Blood Parameters of Broiler Chickens. *Asian J. Anim. Vet. Adv.* **2016**, *11*, 168–177. [CrossRef]

18. Elnany, B.A.; Elkholy, K.H. Effect of natural additive (bee pollen) on immunity and productive and reproductive performances in rabbits. 1- Growth Performance, Digestive and Immune Responses in Growing Rabbits. *Egypt. Poult. Sci. J.* **2014**, *34*, 579–606. [CrossRef]
19. Ray, C.S.; Singh, B.; Jena, I.; Behera, S.; Ray, S. Low Alkaline Phosphatase (ALP) In Adult Population an Indicator of Zinc (Zn) and Magnesium (Mg) Deficiency. *Curr. Res. Nutr. Food Sci. J.* **2017**, *5*, 347–352.
20. Lim, S.M.; Kim, Y.N.; Park, K.H.; Kang, B.; Chon, H.J.; Kim, C.; Kim, J.H.; Rha, S.Y. Bone Alkaline Phosphatase as a Surrogate Marker of Bone Metastasis in Gastric Cancer Patients. *BMC Cancer* **2016**, *16*, 385. [CrossRef]
21. Papadopoulos, M.A.; Papageorgiou, S.N.; Zogakis, I.P. Success rates and risk factors of miniscrew implants used as temporary anchorage devices for orthodontic purposes. In *Skeletal Anchorage in Orthodontic Treatment of Class II Malocclusion*; Papadopoulos, M.A., Ed.; Elsevier: Amsterdam, The Netherlands, 2015; pp. 258–273, ISBN 978-0-7234-3649-2.
22. Yamaguchi, M.; Uchiyama, S.; Nakagawa, T. Anabolic Effects of Bee Pollen Cistus Ladaniferus Extract in Osteoblastic MC3T3-E1 Cells In Vitro. *J. Health Sci.* **2007**, *53*, 625–629. [CrossRef]
23. Zedda, M.; Lepore, G.; Biggio, G.P.; Gadau, S.; Mura, E.; Farina, V. Morphology, Morphometry and Spatial Distribution of Secondary Osteons in Equine Femur. *Anat. Histol. Embryol.* **2015**, *44*, 328–332. [CrossRef]
24. Filipowska, J.; Tomaszewski, K.A.; Niedźwiedzki, Ł.; Walocha, J.A.; Niedźwiedzki, T. The Role of Vasculature in Bone Development, Regeneration and Proper Systemic Functioning. *Angiogenesis* **2017**, *20*, 291–302. [CrossRef]
25. Felder, A.A.; Phillips, C.; Cornish, H.; Cooke, M.; Hutchinson, J.R.; Doube, M. Secondary Osteons Scale Allometrically in Mammalian Humerus and Femur. *R. Soc. Open Sci.* **2017**, *4*, 170431. [CrossRef] [PubMed]

 *animals* 

Article

Comparative Growth and Economic Performances between Indigenous Swamp and Murrah Crossbred Buffaloes in Malaysia

Amirul Faiz Mohd Azmi [1], Hasliza Abu Hassim [1,2,*], Norhariani Mohd Nor [1], Hafandi Ahmad [1], Goh Yong Meng [1], Punimin Abdullah [3], Md Zuki Abu Bakar [1], Jaizurah Vera [1], Nurain Syahida Mohd Deli [1], Annas Salleh [2,4] and Mohd Zamri-Saad [4]

[1] Department of Veterinary Preclinical Sciences, Faculty of Veterinary Medicine, Universiti Putra Malaysia, Serdang 43400, Selangor, Malaysia; amirulfaizazmi@gmail.com (A.F.M.A.); norhariani@upm.edu.my (N.M.N.); hafandi@upm.edu.my (H.A.); ymgoh@upm.edu.my (G.Y.M.); zuki@upm.edu.my (M.Z.A.B.); jaizurahvera@gmail.com (J.V.); NurainSyahida@gmail.com (N.S.M.D.)

[2] Laboratory of Sustainable Animal Production and Biodiversity, Institute of Tropical Agriculture and Food Security, Universiti Putra Malaysia, Serdang 43400, Selangor, Malaysia; annas@upm.edu.my

[3] Faculty of Sustainable Agriculture, Universiti Malaysia Sabah, Sandakan 90509, Sabah, Malaysia; puniminabdullah@gmail.com

[4] Department of Veterinary Laboratory Diagnosis, Faculty of Veterinary Medicine, Universiti Putra Malaysia, Serdang 43400, Selangor, Malaysia; mzamri@upm.edu.my

* Correspondence: haslizaabu@upm.edu.my; Tel.: +603-9769-3417

Citation: Mohd Azmi, A.F.; Abu Hassim, H.; Mohd Nor, N.; Ahmad, H.; Meng, G.Y.; Abdullah, P.; Abu Bakar, M.Z.; Vera, J.; Mohd Deli, N.S.; Salleh, A.; et al. Comparative Growth and Economic Performances between Indigenous Swamp and Murrah Crossbred Buffaloes in Malaysia. *Animals* **2021**, *11*, 957. https://doi.org/10.3390/ani11040957

Academic Editors: María-Luz García, María-José Argente and Maria Caria

Received: 1 February 2021
Accepted: 24 March 2021
Published: 30 March 2021

Simple Summary: A buffalo breeding farm was selected to study the growth performance of Swamp and Murrah crossbred buffaloes. The farm was practicing extensive grazing system without supplementation since 2010 to 2011. In early 2012, the farm had implemented a new intervention to improve the growth performance via improving the feed and the feeding management. Farm records between 2010 to 2015 were analyzed for growth performance and partial budget analysis. So far, there is no comparative study done between Swamp and Murrah crossbred buffaloes in Malaysia. Therefore, in the present study, we aimed to study the differences in the biological and economical performances between Swamp and crossbred buffaloes in Malaysia. With a new intervention, a significant improvement was reported of the number of calves born, average birth weight, and reduced percentage of calf mortality rate, calving interval, and weaning age. Crossbred buffalo showed dominance in biological performance in terms of higher pre- and post-weaning daily weight gain and taking a shorter period to achieve market and breeding weight compared to Swamp buffaloes. Thus, reared Murrah crossbred buffaloes with new intervention management would give a farmer a higher profit return. However, with reared Swamp, the farmer potentially conserves the local indigenous breed of Swamp buffalo.

Abstract: This study was conducted to compare the growth and economic performances between Swamp and Murrah crossbred buffaloes. The records of 108 Swamp and 276 Murrah crossbred buffaloes born between January 2010 and December 2015 were used in this study. The farm was practicing an extensive grazing system without supplementation from January 2010 to December 2011 (pre-intervention) and a new implementation of supplement in the feeding regime from January 2012 to December 2015 (post-intervention). The birth, weaning, and body weight at three monthly intervals, number of calves born, and mortality rate of calves at different years and during pre- and post-intervention were analyzed using a general linear model procedure. The interventions in 2012 had a positive effect on increasing the number of calves born for both breeds, average birth weight, economic performance, and reduce mortality calf rate. As a result, the birth weight of Murrah crossbred buffaloes was higher (36.63 ± 0.50 kg) than Swamp buffaloes (34.69 ± 0.40 kg) ($p < 0.05$). The average pre-weaning daily weight gain for Swamp and Murrah crossbred buffaloes was 0.73 and 0.98 kg/day ($p < 0.05$), while the average post-weaning daily weight gain was 0.39 and 0.44 kg/day, respectively ($p < 0.05$). The Swamp and Murrah crossbred buffaloes achieved the targeted market weight of 250 kg at 18 and 15 months old, respectively, while the targeted breeding weight of 385 kg was achieved at 30 and 26 months old, respectively. In this farm, on average a total of 64 calves were born yearly, with the ratio number of born calves per number of mated dams recorded higher in Murrah crossbred buffaloes as compared to Swamp buffalo (0.64 vs. 0.37) ($p < 0.05$). Furthermore, the average number of calves born in the post-intervention period (January 2012–December 2015) was significantly higher than in the pre-intervention period (January 2010–December 2011), respectively (Swamp: 23 vs. 8 and Murrah crossbred: 53 vs. 31, respectively) ($p < 0.05$). Partial budget method was used to estimate the net gain or loss between the two breeds. The average annual revenue was 2304.14 MYR (566.13 USD) for Swamp buffaloes and 4531.50 MYR (1113.39 USD) for Murrah crossbred buffaloes. The average annual cost saving was 340.02 MYR (83.54 USD) for Swamp and 215.75 MYR (53.01 USD) for Murrah crossbred buffaloes. On the other hand, annual added cost was 84.95 MYR (20.87 USD) for Swamp and 96.76 MYR (23.77 USD) for Murrah crossbred buffaloes. Therefore, the annual net benefit was 2559.21 MYR (628.80 USD) for Swamp and 4650.49 MYR (1142.63 USD) for Murrah crossbred buffaloes. As a conclusion, this study had shown that the higher average daily weight gain contributes to better cost savings, as shown by the crossbred buffaloes.

Keywords: buffaloes; economy; growth; Murrah crossbred; Swamp

1. Introduction

There are two types of domestic water buffaloes or Asian buffaloes, the Swamp and river buffaloes and Murrah buffaloes [1] as shown in Figures 1 and 2. The Swamp buffaloes are found mainly in Southeast Asia and China and are kept mainly for meat and draft power in paddy fields and oil palm plantation. However, Murrah buffaloes are found mainly in South Asia, especially India and Pakistan, and are kept mainly for milk. In the field, river and Swamp buffaloes can be differentiated by their morphology and behavior. Swamp buffaloes are ash or dark grey with a white chevron line on the neck either one or two stripes and have socks, while the tip of the tail and the horns are swept backwards [2,3]. They prefer to wallow in the marshland and mud and have large feet with slow steady movement that make them well suited for paddy land preparation in swampy waterlogged rice fields [4]. On the other hand, Murrah buffaloes have a black body with tightly and forwardly curled horns. They prefer to wallow in clean water [2]. Generally, Swamp buffaloes are smaller than the Murrah buffaloes, while the crossbred buffaloes have the same morphology as Murrah. However, they are smaller than Murrah, but bigger than the Swamp buffaloes.

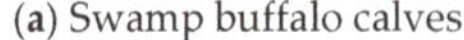

(**a**) Swamp buffalo calves

(**b**) Adult Swamp buffalo

Figure 1. The Swamp buffaloes; (**a**): Swamp buffalo calves; (**b**): Adult Swamp buffalo.

(a) Crossbred buffalo

(b) Pure breed Murrah buffalo

Figure 2. The crossbred and pure breed Murrah buffaloes; (**a**): Crossbred buffalo, (**b**): Pure breed Murrah buffalo.

In trying to improve the milk performance of buffaloes, crossbreeding between the Murrah and the Swamp buffaloes was attempted in many Southeast Asian countries, notably in the Philippines and Malaysia, with the hope that the resulted Murrah crossbred animals are bigger in size to produce better milk yield than the native Swamp buffaloes [5]. In Malaysia, Swamp buffaloes are raised purposely for meat and draught work, while Murrah crossbred buffaloes are raised for milk. Later, it was reported that milk yields from crossbred buffaloes are significantly higher than the local buffaloes [6]. However, the status of buffalo milk production in Malaysia is poorly understood and is currently underdeveloped. Milk production in dairy buffalo in Malaysia varies from 4.00 to 6.00 L/day [7] to approximately 8.00 L/day [8], which is still behind the potential of superior buffaloes at 15.00–20.00 L/day [9]. However, our local Murrah crossbred buffalo can only produce an average of 4.7 L of milk/day/animal per lactation period, producing on average a total of 1000 L of milk annually [9], and this is far from meeting the local milk demand.

In general, buffalo is one of the ruminants with economic importance for small holder farmers in many developing countries in Asia [10]. Furthermore, due to farm record, they have a longer lactation period; hence, farmers potentially harvest more milk and generate more income. In fact, crossbred buffaloes mature earlier and have shorter calving interval, hence producing more calves in their lifetime [6]. Therefore, to estimate the potential improvement in the income from buffaloes, the economic performance between Swamp buffaloes and Murrah crossbred buffaloes should be compared. This will aid future farmers in choosing the beneficial buffalo breed.

According to Nanda et al. [11], reproductive and growth performance of buffaloes are generally poor. This was supported by previous study reports that buffalo management in a buffalo breeding farm in Sabah was practicing the traditional feeding management before 2012 and changed the feeding management in January 2012 [12]. Although there was a retrospective study, comparing the reproductive performance at that farm in 2004 until 2011, there was no comparative study on the differences in the biological and economical performances between Swamp and Murrah crossbred buffaloes in Sabah, Malaysia, during pre- and post-intervention implementation of new feeding regime. Therefore, the objective of this study was to analyze the growth and economic performances between the indigenous Swamp and Murrah crossbred buffaloes kept at a breeding farm in Sabah, Malaysia, and describe their performance following pre- (January 2010–December 2011) and post-interventions (January 2012–December 2015) with an improved rearing system.

2. Materials and Methods

2.1. Statement of Animal Rights

The study was performed and managed according to approval of Animal Utilization Protocol (AUP), Institutional Animal Care and Use Committee (IACUC), Universiti Putra Malaysia (Approval No. UPM/IACUC/AUP-R039/2012, on 2 February 2012).

2.2. Study Area

This study was conducted at the Buffalo Breeding and Research Centre, Telupid, Sabah (5°30′ N, 117°7′ E). The farm consisted of 749 acres of land, and currently with a total of 405 heads of buffaloes. There were two types of buffalo breeds available in this farm, the Swamp and the Murrah × Swamp crossbred, representing 54% and 46% of the total buffalo population, respectively. The sample size in this study was represented by 10% to 15% of the total buffaloes in the farm, according to the Cochran method [13,14]. The sample size reflected proportions of the population in the farm, with the buffaloes selected randomly.

The crossbred animals were selected from the farms that were practicing natural breeding, which involved breeding between the pure Murrah males with Swamp females. The identification of these two subspecies of buffaloes was conducted based on conventional techniques (morphological or physiological characteristics). For genetic predisposition, dominance, and origin of crossbred, karyotypic (based on chromosomes number) and molecular identification techniques were used to identify the phylogeny of the animals or to compare the crossbred buffaloes with their purebred parents [15–17]. In addition, prior study on this farm had reported that phylogenetic tree and mtDNA analysis on Swamp buffaloes were genetically conserved and the crossbred buffaloes were dominantly Swamp according to the maternal lineage using d-loop mtDNA [17].

The 398.5 acres of pastureland were divided into two paddocks with establish pasture (*Brachiaria decumbens*). The Swamp and Murrah crossbred buffaloes in this farm were kept separated. Wallowing areas were available in each paddock and drinking water was available ad libitum. The buffaloes were monitored daily by farm supervisor on their health and nutritional management. The buffaloes were kept extensively and free to graze all day within the paddock enclosed by barbed wire. Even though the pasture in general was poorly maintained, the farm practiced an extensive one-month rotational grazing system without feed supplementation, and the intervals were determined according to the size of each paddock and through the supervision of the farm manager. This farm practiced natural breeding with male to female ratio of 1:20. Breeding season was between November and January each year, and pregnancy diagnosis was carried out every three months following breeding. As routinely practiced by the farm, upon reaching the market weight of 250 kg, the buffaloes were sold, mainly to the surrounding farmers for fattening. The farm practiced minimum breeding weight of 385 kg, and all buffaloes were weighed every three months.

2.3. Pre-Intervention

At the start of the study, two years of farm records between January 2010 and December 2011 were selected and analyzed for growth performance parameters. At the same time, pasture samples were collected at six sites of 1 m^2 area from the paddock for proximate analysis of the nutrient contents [18].

2.4. The Intervention

The intervention was implemented from January 2012 until December 2015 by improving the feeding management. This included the use of fertilizer (20 tons organic fertilizer with 500 kg of urea) and proper management to improve the pasture. Fertilizer was used on the pastureland twice a year. The animals were offered concentrate supplementation at the rate of 1.5 kg/animal/day (moisture: 9.64%, ash: 5.44%, crude fiber: 7.49%, crude fat: 5.46%, crude protein: 18.15%, gross energy: 15.74% MJ/kg). After six months, pasture and concentrate samples were collected from the paddocks and re-analyzed of the nutrient contents (Table 1). In January 2016, all data from 2010 to 2015 were gathered for further analysis.

Table 1. Nutritional composition of diet on pre-intervention (January 2010–December 2011) and post-intervention (January 2012–December 2015).

Nutrient Composition	Pre-Intervention (*Brachiaria decumbens*)	Post-Intervention (*Brachiaria decumbens* + Concentrate)
[1] DM (%)	99.50	99.49
Ash (% DM)	5.09	5.69
Crude fiber (% DM)	26.03	23.73
Ether extract (% DM)	2.03	2.92
Crude protein (% DM)	6.09	8.08
[2] NDF (% DM)	64.27	57.96
[3] ADF (% DM)	33.86	28.70
[4] ADL (% DM)	3.55	3.32
Carbohydrate (% DM)	59.43	61.53
Gross energy (MJ/kg)	11.07	12.1
Hemicellulose (% DM)	30.41	29.25
Cellulose (% DM)	30.32	25.38

The data are the mean of triplicate analyses of each diet.[1] DM: dry matter; [2] NDF: neutral detergent fiber, [3] ADF: acid detergent fiber, [4] ADL: acid detergent lignin.

2.5. *Analysis of Growth Performance*

A total of 108 Swamp (female: 60 and male: 48) and 276 Murrah crossbred (female: 152 and male: 124) buffaloes that were born between January 2010 and December 2015 were included in this study. The average number of female's breeders also was recorded in this study since 2010 to 2015 (Swamp: 49 and Murrah crossbred: 72). The average number of calving rates per year, mortality rate, and average birth weight were recorded in this study. The number of calving rates per year was calculated by the average number of calves born per year divided with the number of mating female per year. All selected buffaloes had gone through the same management system described previously. Records of the selected animals in both sexes, which included the animal identification and breed, birth weight, weaning weight (at three months old), and the three-monthly body weights were further analyzed using statistical software. The average daily weight gain and the period taken to reach the targeted 250 kg market weight (taking on average 12 to 18 months) and the 385 kg breeding weight (taking on average 24 to 30 months) were calculated.

The average weight gain was determined as the weight difference between the final weight and initial weight divided by the number of days between the final and the initial weights. Similarly, records of the average body weight and differences in the three-monthly body weight of the breeder Swamp and Murrah crossbred buffaloes were analyzed.

2.6. *Statistical Analysis*

All data were collected and subjected using Microsoft Excel and were analyzed using software package SPSS (Statistical Package for the Social Science 25.0, Inc., Chicago, IL, USA). Comparisons between breeds, months, years, and pre- and post-intervention and their interaction for parameters body weight, average daily weight gains, calving rate per year, mortality rate of calves, and average birth weight of calves from 2010 until 2015, were performed using the general linear model (GLM) procedures, and the resulting P-values were corrected by Bonferroni's method for multiple comparisons. Linear mixed effects models were utilized with both 'Diets' and 'Breed' as fixed effects to capture the appropriate structure for GLM, while the months, years, and pre- and post-intervention were considered as a random effect. For all the statistical tests used, results were considered significant at $p \leq 0.05$.

2.7. Partial Budget Analysis

The economic analysis was done to compare the differences between pre- and post-intervention using partial budget where net gain or loss was estimated by subtracting total loss from the total gain [19]. The assumption of total gain included the sum of increased revenue and reduced costs, while total loss included the sum of decreased revenue and added costs. Increased revenue included the increased values in improved birth weight, increased number of calves, improvement number survival rate of calves, and sales calves per year. Decreased revenue included reduced values in shorter first calving age cost and shorter calving interval cost. Added costs/losses were the costs of fertilizer, flushing, cost of calf supplemental feed, cost of feed for increase total population of calf heifer, deworming drugs, and ID tag.

The biological input was collected from the farm records, which included the birth, death, weaning, and breeding records reviewed as the data for the study. Other biological inputs, including the average increase in the number of calves born, average increase in birth weight, average decrease in calf mortality, and average age at first calving, were collected from the farm record books and by interviewing the farm manager. The calving interval was calculated as the interval period between two successive calvings of each buffalo divided by the number of buffaloes in calving group, whereas the calving rate was recorded as the ratio number of born calves per number of mated dams. In this calculation, the first 12 weeks following calving were deducted from the calculation, assuming that 12 weeks were required for recovery in which the females were not used for breeding. The biological inputs collected are as in Table 2. Early age at first calving was assumed to be at 33 months old (assuming 24 months to reach 350 kg from birth was adding with nine months for successful pregnancy).

Table 2. Biological parameters used in the partial budget analysis.

Variables	Inputs	
	Swamp Buffalo	Murrah Crossbred
Difference ratio number of calves per number of mated dams between post- and pre-intervention	0.05	0.11
Difference in number of calves born between post- and pre-intervention (head)	15	20
Difference in calves birth weight between post- and pre-intervention (kg)	6.76	5.72
Difference in calve survival rate between post- and pre-intervention (%) [1]	7.35	8.17
Difference in first calving age between post- and pre-intervention (months)	12.5	6.5
Difference in calving interval between post- and pre-intervention (months)	33.17	21.91
Average sales of buffalo calves (2010–2011)	8647.25	26,247
Average sales of buffalo calves (2012–2015)	10,943.38	30,758

All the calculation had been done between post- (2012–2015) and pre-intervention (2010–2011) period. [1] Survival rate: 100—mortality rate.

For the economical inputs, the prices of live buffalo per kg and feed per kg, and the costs of feed per week per animal, the fertilizer, the deworming drugs, and the ID tag per animal were obtained from the farm records. Average value of sold animals per year was obtained from the buffalo sales record book. The cost of treatment involved only the deworming drugs. The economic input is as shown in Table 3.

Table 3. Economic parameters used in the partial budget analysis.

Variables	Price (MYR) [1]
Buffalo calves per kg [2]	5.50
Price of PKC per kg	0.80
Feed cost per week (MYR/animal)	7.20
Feed cost per year calf heifer (MYR/animal)	49.00
Fertilizer cost (MYR/animal)	64.79
Deworming cost (MYR/animal)	0.50
ID tag cost (MYR/animal)	2.00
Flushing cost (MYR/animal)	5.60

[1] All amounts are in Malaysian Ringgit (MYR); [2] The price of live buffalo based on average market price since 2010 to 2015.

There was no foregone revenue, as Murrah crossbred and Swamp breeds were assumed to be similar in aesthetic value. To complete the calculation for the items, in total 12 formulas (1–12) were derived followed the method by Moran [20] as below:

Increased Revenues:

$$\text{Improved birth weight} = \text{Estimation value of improved birth weight per year (MYR)} = BW_{mean} \times \Delta R \times \$_{\text{per kg calf}} \quad (1)$$

BW_{mean} = Improved mean calf body weight; ΔR = Change ratio number born calves per number of mated dams; $\$_{\text{per kg calf}}$ = Price of calf per kg

$$\text{Increased number of calves} = \text{Estimation value of increase number calf per year (MYR)} = \Delta n_{\text{improved number of calves born}} \times \Delta R \times \$_{\text{per kg calf}} \quad (2)$$

$\Delta n_{\text{improved number of calves born}}$ = Improved number of calves born; ΔR = Change ratio number born calves per number of mated dams; $\$_{\text{per kg calf}}$ = Price of calf per kg

$$\text{Increased number of survival calves} = \text{Estimation number of survival calves per year (MYR)} = \Delta SR \times \Delta R \times \$_{\text{per kg calf}} \quad (3)$$

ΔSR = Increased in survival rate; ΔR = Change ratio number born calves per number of mated dams; $\$_{\text{per kg calf}}$ = Price of calf per kg

$$\text{Increased sales of buffaloes} = \text{Estimation number of calves sales (MYR)} = AS_{2012\text{-}2015} - AS_{2010\text{-}2011} \quad (4)$$

$AS_{2012–2015}$ = Average sales of buffaloes in 2012 to 2015; $AS_{2010–2011}$ = Average sales of buffaloes in 2010 to 2011

Reduced cost:

$$\text{Shorter in first calving age (MYR)} = \text{Estimation of shorter calving interval after new feeding management cost (MYR)} = \Delta FCA \times FC_{week} \quad (5)$$

ΔFCA = Change in first calving age; FC_{week} = Feed cost per week

$$\text{Shorter in calving interval cost} = \text{Estimation of shorter first calving age after new feeding management cost (MYR)} = (\Delta CI \times FC_{week}) + (\$_{flushing} \times 2) \quad (6)$$

ΔCI = Change in calving interval; FC_{week} = Feed cost per week; $\$_{flushing}$ = Cost of flushing per animal; 2 = number of flushing frequencies

Increased cost:

$$\text{Cost of organic fertilizer per year} = \text{Estimation of organic fertilizer per year (MYR)} = \$_{\text{fertilizer per animal}} \quad (7)$$

$\$_{\text{fertilizer per animal}}$ = Cost of fertilizer per animal

$$\text{Flushing cost per year} = \text{Estimation of flushing cost per year (MYR)} = n_{\text{flushing days}} \times n_{\text{amount PKC (kg)}} \times \$_{\text{PKC/kg}} \times 3 \quad (8)$$

$n_{\text{flushing days}}$ = Number of flushing days; $n_{\text{amount PKC (kg)}}$ = Amount of PKC used (kg); $\$_{\text{PKC/kg}}$ = PKC price per kg (MYR); 3 = number of PKC given to animals' frequency

$$\text{Calf feed cost (Supplemented feed)} = \text{Estimation of additional calf feed cost per year (MYR)} = n_{\text{days (within 3months)}} \times \$_{\text{feed animals/week}} \times \Delta R \quad (9)$$

$n_{\text{days (within 3months)}}$ = considering supplement given once a week with duration of 3 month (12 days); $\$_{\text{feed/animals/week}}$ = Cost of animal feed per week; ΔR = Change ratio number born calves per number of mated dams

$$\text{Feed cost for increase in total calf-heifer population} = \text{Estimation of additional feed cost for total calf-heifer population per year (MYR)} = \Delta R \times \$_{\text{feed/animal/year}} \quad (10)$$

ΔR = Change ratio number born calves per number of mated dams; $\$_{\text{feed/animal/year}}$ = Cost of feed calf-heifer per year

$$\text{Cost for deworming} = \text{Estimation of additional in deworming cost per year (MYR)} = \Delta R \times \$_{\text{deworming/animal}} \quad (11)$$

ΔR = Change ratio number born calves per number of mated dams; $\$_{\text{deworming/animal}}$ = Deworming cost per animal

$$\text{Cost for ID tag} = \text{Estimation of additional cost for ID tag per year (MYR)} = \Delta R \times \$_{\text{ID tag cost/animal}} \quad (12)$$

ΔR = Change ratio number born calves per number of mated dams; $\$_{\text{deworming/animal}}$ = ID tag cost per animal

Forgone Revenue:

Assume similar quality present in both breed = The value given for both breeds were 0

The sum of the new revenue and the costs saved gave the value of total additional gains due to the intervention. The sum of revenue forgone and the new costs gave the value of total additional costs due to the intervention. The net benefit due to the intervention was the difference between total additional gains and total additional costs. The formula for the calculation is as below:

$$\text{Net benefit} = (\text{Increased revenue} + \text{Decreased cost}) - (\text{Revenue foregone} + \text{Increased costs})$$

3. Results

3.1. Growth Performance

In general, the mean bodyweight patterns of Swamp and Murrah crossbred buffaloes from birth until 24 months old are shown in Table 2. There was significant ($p < 0.05$) difference in the birth weight between Swamp (34.69 ± 0.50 kg) and Murrah crossbred (36.63 ± 0.40 kg) buffaloes. Thereafter, the crossbred buffaloes showed significantly

($p < 0.05$) higher body weights at each of the three-monthly intervals. Interaction among breeds and three-month intervals were significantly ($p < 0.05$) different. In this farm, the weaning age was practiced at three months old. Thus, the body weight at three months old showed a significant higher in Murrah crossbred (124.7 ± 2.80 kg) as compared to Swamp (99.9 ± 3.10 kg) buffaloes. At 24 months old, Swamp showed significantly ($p < 0.05$) lower body weight of 320.70 ± 5.60 kg compared to 356.60 ± 5.90 kg for Murrah crossbred buffaloes (Table 4). The targeted 250 kg market weight was achieved at 18 months old for Swamp and at 15 months old for Murrah crossbred buffaloes (Figure 3).

Table 4. The body weight of Swamp and Murrah crossbred buffaloes for both sexes at three-month intervals (Mean ± SEM).

	Body Weight (kg)								
Month	**Birth**	**3 ***	**6**	**9**	**12**	**15**	**18**	**21**	**24**
Swamp (n = 35)	34.7 ± 0.50 ^{a}	99.9 ± 3.10 ^{bc}	132.6±3.50 ^{c}	163.6 ± 4.80 ^{d}	189.8 ± 4.00 ^{e}	235.3 ± 5.90 ^{f}	271.1 ± 4.60 ^{g}	307.9 ± 6.50 ^{h}	320.7 ± 5.60 ^{i}
Murrah Crossbred (n = 35)	36.6 ± 0.40 ^{b}	124.7 ± 2.80 ^{bc}	160.5±2.70 ^{d}	198.8 ± 3.00 ^{e}	232.6 ± 3.40 ^{f}	276.6 ± 3.30 ^{g}	302.1 ± 3.90 ^{h}	325.8 ± 4.20 ^{i}	356.6 ± 5.90 ^{j}
p-Value									
Breeds	<0.05								
Months	<0.05								
Interactions	<0.05								

^{a,b,c,d,e,f,g,h,i,j} Different superscripts within the same column indicate significant difference $p < 0.05$, * calves weaned at 3 months old.

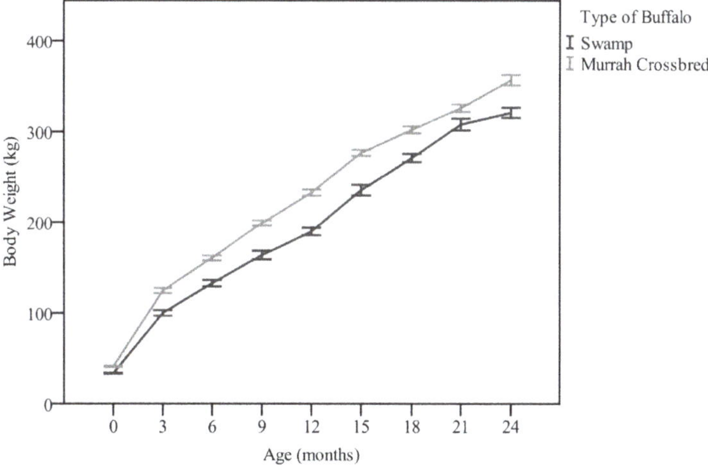

Figure 3. The body weight patterns of Swamp and Murrah crossbred buffaloes for both sexes.

3.2. Average Daily Weight Gain

The mean (±SEM) average daily weight gain (ADG) for Swamp and Murrah crossbred buffaloes are summarized in Table 5. In this study, Murrah crossbred buffaloes showed significant ($p < 0.05$) higher in average daily gain compared to Swamp. Highest average daily weight gain was observed during the pre-weaning period, while the lowest was in the period between 12 and 24 months old. The pre-weaning ADG was 0.73 ± 0.03 kg and 0.98 ± 0.03 kg ($p < 0.05$) for Swamp and Murrah crossbred buffaloes, respectively. However, between 12 and 24 months old, the average daily weight gain did not differ significantly ($p > 0.05$) with 0.30 ± 0.01 and 0.29 ± 0.01 kg/day for Swamp and Murrah crossbred buffaloes, respectively. Moreover, when buffaloes reach more than 24 to 30 months old, the average daily weight gain by Murrah crossbred buffaloes (0.44 kg/day) was significantly

($p < 0.05$) higher compared with to the 0.39 kg/day for Swamp buffaloes. Thus, the Murrah crossbred buffaloes reached the targeted breeding weight of 385 kg by 26 months old, while the Swamp buffalo reached it by 30 months old. However, there was no significant record on the month intervals and interaction between months and breeds.

Table 5. Average daily weight gain (kg) of Swamp and Murrah crossbred buffaloes for both sexes (Mean ± SEM).

Month	0–3 *	3–12	12–24	24–30
Swamp (n = 35)	0.73 ± 0.03 ^{a}	0.32 ± 0.02 ^{b}	0.30 ± 0.01 ^{b}	0.39 ± 0.07 ^{b}
Murrah crossbred (n = 35)	0.98 ± 0.03 ^{c}	0.37 ± 0.01 ^{bd}	0.29 ± 0.01 ^{be}	0.44 ± 0.08 ^{bf}
p-Value				
Breeds	<0.001			
Months	0.304			
Interactions	0.108			

^{a,b,c,d,e,f} Different superscripts within the same column indicate significant difference $p < 0.05$, * calves weaned at three months old.

3.3. *Number of Calves and Calf Quality*

In general, the post-intervention period showed significant ($p < 0.05$) improvement in number of calves born, average birth weight, and reduced mortality calf rate compared to the pre-intervention period. The number of calves born was taken from the farm record, where every single female produced calving annually, considering the period for pregnancy was nine months and 12 months of resting period. The numbers of calves born per year for Swamp and Murrah crossbred buffaloes are shown in Figure 4. In this farm, on average a total of 64 calves were born yearly with the ratio number born calves per number of mated dams recorded higher in Murrah crossbred buffaloes compared to Swamp buffalo (0.64 vs. 0.37) ($p < 0.05$). Over the study period, both Swamp and Murrah crossbred buffaloes showed an increasing trend of calves' birth with an average annual increase of 18 calves for Swamp and 46 calves for the Murrah crossbred buffaloes; significantly ($p < 0.05$), calves' birth was more than the Swamp buffaloes. During pre-intervention (2010–2011), the average number of calves born recorded significantly lower compared to post-intervention period (2012–2015) (Murrah crossbred: 31 vs. 53, Swamp: 8 vs. 23, respectively) ($p < 0.05$). Furthermore, the ratio number of calves born per number of mating females during pre-intervention (2010–2011) showed significant lower compared to post-intervention period (2012–2015) (Murrah crossbred: 0.52 vs. 0.63, Swamp: 0.40 vs. 0.45) ($p < 0.05$). However, there were no interaction ($p > 0.05$) showed between number of calves born with different years and pre- and post-interventions.

The improvement in calf quality was measured through the rate of calf mortality and the average birth weight. Figure 5 shows the rate of calf mortality in which the Swamp and Murrah crossbred buffaloes showed the average decreasing pattern of 7.4% and 8.1%, respectively, and significantly ($p < 0.05$) showed better improvement than the Swamp buffaloes. However, no interaction ($p > 0.05$) was recorded between calf mortality rate with different years and pre- and post-interventions. Other than calf mortality, improving birth weight was also considered as an increased in calf quality. Figure 6 shows the average birth weight of calves for Swamp and Murrah crossbred buffaloes. During the study period, Swamp buffaloes showed an average increment of 9.15 kg, which was significantly ($p < 0.05$) better than the 7.4 kg for the Murrah crossbred buffaloes. Furthermore, there were significant differences ($p < 0.05$) on average birth weight, years of birth, and interaction between average birth weight with years of birth and pre- and post-interventions. These data were included as increased revenue to the farm.

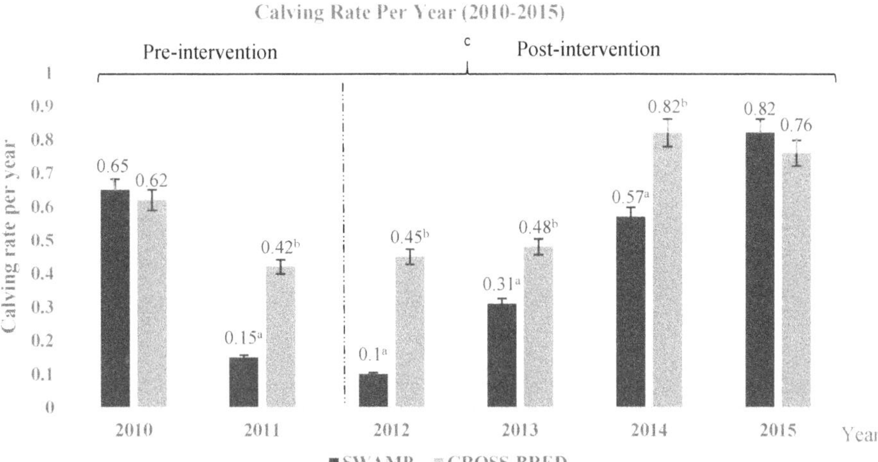

Figure 4. Total number of calving rates per year of Swamp and Murrah crossbred buffaloes between 2010 to 2015. There are increasing patterns for both breeds, and the crossbred buffaloes show significantly ($p < 0.05$) higher number than the Swamp buffaloes. [a,b] Different superscripts indicate significant difference of calving rate at $p < 0.05$; [c] indicating significant ($p < 0.05$) difference comparing between pre- and post-intervention. There is no significantly different ($p > 0.05$) interaction between number of calves born with different years of birth and pre- and post-intervention.

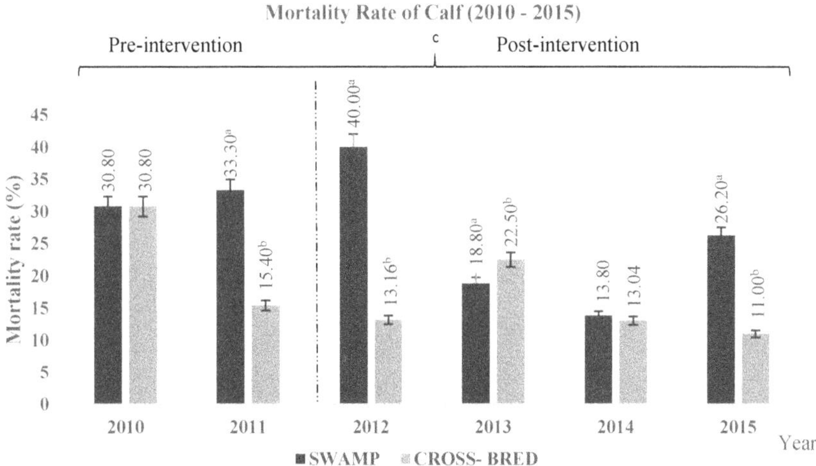

Figure 5. The rate of calf mortality among Swamp and Murrah crossbred buffaloes between 2010 to 2015. Based on Figure 5, the Murrah crossbred buffaloes show a significantly ($p < 0.05$) better improvement in the rate of calf mortality than the Swamp buffaloes. [a,b] Different superscripts indicate significant difference of mortality rate at $p < 0.05$; [c] indicating significant ($p < 0.05$) difference comparing between pre- and post-intervention. There is no significantly different ($p > 0.05$) interaction between number of calves born with different years of birth and pre- and post-intervention.

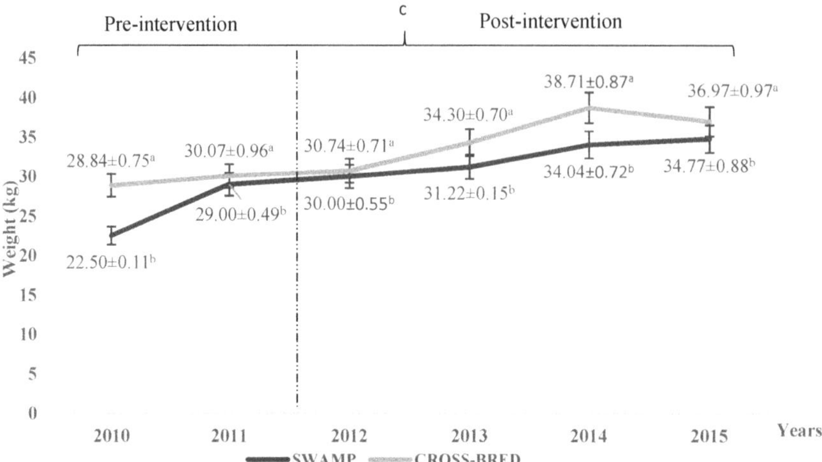

Figure 6. Average birth weight pattern of the Swamp and Murrah crossbred buffaloes between 2010 to 2015. Based on the Figure 6, the Swamp shows significantly ($p < 0.05$) better improvement than the Murrah crossbred buffaloes. ^{a,b} Different superscripts within the same year indicate significant difference of birth weight at $p < 0.05$; ^{c} indicating significant ($p < 0.05$) difference comparing pre- and post-intervention. There is a significantly different ($p < 0.05$) interaction between average birth weight with different years of birth and pre- and post-intervention.

3.4. Calving Interval and First Calving Age

The average calving interval for Swamp buffaloes was 398 days (13 months) and 460 days (15 months) for the Murrah crossbred ($p < 0.05$). Both breeds showed the calving age at 33 months in 2015, reduction of 12.5 months for Swamp, and 6.5 months for Murrah crossbred buffaloes from 2010.

3.5. Partial Budget

Table 6 shows the increased revenue for Swamp and Murrah crossbred buffaloes. The increased revenue for both breeds came from the increased number of animals due to the increased number of calves being born and the improvements in mortality rate, the improved quality of the animals following improvement in birth weight, and the increased sales per year. The improved birth weight per animal of Swamp buffaloes was valued at 1.86 MYR (0.46 USD), while the Murrah crossbred buffaloes was 3.46 MYR (0.85 USD). The increased numbers of calves were valued at 4.13 MYR (1.02 USD) for Swamp and 12.10 MYR (2.97 USD) for Murrah crossbred buffaloes. The improvement number survival rate of Swamp and Murrah crossbred buffaloes were 2.02 MYR (0.50 USD) and 4.94 MYR (1.21 USD), respectively. The increased sales per year for Swamp and Murrah crossbred buffaloes were 2296.13 MYR (564.16 USD) and 4511.00 MYR (1108.35 USD), respectively. Therefore, the overall increase in revenue for Swamp buffaloes was 2,304.14 MYR (566.13 USD), while the Murrah crossbred buffalo was 4531.50 MYR (1113.39 USD). The Murrah crossbred buffalo showed higher scale of increased revenue than the Swamp buffaloes.

Table 6. Partial budget analysis on intervention of feed supplementation for Swamp and Murrah crossbred (per female breeder per year).

Increases in Net Income			Decreases in Net Income		
Breed	**Swamp**	**Murrah Crossbred**	**Breed**	**Swamp**	**Murrah Crossbred**
Added Income Due to Change	MYR	MYR	Added Cost Due to Change	MYR	MYR
[1] Improved birth weight	1.86	3.46	[7] Fertilizer	64.79	64.79
[2] Increased number of calves	4.13	12.10	[8] Flushing	13.26	16.80
[3] Improvement number survival rate of calves	2.02	4.94	[9] Calf supplemental feed cost	4.32	9.50
[4] Increased sales of calves per year	2296.13	4,511.00	[10] Feed cost for increase total population of calf-heifer	2.45	5.39
			[11] Deworming	0.03	0.06
			[12] ID tag	0.10	0.22
(A) Total increase revenue	2304.14	4,531.50	(B) Total increased cost	84.95	96.76
Reduced cost due to change			Reduced income due to change		
[5] Shorter calving interval cost	90.00	46.80	Assume similar quality	0	0
[6] Shorter first calving age cost	250.02	168.95			
(C) Total decrease cost	340.02	215.75	(D) Total Foregone Revenue	0	0
E. Subtotal added gains (A + C)	2644.16	4,747.25	F. Subtotal added costs (B + D)	84.95	96.76
[13] New Benefit (E − F)	Swamp = MYR 2644.16 − MYR 84.95 = MYR 2559.21 (USD 628.80) Murrah cross = MYR 4747.25 − MYR 96.76 = MYR 4650.49 (USD 1142.63)				

[1] USD = 4.07 MYR Currency Conversion 5 March 2021, MYR Malaysian Ringgit. The status quo of the partial budget analysis of the buffalo farm management is an extensive system. The change of the farm management system is the feed supplementation for Swamp and Murrah crossbred. The change that occurs reflects a semi-intensive farm management system. The additional cost to the farm due to change in farm management (pre- to post-intervention) increased. In return, the farm has gained their income due to improvement of birth weight, survival rate of calves, and increased number of calves and sales of calves per year. The reduced cost due to change of management also has increased when shorter calving interval period and first calving age are taken after new intervention. [1–13]: [1] Estimation value of improved birth weight per year (MYR); [2] Estimation value of increase number calf per year (MYR); [3] Estimation value of selling calves to other farms for fattening per year (MYR); [4] Estimation number of calves sales; [5] Estimation of shorter calving interval after new feeding management cost (MYR); [6] Estimation of shorter first calving age after new feeding management cost (RM); [7] Estimation of organic fertilizer per year (RM); [8] Estimation of flushing cost per year (MYR); [9] Estimation of additional calf feed cost per year (MYR); [10] Estimation of additional feed cost of total population of calf-heifer per year (MYR); [11] Estimation of additional in deworming cost per year (MYR); [12] Estimation of additional cost for ID tag per year (MYR); [13] Estimation of net gain or loss per year (MYR).

Another input that was considered under gains was the decreased cost of production, which included the shorter calving interval cost and shorter first calving age cost. The shorter calving interval after new feeding management cost for Swamp buffaloes was valued at 90.00 MYR (22.11 USD), while for Murrah crossbred buffaloes it was at 46.80 MYR (11.50 USD). The shorter first calving age after the new feeding management cost was valued at 250.02 MYR (61.43 USD) for Swamp and 168.95 MYR (41.51 USD) for Murrah crossbred, respectively. Therefore, the total decreased cost for Swamp buffaloes was 340.02 MYR (83.54 USD) and for Murrah crossbred buffaloes was 215.75 MYR (53.01 USD).

Total gains were calculated by the sum of total increased revenue and decreased cost. Therefore, total gain for Swamp buffalo was 2644.16 MYR (649.67 USD), while the Murrah crossbred buffalo was 4747.25 MYR (1166.40 USD), higher than the Swamp buffaloes.

In this study, the losses included the increased costs of production, which consisted of costs of fertilizer, flushing, supplemented feed, feed cost for increase total population of calf heifer, deworming, and ID tags. Table 6 also summarizes the increased costs for rearing

Swamp and Murrah crossbred buffaloes. The cost of fertilizer for per animal Swamp and Murrah crossbred buffaloes was 64.79 MYR (15.92 USD). The cost of flushing the Swamp buffaloes was 13.26 MYR (3.26 USD), while the crossbred buffaloes was 16.80 MYR (4.13 USD). The cost of supplemented feed due to the increase in a number of calves was valued at 4.32 MYR (1.06 USD) for Swamp and 9.50 MYR (2.33 USD) for Murrah crossbred buffaloes. The increased total population calf heifer of Swamp buffaloes was valued at 2.45 MYR (0.60 USD), while the Murrah crossbred buffaloes was 5.39 MYR (1.32 USD). The costs of deworming and ID tags were 0.03 MYR (0.01 USD) and 0.10 MYR (0.03 USD), respectively, for Swamp, while the Murrah crossbred buffaloes were 0.10 MYR (0.03 USD) and 0.22 MYR (0.05 USD), respectively. In total, the additional cost for Swamp buffalo was 84.95 MYR (20.87 USD) and for the Murrah crossbred buffalo was 96.76 MYR (23.77 USD). The Murrah crossbred buffalo showed slightly higher additional cost.

3.6. Net Gains/Loss

The difference between total gain and total loss was the net change. The net change for Swamp buffalo was 2559.21 MYR (628.80 USD), while the Murrah crossbred buffalo was 4650.49 MYR (1142.63 USD).

4. Discussion

The birth weight of Murrah crossbred buffaloes, as shown earlier, was heavier than that of the Swamp buffaloes [5,21,22]. However, this study reported much heavier birth weights for both Swamp and Murrah crossbred buffaloes compared to the previous reports. In fact, the average of birth weights of Swamp buffaloes reported in Malaysia [23] and Thailand [24] were lower compared to the present study. Similarly, the average of birth weights of Murrah crossbred buffaloes reported in Sri Lanka [25] was also lower than the present study. In this study, the average of birth weight showed significant ($p < 0.05$) interaction with the year of birth. According to previous research, the birth weight is influenced by the season of birth, year of birth, sex of the calf, and parity of the dam [24,26,27]. However, it is believed that the better birth weights observed in this study might be due to the good feeding and husbandry managements [12,28]. Thus, this study was in agreement with our findings where the new feeding regime potentially improved average birth weight.

Following birth, both Swamp and Murrah crossbred buffaloes showed rapid pre-weaning growth, but the crossbred buffaloes showed significantly better rate, leading to a better overall daily weight gain of 0.86 kg/day compared to 0.65 and 0.74 kg/day at three months old reported by [29,30]. As a result, the average body weight of three-month-old buffaloes in the present study (112.29 kg) was higher than the 86.5 kg reported earlier [29]. A similar study revealed that approximately nine months were needed for the calves to achieve body weight of more than 100 kg, irrespective of the breed of the buffaloes [5].

In this study, it was observed that ADG at birth to weaning age (zero to three months) was higher than at four to 30 months old. Comparing the performance ADG of buffalo between 0–3 months up to 24 to 30 months, it is revealed that weight gains and growth rates were higher during the periods when the calves were offered milk than when offered solid feed. In addition, the higher ADG before weaning age could be due to underdevelopment of the rumen buffalo calves at an early age, as observed and reported by Abbas et al. [30].

In general, the Murrah crossbred buffaloes had significantly heavier bodyweight than Swamp buffaloes from birth until 24 months old, which is in line with buffaloes in Indonesia [31], Thailand [22], and the Philippines [5,21,32]. The body weights of the buffalo calves can be affected by many factors, such as the feeding management [33–35], breeds of buffalo [5,25,27], and environmental factors [24,36,37].

However, there were several factors contributing to the poor performance of animal growth at farm levels, namely improper nutritional management, climate change, seasonal stress, metabolic diseases, and mismanagement of farm [12,38]. Prior to intervention, nutritional analysis of the pasture before 2012 showed the average of crude protein, carbohydrates, ash, ether extract, and gross energy were low in value contents. The grass used by this farm had been the primary diet to boost performance of the animals. Additionally, another study has shown that *Brachiaria decumbens* were able to give the animals with nutritionally inferiority [39] when farms have proper pasture management. Thus, managing the farm with a proper management of pasture and supplementation diet in future will potentially make the buffalo have a better performance in growth, reproduction, and high production of quality milk to calves and industry over a long productive life [38,40]. Meanwhile, after the implementation of the intervention (2012–2015) via a new feeding regime by supplementing concentrate in the basal diet and improvement of pasture management, the nutritional contents of the buffalo's diets were improved.

Since the improvement of the diet due to post-intervention potentially improves the growth performance of both breeds, calves after weaning must be allowed to graze on improved pasture, be provided with a proper ratio of concentrate supplement to improve the growth performance, and be able to reach the breeding age in a shorter period [12,40]. This study also was in agreement with Vendramini et al. [41], where the study revealed a positive linear relationship between concentrate supplementation and ADG, liveweight gain, and stocking rate of early weaned calves of *Bos sp.* In addition, the new intervention revealed a better impact on the improvement of average birth weight, number of calving rate, and reduce calf mortality rate. Indeed, the number of calving rates in this study was in agreement with Wahid and Rosnina's [9] findings, which indicated that the buffalo in Asia undergo year-round breeding and produces two calves every three years on average.

The reduction of calf mortality rate was shown before and after intervention from 26.83% to 19.81%, respectively. However, a study in Pakistan revealed an average of 17.98% mortality rate among buffalo calves [42], which was lower than this study. Other studies from India recorded that the mortality rate among dairy buffalo farms was 81.09% [43]. According to Panchasara et al. [42] and Othman et al. [12], the death of buffalo calves was usually reported during the third to fourth weeks of age, during certain seasons (monsoon, winter, or during heavy rainfall season) and improper feeding management, which had led to metabolic disease such as ketosis.

Between 2011 and 2015, improved farm husbandry practices have resulted in an increased number of new-born calves and low rate of calf mortality, which had contributed to the increased revenue for both Swamp and Murrah crossbred buffaloes [12,28]. However, the Murrah crossbred buffaloes showed higher value due to the marked increase in the number of new-born calves and lower rate of calf mortality, making the population of crossbred calves much higher, which had brought more income to the farm. In addition, the Murrah crossbred buffaloes showed heavier birth weight; thus, they tended to reach weaning weight much earlier than the Swamp buffaloes.

It was reported in this study that the reduction of calving intervals had occurred after implementing the intervention. The data on calving interval recorded in Murrah crossbred (13 months) and Swamp buffalo (15 months) were lower compared to what was reported in India, where Murrah buffaloes had an average of 15 months [44] and Swamp buffaloes had an average of 18 months [4].

Partial budgeting is an economic model providing information on consequences of certain change in farm procedure without any specific time pattern [45]. The data retrieved from partial budget allows the farmer to understand the cost of production changes that determine the profit margins, and it is critical to ensure the sustainable of farm industry [46]. In this study, early weaning is costlier for the farm due to the longer weaning to production period. This leads to a slightly higher additional cost of the Murrah crossbred buffalo. On the other hand, the birth weight could also affect the reproduction maturity of females [47], when the Murrah crossbred buffalo tends to reach the age at first calving earlier than the Swamp buffaloes. Thus, the females can reproduce earlier and for a longer period, which brings more economic benefit to the farm. Earlier age of first calving also reduces the cost of rearing heifers.

Although better birth weight was beneficial to the farm economically, it also brings extra cost, especially in terms of feed consumption. As Murrah crossbred buffaloes were born bigger, their feed consumption was also higher than the Swamp buffaloes. Thus, they need more feed to get enough energy and nutrition for maintenance and growth, and this adds cost to the farm. Nevertheless, the Murrah crossbred buffaloes produced significantly higher overall net gains than the Swamp buffaloes. Therefore, farmers can reduce the rearing cost and earn more profit by rearing Murrah crossbred buffaloes and selling them at an earlier age either for slaughter or fattening purposes. Estimating costs via partial budget for rearing young buffalo could improve the farmer's awareness and improve decision making in either rearing Murrah crossbred or Swamp buffalo, which could increase economic sustainability, meat, and milk production, as well as conserve the local indigenous breed in Malaysia.

5. Conclusions

In conclusion, the use of supplementation is a good practice with favorable findings when the system of buffalo rearing is addressed for meat and milk purposes. The early weaning of the buffalo calves, calving interval, reducing calf mortality rate, and improved calf birth weight may be a successful practice when having feed of good nutritional quality and proper pasture management. In future, this allows the calves to have early development of rumen physiology and metabolic function. Murrah crossbred buffaloes showed significantly heavier birth weight and better daily weight gains and thus reached market and breeding weights at a significantly younger age as compared to Swamp buffaloes. Subsequently, they resulted in better income gain for the farm when rearing Murrah crossbred buffaloes (4650.49 MYR) as compared to Swamp buffaloes (2559.21 MYR), whose crossbred buffaloes contribute to better cost savings and high net gains. The use of new intervention practice toward buffalo calves according to the rearing of both breeds has given a better extent of efficiency in the growth performance of the stock.

Author Contributions: Conceptualization: H.A.H., P.A., M.Z.A.B. and M.Z.-S.; data curation: N.M.N., M.Z.A.B., J.V. and N.S.M.D.; formal analysis: A.F.M.A., P.A., J.V. and N.S.M.D.; investigation: H.A., G.Y.M. and A.S.; methodology: A.F.M.A., N.M.N., P.A., M.Z.A.B., J.V. and N.S.M.D.; project administration: M.Z.-S.; resources: H.A., G.Y.M., P.A. and A.S.; supervision: H.A.H., H.A., G.Y.M., A.S. and M.Z.-S.; validation: H.A., G.Y.M. and A.S.; writing—original draft: N.S.M.D.; writing—review and editing: A.F.M.A., H.A.H., N.M.N., M.Z.A.B., J.V. and M.Z.-S. All authors have read and agreed to the published version of the manuscript.

Funding: The whole research work and manuscript writing were fully funded by Ministry of Science, Technology, and Innovation (MOSTI), Malaysia, under the FLAGSHIP Research Grant Scheme (Project Title: Enhancing the Productivity of Farm Buffalo (*Bubalus bubalis*) Through Modification of Rearing Practices; Reference number: FP05514B0020-2(DSTIN)/6300858). Amirul Faiz M. A. was a recipient of Graduate Research Fellowship from the Universiti Putra Malaysia.

Institutional Review Board Statement: Not applicable.

Data Availability Statement: The datasets used and/or analyzed during the current study are available from the corresponding author on reasonable request.

Acknowledgments: The authors would like to thank the staff members of Department of Veterinary Services Sabah and Buffalo Breeding and Research Centre, Telupid Sabah, as well as all postgraduate and staff from Nutrition Lab, Faculty of Veterinary Medicine, Universiti Putra Malaysia, for their assistance in this study. This study was financially supported by the Flagship Research grant ABI/FS/2016/01 of the Ministry of Science, Technology, and Innovation, Malaysia and Ministry of Higher Education Malaysia through Higher Institutions Centres of Excellence (HiCoE) and Universiti Putra Malaysia.

Conflicts of Interest: The authors declare that they have no conflict of interest.

Abbreviations

MYR	Malaysian Ringgit
USD	United States Dollar
L	Liter
m	meter
kg	Kilogram
MJ	Millijoule
DM	Dry matter
ADF	Acid detergent fiber
NDF	Neutral detergent fiber
ADL	Acid detergent lignin
SEM	Standard error of mean
PKC	Palm kernel cake

References

1. Johari, J.A. Conservation and utilization of animal genetic resources in Malaysia. In Proceedings of the 6th National Congress on Genetics, Kuala Lumpur, Malaysia, 12–14 May 2005; pp. 30–36.
2. Triwitayakorn, K.; Moolmuang, B.; Sraphet, S.; Panyim, S.; Na-Chiangmai, A.; Smith, D.R. Analysis of genetic diversity of the Thai swamp buffalo (Bubalus bubalis) using cattle microsatellite DNA markers. *Asian Australas. J. Anim. Sci.* **2006**, *19*, 617–621. [CrossRef]
3. Zhang, Y.; Colli, L.; Barker, J.S.F. Asian water buffalo: Domestication, history and genetics. *Anim. Genet.* **2020**, *51*, 177–191. [CrossRef] [PubMed]
4. Tuyen, D.K.; Ly, N.V. The role of swamp buffalo in agricultural production of small farm holder. In Proceedings of the Workshop on Swamp Buffalo Development, Bangkok, Thailand, 17–18 December 2001; Volume 17.
5. Salas, R.C.; van der Lende, T.; Udo, H.M.; Mamuad, F.V.; Garillo, E.P.; Cruz, L.C. Comparison of growth, milk yield and draughtability of murrah-Philippine crossbred and Philippine native buffaloes. *Asian Australas. J. Anim. Sci.* **2000**, *13*, 580–586. [CrossRef]
6. Timsina, M.P.; Tamang, N.B.; Rai, D.B.; Siddiky, M.A. Comparative production and reproduction performances of local and murrah-cross buffaloes managed by smallholder farmers in Bhutan. *SAARC J. Agric.* **2015**, *13*, 200–206. [CrossRef]
7. Zainuddin, N. Evaluation of Milk Production in Murrah Buffalo Cows under Different Farm Management System. Ph.D. Thesis, Universiti Putra Malaysia, Selangor, Malaysia, 2017.
8. Feroz, K.; Owner of Buffalo Farms, Beranang, Selangor, Malaysia. Personal Communication, 19 April 2020.
9. Wahid, H.; Rosnina, Y. Buffalo: Asia. *Ref. Modul. Food Sci.* **2016**. [CrossRef]
10. Cruz, L.C. Recent Developments in the Buffalo Industry of Asia. *Rev. Vet.* **2010**, *21*, 7–19.
11. Nanda, A.S.; Brar, P.S.; Prabhakar, S. Enhancing reproductive performance in dairy buffalo: Major constraints and achievements. *Reprod. Camb. Suppl.* **2003**, *61*, 27–36. [CrossRef]
12. Othman, R.; Bakar, M.Z.A.; Kasim, A.; Zamri-Saad, M. Improving the reproductive performance of buffaloes in Sabah, Malaysia. *J. Anim. Health Prod.* **2014**, *2*, 1–4. [CrossRef]
13. Van Os, J.M.; Weary, D.M.; Costa, J.H.; Hötzel, M.J.; von Keyserlingk, M.A. Sampling strategies for assessing lameness, injuries, and body condition score on dairy farms. *J. Dairy Sci.* **2019**, *102*, 8290–8304. [CrossRef] [PubMed]
14. Cochran, W.G. *Sampling Techniques*; John Wiley & Sons: Hoboken, NJ, USA, 2007.
15. Fischer, H.; Ulbrich, F. Chromosomes of the Murrah buffalo and its crossbreds with the Asiatic swamp buffalo (Bubalus bubalis). *J. Anim. Breed. Breed. Biol.* **1967**, *84*, 110–114. [CrossRef]
16. Udroiu, I.; Sgura, A. Cytogenetic tests for animal production: State of the art and perspectives. *Anim. Genet.* **2017**, *48*, 505–515. [CrossRef]
17. Shaari, N.A.L.; Jaoi-Edward, M.; San Loo, S.; Salisi, M.S.; Yusoff, R.; Ab Ghani, N.I.; Ahmad, H. Karyotypic and mtDNA based characterization of Malaysian water buffalo. *BMC Genet.* **2019**, *20*, 1–6. [CrossRef]
18. Galyean, M.M. *Laboratory Procedures in Animal Nutrition Research*; Department of Animal and Food Science, Texas Tech University: Lubbock, TX, USA, 2010.
19. Rushton, J.; Ahuja, V.; Beck, H.; Bonnet, P.; Devendra, C.; James, A.; Leonard, D.; Lesnoff, M.; Otte, J.; Pica-Ciamarra, U.; et al. *The Economics of Animal Health and Production*; CABI: London, UK, 2009.
20. Moran, J. *Business Management for Tropical Dairy Farmers*; Landlinks Press: Collingwood, Australia, 2009.
21. Momongan, V.G.; Sarabia, A.S.; Roxas, N.P.; Palad, O.A.; Obsioma, A.R.; Nava, Z.M.; del Barrio, A.N. Increasing the productive efficiency of carabaos under smallholder farming systems. In *Domestic Buffalo Production in Asia, Proceedings of the Final Research Co-Ordination Meeting on the Use of Nuclear Techniques to Improve Domestic Buffalo Production in Asia-Phase II, Rockhampton, Australia, 20–24 February 1989*; International Atomic Energy Agency: Vienna, Austria, 1990; pp. 167–178.

22. Kamonpatana, M.; Sophon, S.; Jetana, T.; Sravasi, S.; Tongpan, R.; Thasipoo, K. A Comparative Study of Reproductive performance and Growth of Female Swamp and Murrah × Swamp Buffalo under Village Conditions. In Proceedings of the Buffalo and Goats in Asia, Genetic Diversity and Its Application, Kuala Lumpur, Malaysia, 10–14 February 1991; Volume 76.
23. Nordin, Y.; Sani, R.A.; Rosalan, A.M.; Ghadzi, M.S. Productivity of swamp buffaloes under three production systems. *J. Trop. Agric. Food Sci.* **2004**, *32*, 229.
24. Thevamanoharan, K.; Vandepitte, W.; Mohiuddin, G.; Chantalakhana, C. Environmental factors affecting various growth traits of swamp buffalo calves. *Pak. J. Agric. Sci.* **2001**, *38*, 5–10.
25. Charlini, B.C.; Sinniah, J. Performance of Murrah, Surti, Nili-Ravi buffaloes and their crosses in the intermediate zone of Sri Lanka. *Development* **2015**, *27*, 1–7.
26. Thiruvenkadan, A.K.; Panneerselvam, S.; Rajendran, R. Non-genetic and genetic factors influencing growth performance in. *S. Afr. J. Anim. Sci.* **2009**, *39*, 102–106. [CrossRef]
27. Pandya, G.M.; Joshi, C.G.; Rank, D.N.; Kharadi, V.B.; Bramkshtri, B.P.; Vataliya, P.H.; Solanki, J.V. Genetic analysis of body weight traits of Surti buffalo. *Buffalo Bull.* **2015**, *34*, 189–195.
28. Zamri-Saad, M.; Azhar, K.; Zuki, A.B.; Punimin, A.; Hassim, H.A. Enhancement of Performance of Farmed Buffaloes Pasture Management and Feed Supplementation in Sabah, Malaysia. *Pertanika J. Trop. Agric. Sci.* **2017**, *40*, 553–564.
29. Vaz, R.Z.; Lobato, J.F.P.; Restle, J. Influence of weaning age on the reproductive efficiency of primiparous cows. *Rev. Bras. Zootec.* **2010**, *39*, 299–307. [CrossRef]
30. Abbas, W.; Bhatti, S.A.; Khan, M.S.; Saeed, N.; Warriach, H.M.; Wynn, P.; McGill, D. Effect of weaning age and milk feeding volume on growth performance of Nili-Ravi buffalo calves. *Ital. J. Anim. Sci.* **2017**, *16*, 490–499. [CrossRef]
31. Situmorang, P.; Sitepu, P. Comparative growth performance, semen quality and draught capacity of Indonesian swamp buffalo and its crosses. *ACIAR Proc.* **1991**, *34*, 102–112.
32. Parker, B.A.; Nava, Z.M.; Lapitan, R.M.; del Barrio, A.N.; Momongan, V.G. Comparative growth performance of carabao and carabao crosses under controlled (institutional) and smallholder farms. In Proceedings of the Buffalo and Goats in Asia, Genetic Diversity and Its Application, Kuala Lumpur, Malaysia, 10–14 February 1991; pp. 83–92.
33. Burque, A.R.; Abdullah, M.; Babar, M.E.; Javed, K.; Nawaz, H. Effect of urea feeding on feed intake and performance of male buffalo calves. *J. Anim. Pl. Sci.* **2008**, *18*, 1–6.
34. Mahmoudzadeh, H.; Fazaeli, H. Growth response of yearling buffalo male calves to different dietary energy levels. *Turk. J. Vet. Anim. Sci.* **2009**, *33*, 447–454.
35. Peangkoum, P.; Tatsapong, P.; Pimpa, O.; Hare, M.D.; Paengkoum, S. Effect of dietary protein on nutrients intake, nitrogen balance and crude protein for maintenance in swamp buffalo. *Khon Kaen Agri. J.* **2010**, *38*, 112–114.
36. Bondoc, O.L.; Flor, M.C.; Rebollos, S.D.; Albarace, A.G. Variations in karyotypic characteristics of different breed groups of water buffaloes (Bubalus bubalis). *Asian Australas. J. Anim. Sci.* **2002**, *15*, 321–325. [CrossRef]
37. Javed, K.; Mirza, R.H.; Abdullah, M.; Akhtar, M. Environmental factors affecting live weight and morphological traits in Nili Ravi buffaloes of Pakistan. *Editor. Board* **2013**, *32*, 1161–1164.
38. Fukumoto, G.K.; Lee, C.N. Signalgrass for Forage. Available online: http://www2.ctahr.hawaii.edu/oc/freepubs/pdf/lm--3.pdf (accessed on 28 December 2020).
39. Campanile, G.; Di Palo, R.; Infascelli, F.; Gasparrini, B.; Neglia, G.; Zicarelli, F.; D'Occhio, M.J. Influence of rumen protein degradability on productive and reproductive performance in buffalo cows. *Reprod. Nutr. Dev.* **2003**, *43*, 557–566. [CrossRef]
40. Perera, B.M. Reproduction in domestic buffalo. *Reprod. Domest. Anim.* **2008**, *43*, 200–206. [CrossRef]
41. Vendramini, J.M.; Sollenberger, L.E.; Dubeux, J.C., Jr.; Interrante, S.M.; Stewart, R.L., Jr.; Arthington, J.D. Concentrate supplementation effects on the performance of early weaned calves grazing Tifton 85 bermudagrass. *Agron. J.* **2007**, *99*, 399–404. [CrossRef]
42. Panchasara, H.H.; Sutaria, T.V.; Shah, R.R. Factors affecting mortality of Mehsana buffalo calves. *Intas Polivet* **2009**, *10*, 170–173.
43. Tiwari, R.; Sharma, M.C.; Singh, B.P. Buffalo calf health care in commercial dairy farms: A field study in Uttar Pradesh (India). *Livest. Res. Rural Dev.* **2007**, *19*, 38.
44. Khan, Z.U.; Khan, S.; Ahmad, N.; Raziq, A. Investigation of mortality incidence and managemental practices in buffalo calves at commercial dairy farms in Peshawar City. *J. Agric. Biol. Sci.* **2007**, *2*, 16–21.
45. Dijkhuizen, A.A.; Huirne, R.B.; Jalvingh, A.W. Economic analysis of animal diseases and their control. *Prev. Vet. Med.* **1995**, *25*, 135–149. [CrossRef]
46. Tong, A.X.; Nor, N.M.; Indrawan, D.; Suhaimi, N.A. A Review: Should small holder dairy farms rear own young stock? *Malays. J. Anim. Sci.* **2018**, *21*, 1–6.
47. Naqvi, A.; Shami, S.A. Factors affecting birth weight in Nili-Ravi buffalo calves. *Pak. Vet. J.* **1999**, *19*, 119–122.

 MDPI

Review

The Impact of Feed Supplementations on Asian Buffaloes: A Review

Amirul Faiz Mohd Azmi [1], Hafandi Ahmad [1], Norhariani Mohd Nor [1], Yong-Meng Goh [1], Mohd Zamri-Saad [2], Md Zuki Abu Bakar [1], Annas Salleh [2], Punimin Abdullah [3], Anuraga Jayanegara [4,5] and Hasliza Abu Hassim [1,5,6,*]

1 Department of Veterinary Preclinical Sciences, Faculty of Veterinary Medicine, Universiti Putra Malaysia, UPM Serdang, Serdang 43400, Malaysia; amirulfaizazmi@gmail.com (A.F.M.A.); hafandi@upm.edu.my (H.A.); norhariani@upm.edu.my (N.M.N.); ymgoh@upm.edu.my (Y.-M.G.); zuki@upm.edu.my (M.Z.A.B.)
2 Department of Veterinary Laboratory Diagnosis, Faculty of Veterinary Medicine, Universiti Putra Malaysia, UPM Serdang, Serdang 43400, Malaysia; mzamri@upm.edu.my (M.Z.-S.); annas@upm.edu.my (A.S.)
3 Faculty of Science and Natural Resources, Universiti Malaysia Sabah, Jalan UMS, Kota Kinabalu 88400, Malaysia; puniminabdullah@ums.edu.my
4 Department of Nutrition and Feed Technology, Faculty of Animal Science, IPB University, Bogor 16680, Indonesia; anuraga.jayanegara@gmail.com
5 Animal Feed and Nutrition Modelling (AFENUE) Research Group, Department of Nutrition and Feed Technology, Faculty of Animal Science, IPB University, Bogor 16680, Indonesia
6 Laboratory of Sustainable Animal Production and Biodiversity, Institute of Tropical Agriculture and Food Security, Universiti Putra Malaysia, UPM Serdang, Serdang 43400, Malaysia
* Correspondence: haslizaabu@upm.edu.my; Tel.: +60-3-9769-3417

Citation: Mohd Azmi, A.F.; Ahmad, H.; Mohd Nor, N.; Goh, Y.-M.; Zamri-Saad, M.; Abu Bakar, M.Z.; Salleh, A.; Abdullah, P.; Jayanegara, A.; Abu Hassim, H. The Impact of Feed Supplementations on Asian Buffaloes: A Review. *Animals* **2021**, *11*, 2033. https://doi.org/10.3390/ani11072033

Academic Editors: Maria de la Luz Garcia, María-José Argente and Frank van Eerdenburg

Received: 4 May 2021
Accepted: 3 June 2021
Published: 7 July 2021

Simple Summary: Apart from feeding with forages, dietary supplementation with concentrate and rumen bypass fat is one of the feeding strategies to enhance nutrient availability and improve buffalo performance and productivity. This review paper thoroughly discussed the utilization of concentrate and bypass fat as dietary supplementation in buffalo feeding, and discussed the effects on performance, fermentation characteristics and general health of buffaloes to give better insight about the potential and challenges of dietary supplementation in buffalo diet. Based on the literature studies, it can be summarized that supplementation of concentrate and bypass fat in buffaloes may overcome the nutritional problems and improve the growth performance, health status, rumen environment and carcass traits.

Abstract: With the increase in the global buffalo herd, the use of supplementation in the ruminant feeding has become an important area for many researchers who are looking for an isocaloric and isonitrogenous diet to improve production parameters. In order to improve the performance of the Asian water buffalo, the optimal balance of all nutrients, including energy and protein, are important as macronutrients. Dietary supplementation is one of the alternatives to enhance the essential nutrient content in the buffalo diet and to improve the rumen metabolism of the animal. Researchers have found that supplementation of concentrate and rumen bypass fat could change growth performance and carcass traits without causing any adverse effects on the buffalo growth. Some studies showed that dry matter intake, body condition score and some blood parameters and hormones related to growth responded positively to concentrate and rumen bypass fat supplementation. In addition, changes of feeding management by adding the supplement to the ruminant basal diet helped to increase the profit of the local farmers due to the increased performance and productivity of the animals. Nevertheless, the effects of dietary supplementation on the performance of ruminants are inconsistent. Thus, its long-term effects on the health and productivity of buffaloes still need to be further investigated.

Keywords: bypass fat; buffalo; concentrate; performance; supplementation

1. Introduction

The Asian water buffalo (*Bubalus bubalis*) is an important animal resource in a minimum of 67 developing countries. Many people rely on this species for their livelihoods [1]. It provides economic value from its meat, milk and leather, and especially draft power [2]. An estimation by Scherf [3] revealed that more than 2 billion people depend on approximately 200 million heads of buffaloes, which is more than any other domesticated animal. The majority of buffaloes operate in close association with humans on small farms, for whom these animals are also their largest capital asset. Based on morphological and behavioral criteria, there are two subspecies of domesticated water buffaloes, the swamp (*Bubalus bubalis carabanesis*) and the river or Murrah (*Bubalus bubalis bubalis*) buffaloes [4,5]. Crossbreds between Murrah and Swamp (*Bubalus sp.*) buffaloes are found in some parts of Asia, especially in China, Indonesia, the Philippines, Thailand, Vietnam and Malaysia [6].

Currently, buffalo management practices are intrinsically correlated to the status of buffalo health, production and welfare. In order to ensure the optimum health of the animals and the organized production of high-quality and safe animal products (meat, milk and dairy products), a proper diet ration arguably represents the easiest strategy that can be implemented by farmers at the farm level. Both the meat and the dairy buffalos' industries have made significant advances in animal management, husbandry, genetics and nutrition. However, the current climate change phenomenon has caused a reduction of the availability of rangeland pastures and forages, especially for animals reared under the free-grazing system [7]. According to Henry et al. [8], extreme and fluctuating temperature may impair the optimum health and quality of life of animals. Furthermore, changes of the economic patterns around the world due to COVID-19 issues and the increasing demand for livestock as well as animal products from the developing countries trigger all the stakeholders involved in ruminant production to reconsider the strategic use of nutrition for enhancing animal health and production. Indeed, the livestock industry must find alternative nutritional strategies which meet the demand of consumers for economical animal products that are produced in clean, halal, green and ethical manners [9]. From our literature review, we found an abundance of scientific trial studies on animals (in vitro and in vivo) that indicated reliable and cost-effective approaches for increasing ruminant profitability through optimizing the composition of the feed nutrients and with the addition of supplementation [7].

Nowadays, the buffalo industry in Malaysia is starting to gain some attention as there are currently more studies on buffalo production, especially on how to improve it. This is in line with what has been carried out elsewhere, especially in other Southeast Asian countries. We therefore propose an approach using a suitable diet of a combination of basal diet with supplementation aimed for growing buffaloes. Hence, in this review article, we discussed the challenges of the buffalo industry, the requirements for energy and protein for buffalo growth, the potential of supplementation in the buffalo diet and the effects of the supplemented diet on rumen fermentation characteristics, growth performance, quality of the buffalo meat and the economics of the feed ration in ruminants. Our findings would allow specific dietary supplementation-based strategies to be established, which could efficiently enhance the health, welfare and longevity of the buffalo.

2. Characteristics of Asian Water Buffalo

2.1. Morphology and Genetics

In the field, river and swamp buffaloes can be differentiated based on their morphology and behavior. Swamp buffaloes are ash or dark grey with a white chevron line on the neck, either one or two stripes, and have socks, while the tip of the tails and the horns are swept backwards [10,11]. They prefer to wallow in the marshland and mud, and have large feet with slow steady movement that make them well-suited for paddy land preparation in swampy waterlogged rice fields [12]. On the other hand, river (Murrah) buffaloes have black bodies with tightly and forwardly curled horns. They prefer to wallow in clean water [10]. The crossbred buffaloes have the same morphology as the Murrah but are smaller than the Murrah and bigger than the Swamp buffaloes.

The chromosome number also differs between Swamp and Murrah, with river 2n = 50 and swamp 2n = 48 [11], owing to the fusion of chromosomes 4p and 9 [13]. The river and swamp buffaloes are also genetically distinct, as confirmed by the variations in allozymes, sex-linked, autosomal DNA markers, mitochondrial DNA sequences, microsatellite variation and a comparison with protein-coding loci [11,14,15]. Malaysian swamp buffaloes from the peninsula and Borneo region such as Sabah and Sarawak states were studied in this work. The peninsular and the Borneo swamp buffaloes were found to be genetically distinct, hence paving the way for the possibility of crosses between them to improve the Malaysian swamp buffaloes [6]. In addition, a prior study in Malaysia had reported that phylogenetic tree and mtDNA analysis on Swamp buffaloes were genetically conserved and the crossbreds were dominantly Swamp according to the maternal lineage using d-loop mtDNA [6].

On the other hand, crossbred buffaloes are the result of a combination of the two genetic types, but showing 80% dominant characteristics of the Murrah breed [6]. Crosses between river and swamp buffalo are fertile, despite having 2n = 49 chromosomes in F_1 and F_2 offspring. The crossbred buffaloes show better growth, larger body size and much improved draft power compared to the local swamp buffaloes [16]. Furthermore, the crossbred buffaloes are capable of yielding extra milk production, with an average of between 4.0 and 1.94 kg/day [17]. In fact, crossbred buffaloes are much more improved in terms of birthweight, age at maturity and first calving, duration of heat and period of inter-calving [18]. Indeed, with improved feeding, the crosses were recorded to grow 40% faster, with significantly improved meat quality than the Swamp buffaloes [19].

2.2. Distribution and Use

The indigenous breed of buffalo in East and Southeast Asia is the Swamp buffalo (*Bubalus bubalis carabanesis*). It is largely concentrated in Southeast Asia and Southern China (Figure 1). In China, the Swamp buffaloes have adapted to a range of climates, altitudes and temperatures [20]. Therefore, the Swamp buffaloes could be found in both low- and high-lands. Traditionally, the Swamp buffalo is used mainly for draught power, particularly for ploughing paddy fields and transportation, but it is also used to supply meat for human consumption. Until now, some oil palm estates are still using Swamp buffaloes as draught animal to pull carts carrying oil palm bunches. Occasionally, Swamp buffaloes' milk is used to make dairy products such as yoghurt and mozzarella cheese.

Figure 1. The swamp buffaloes. (**a**) Calves Swamp buffalo, (**b**) adult Swamp buffalo. Reprinted with permission from ref. [21]. Copyright 2021 Mohd Azmi et al.

River or Murrah buffaloes (*Bubalus bubalis bubalis*) are mainly found in South Asia, with the highest distributions in Pakistan, India, the Middle East and Italy (Figure 2). They are primarily reared for their milk typified by the high contents of fat and dry matter. Furthermore, buffalo milk has lower cholesterol content, but compared to cow's milk, it has more calories and fat. Thus, it is used to produce high-value thick and creamy cheese, yoghurt and ghee [20]. Murrah buffaloes are also used to improve buffalo milk production in many other countries such as Egypt and Bulgaria [22]. According to Hamid et al. [23], the Murrah buffaloes have been described as the "Asian tractors" and serve the purpose of meat, milk and work. In contrast to the Swamp buffaloes, river buffaloes are more aggressive, with temperamental instability [24].

Figure 2. The crossbred and pure-bred Murrah buffaloes. (**a**) Crossbred buffalo and (**b**) pure-bred Murrah buffalo. Reprinted with permission from ref. [21]. Copyright 2021 Mohd Azmi et al.

The multipurpose crossbred buffaloes are mainly found in several parts of Asia, especially in China, Indonesia, the Philippines, Thailand and Vietnam. They are used to provide draught power and meat in rice-growing areas and milk in other regions [25]. Now, crossbreeding of buffaloes is practiced in almost all countries where Swamp buffaloes are found, such as China, Burma, Thailand, the Philippines, Malaysia and Sri Lanka, with the aim of improving the milk yield and the animal size for work in the field [26].

3. Challenges in Buffalo Production Systems

An alarming decline in the buffalo population of Southeast Asia has happened over the past two decades at an average rate of 1.2 percent per year. This is due to several factors: (i) poor market demand for buffalo products [27], (ii) high rate of slaughter coupled with insufficient input for research and development [28], (iii) increased agricultural mechanization that made the Swamp buffalo redundant, (iv) a myth against the quality of buffalo meat [29], (v) poor reproductive performance coupled with poor responses to the biotechnology currently available, such as embryo transfer technology and artificial insemination, which prevented sufficient proliferation of the buffalo [17], and above all, (vi) a lack of knowledge on farm and feed management, resulting in a rapid decline in the number of buffaloes.

Furthermore, a few other factors have been cited as possible constrains that had contributed to the low output from beef and mutton producers. These factors include the inadequacy of land suitable for grazing to sustain significant livestock breeding populations, low supply of quality breeding stocks, erratic supply of high nutritional value feed and lack of an effective system of marketing. Taking cues from the poultry industry, the beef industry must promote ready supplies for the breeding and fattening of the production stock, ensuring the continuous supply of reasonably priced buffalo feed and creating an effective marketing network [30].

In Asia, in accordance with the environment, soil and socioeconomic opportunities, buffalo production systems vary widely [31]. The semi-intensive production system practiced by smallholder farmers is currently focused upon by many developing countries for the ruminant industry, mainly for cattle and buffalo. According to Saadullah [31], buffaloes in Asia that are mostly under the semi-intensive system are kept mainly for specific purposes, such as for meat and milk production. Approximately, 11%, 5% and 84% of the smallholder farmers reared buffalo for milk, meat and for both milk and meat, respectively [32]. Recently, the extensive production system is used through integrating ruminants, including buffaloes with oil palm or rubber plantations [30], where large-scale pasture lands and green forages or grasses are available [33] that allow the animals to graze for an average of 6 to 8 h daily [34]. However, most buffalo farms practice the semi-extensive grazing system in oil palm or paddy field areas without supplementation [35].

South and Southeast Asia have many marshlands and rivers which are suitable for raising buffaloes. Moreover, improvements in feeding management have influenced the growth performances of buffaloes in countries such as India, Brazil, the Philippines and Malaysia [35,36]. In general, indigenous Swamp and crossbred buffaloes in Southeast Asia are low in numbers, and this has affected the production of dairy and meat. Furthermore, longer puberty age, seasonality of breeding, longer calving interval, high calf mortality and poor genetic pool, nutrition and management practices [37–39] have all influenced the productivity.

4. Feeding and Nutritional Management for Buffalo Production

4.1. Nutrient Requirements and Utilization of Buffalo

Like other ruminants, buffaloes obtain their energy and protein in the form of volatile fatty acids and microbial proteins from fermentation end products. Based on research studies on the nutritional requirements and digestive physiology of buffaloes, it was concluded that buffaloes underwent relatively higher ruminal degradation of both protein and fiber [40] as compared to cattle [41]. This unique ability, particularly the better fermentation of fibers in buffaloes in temperate countries, is the consequence of adaptation following years of being fed with low-quality and highly fibrous diets [42]. According to Manish [43], the protein and energy demands of buffaloes in Asia were being met by feeding roughages containing high levels of lignocellulosic materials, namely cellulose, hemicellulose, lignin and low levels of fermentable carbohydrates and proteins. Indeed, the irregular and inadequate availability of quality feedstuffs and their utilization had been reported as the main causes of the poor performance of buffaloes in Asia [44]. In contrast, developed countries that produce a large number of meat and dairy animals place much emphasis on improving energy and protein levels of animal feed as well as on developing a specific model of nutrient requirements in buffaloes. These were reported by various studies, where the levels of energy and protein requirements varied in buffalo diets during lactation and growth [41,42,45]. Nevertheless, limited strategic studies were performed to establish the protein and energy needs of buffaloes in Asia, particularly on assessing their effects on different physiological stages and on the performances of buffaloes of various breeds.

In general, buffaloes need optimal nutrient requirements such as protein, fat, vitamins, minerals and water in order to maintain life, to reproduce and to enhance productivity. These requirements are influenced by many factors [46,47]. They include:

1. Animal-based factors, such as the physiological status (growth), age and body weight of animal, production line, traits of digestive system and health status [48,49].
2. Ration-associated factors, including the feeds that are used in the ration and the nutritional, physical and chemical composition status of the feed [50].
3. Environmental-related factors, particularly the climate, air temperature and feeding system [51].

Currently, there is paucity of information regarding the energy requirements of growing, breeding, lactating or working buffaloes [52]. Therefore, the nutrient requirements for cattle have always been adopted for buffalo feeding following the recommendations by NRC [53] and AFRC [54].

Energy is typically derived from carbohydrates, namely starch, cellulose and fat [47]. The nature of the buffalo digestive system physiology makes the cellulose in roughages, a rather cheap energy source, important [55]. The recommended values of energy requirements for buffaloes differ depending on their stages and physiological conditions. Estimations of the energy requirements of buffalo gain in the literature varied from 0.78 to 2.23 TDN/g gain [52], and for the lactating stage it was 1.97 TDN/g gain [48]. In addition, the amounts of feedstuffs offered to buffaloes to meet their requirements were also linked to various characteristics, such as the type, quantity, quality and presentation of the feed [52]. Therefore, besides feeding with low-quality roughage, the high demand for energy by growing buffaloes should be fulfilled by supplementing with a mixture of quality roughage and grains that contained an abundance of energy [56]. Furthermore, adequate supplementation of fat was capable of increasing the concentration of energy in the animals, which could also enhance the percentage of fat in the milk as well as the quality of the meat [57].

Protein is an important substance for growth and development of muscles, nerves and other tissues. It is also important for the repair of aged tissues, fetus development and the production of meat and milk [58]. Ammonia is required for the growth of rumen microorganisms before optimal microbial protein synthesis is achieved by supplementation with adequate levels of protein and non-protein compounds [59]. Other studies reported earlier that there was a large variation in the values for protein requirements in buffaloes [48,52]. The estimated range of digestible crude proteins needed by buffalo for maintenance and growth was 1.11 to 5.05 g/kg metabolic body size $^{0.75}$ and 0.30 to 0.45 g/g gain, respectively [52]. Following a low level of protein or energy supply to buffaloes, the demand for proteins was met by a low supply of medium-quality pastures and fodder. Therefore, growing, pregnant and lactating buffaloes should be fed with meadow grass and leguminous forage supplemented with concentrate, grain or oil seed cakes [46]. This could prevent protein insufficiency that might lead to declines in appetite and feed consumption of the animals, a negative utilization of feed and a reduction in cellulose digestion [55].

4.2. Roughage Feeding

It is clear that buffaloes' main diet consists of roughages such as grass, legumes and straw. Most of the buffaloes are fed on tropical grasses as the primary source of nutrients. The roughages are fed either fresh as the grazing pasture or dried such as in a cut and carry system or conserved as hay or silage. Thus, a relatively low cost of digestible energy from grasses is provided to the buffaloes, mainly in the form of fibrous compounds [60]. Nevertheless, the roughage that forms the basis of a feed ration should be of good quality, have both nutritional and hygienic qualities and be capable of meeting at least the total maintenance requirements. Unfortunately, tropical grasses for grazing buffaloes infrequently represent a balanced diet, since they have constraints on one or more nutrients that may limit the intake of forage, digestibility or the metabolism of the absorbed substrates [61]. Therefore, an addition of concentrate in the diet should be practiced in buffaloes' feeding to ensure a balanced ration, and nutrients are provided to meet the buffaloes' requirements [62].

To improve buffalo performance, the utilization of pastures in any season as the main nutrients is not considered to be optimal [61]. Thus, selecting mature, dried foliage and stems of grasses ensures supply of a low protein level of less than three percent of crude protein. Furthermore, the grasses have been variably leached of soluble components, including minerals, proteins, sugar and starchy carbohydrates that are needed for efficient fermentative digestion in ruminants. On the other hand, too much intake of non-fibrous feed can change the environment of the rumen, and in the long term, leads to serious feed digestion problems, including reducing feed intake, which leads to ketosis, weight loss and a reduction in milk yield. According to Figueiras [61], unbalanced energy to protein ratios were recorded in tropical grasses, with relative energy surplus. Thus, the use of this poor-quality forage without a supplementary diet contributed to the low production of the ruminant meat industry in developing countries. In fact, a study in Brazil and Australia revealed that using a proper feed supplementation program improved the body weights of ruminants [60]. For this reason, there is a need to recognize tropical pastures as having limitations of nutrients and to solve this or to change to feeding regimes that may lead to improved performances of the animals and the overall production efficiency [63].

4.3. *Supplementation Strategy*

Low growth performance, poor reproductive performance and milk yield have been reported in buffaloes in other studies [64,65]. Indeed, the poor performance reported by those studies correlated with poor dry matter intake and weight gain and longer calving intervals [64,65]. In many parts of Asia, inadequate and irregular availability of quality feedstuffs and their utilization are the main causes of the poor performance of buffaloes [44]. A report stated that buffaloes in Southeast Asia were mainly fed with hay or forages high in lignin and low in fermentable protein and carbohydrate contents [44]. Many strategies have been trialed in order to improve the nutrient supply and utilization in buffaloes, with varying degrees of success. These are: (1) utilization of available feed resources such as local plants that have high crude protein or energy content, (2) use of industrial and agricultural by products, (3) dietary addition of concentrate, fermentation modifiers and vitamins and (4) usage of ruminally protected dietary fat and protein sources, which have shown significant potentials to improve growth, reproduction and milk yield of buffaloes [44]. However, in order to choose the best strategy to improve buffalo performance, the farmer should identify the main problem causing the low growth performance of the animal.

In the buffaloes' diet, the roughage should often be complemented with concentrate, grains or agro-industrial products as supplements. The supplements should be fed only to fulfil the additional requirements for improving growth, pregnancy and milk production. Therefore, feed supplementation programs should concentrate mainly on the establishment of a diet that contains balanced nutrients by increasing the energy content in the diet as well as by increasing the dry matter intake through the addition of supplements. Indeed, a supplemented diet in the tropics should focus primarily on protein and fat supplementations in order to provide optimum energy for better growth performance of the animals [63,66]. This would allow for utilization of the relative excess of energy from the supplemented diet to be converted into body weight gain [67]. Balancing the fat and protein nutrition through supplementation is one of the strategies for increasing production in ruminants with high-energy diet requirements, such as young post-weaning animals, animals in the last trimester of pregnancy and lactating animals.

In most regions of developing countries, it is not practical to identify the deficient micronutrients and macronutrients in pastures or other forages, as these vary from site to site and year to year. Furthermore, they are also influenced by the pattern of fertilizer application and the weather conditions. The alternative is through providing buffaloes with a supplemented diet, and the supplementation involves providing energy and protein to satisfy the requirements for efficient digestion in the rumen. Diet with a proper supplementation is able to provide optimal ammonia nitrogen and good fatty acid in the rumen. Palatable and tasty feed with a well-balanced ratio of protein and fat as additional energy sources is the best way to increase milk production and live weight, maintaining health and enhancing fertility. This has influenced the changes in supplementation utilization, since a lack of knowledge on feeding management would affect animal production and nutritional performance [63,66]. However, comparative analyses regarding the combination of energy and protein supplementation on buffalo performance in the tropics remain scarce. Therefore, extensive studies are needed to assess the impacts of supplements on grazing buffaloes in the tropics, particularly on the intake, ruminal fermentation pattern and the quality of the meat.

4.4. Types of Supplementation in Buffalo Diets

Supplemented feed offers a promising way to enhance the overall health and performance of buffaloes. In contrast to a nutritionally complete ration, supplemental diets are intended to provide an additional source of energy and protein when forage quantity and quality are inadequate to meet the desired performance [68]. Nutritional husbandry of domestic buffalo often contains high energy and protein supplements in combination with roughage to increase the growth rate of sub-adult animals [69] and to enhance the digestibility of forage diets [70]. Supplementation strategies are essential in designing the feeding programs for this species. In fact, supplementation with larger amounts of energy-rich feeds with a source of protein and fat could reduce the time taken to prepare buffaloes for the market, thus increasing profitability, such as those reported for cattle that consumed low-digestibility forages with energy and protein supplements [71].

Concentrate is one of the dietary supplements that supply protein to animals. Many experiments have demonstrated the benefits of supplementing dietary protein meals or concentrate to ruminants that are fed poor-quality forage [72–74]. In fact, in developing countries, low cattle and buffalo productions are primarily due to limited supplies of nutrients in high forage-based rations [75]. Therefore, it is important to identify new feed sources and technologies for cattle and buffalo production systems. Recently, there are several varieties of concentrate ingredients that have been used by smallholder farmers, namely maize meal, cassava powder, rice bran and mixtures of these feedstuffs in ruminant production [75], such as cattle and buffalo. However, due to inadequate information on the nutritional value, digestibility and characteristics of the rumen fermentation, the benefits of using these feeds for buffalo are not well-understood. This is especially true since previous studies had reported that different proportions of concentrate feeds differed substantially in their rumen fermentation characteristics [76]. Nevertheless, a high starch content in the concentrate is important in ruminant nutrition because it is cost-effective, contains protein sources and has been proven able to influence rumen function and digestion of nutrients [77,78].

The technology of bypass nutrients such as rumen bypass fat has been implemented in feed management through passive rumen manipulation [79]. It is also known as rumen-protected fat, inert fat or rumen bypass fat [80]. Bypass fat is the supplement that escapes rumen degradation as it is being protected from microbial fermentation, biohydrogenation and remains insoluble at normal rumen pH. However, it is easily digested and absorbed in the lower gastrointestinal tract at the abomasum and small intestine, respectively [81]. Therefore, it might be beneficial for ruminants to be fed with rumen bypass fat as a supplementary diet that is absorbed readily from the lower digestive tract. There are two types of rumen bypass fat supplements commercially used by farmers, namely natural bypass fat (e.g., cotton, roasted soybeans, sunflower and canola whole oil seeds) and chemically prepared bypass fat (e.g., crystalline or prilled fatty acids, formaldehyde-treated protein-encapsulated fatty acid, fatty acyl amide and calcium salts of long-chain fatty acids) [82]. The calcium salt-coated method on fatty acid of vegetables has been reported to be a more accessible bypass fat to all types of farmers. It has also been proven to be a cost-effective technology compared to other rumen bypass fats [83]. A few studies reported improvements on the growth and nutrient utilization of buffalo calves after being fed with bypass nutrients [84,85]. Integration of fat supplement into the diet enhances the growth potential of buffalo calves [79]. Studies have reported that high fat supplementation, such as RPF, could also enhance fiber digestibility of various fibrous feedstuff due to the high hydrolysis rate in rumen (85% to 95%) [81,82]. This supplement can also improve energy efficiency, as a result of reduced production of methane from the rumen and direct use of long-chain fatty acids [86]. Many studies recommended that the ration of high producing and growing animals should contain between 3% to 6% fat in the total ration DM, in order to obtain the beneficial effects [82]. Meanwhile, feeding more than 9% rumen bypass fat would not be beneficial for the growth (feed intake) and lactation (milk yield) of the animals [82].

5. Effect of Dietary Supplementation on Buffalo Production

The advantages of high energy and protein in dietary supplements can be the explanation for the improvements in the growth and fattening of ruminants. Indeed, it has been shown that apart from improvement of the growth performance upon supplementation of concentrate and bypass fat, the improvement in dry matter intake may further enhance the average daily gain, growth hormones (e.g., GH and IGF-I), body condition score, carcass traits, meat quality and breeding performance of buffaloes [21,79,87]. In addition, the supplementation of both concentrates and bypass fat did not cause any adverse effects on serum biochemical profiles, rumen fermentation and microbial population [21,28,33–35]. An overview of the role of these concentrate and bypass fat supplementations in buffalo nutrition is presented in Figure 3.

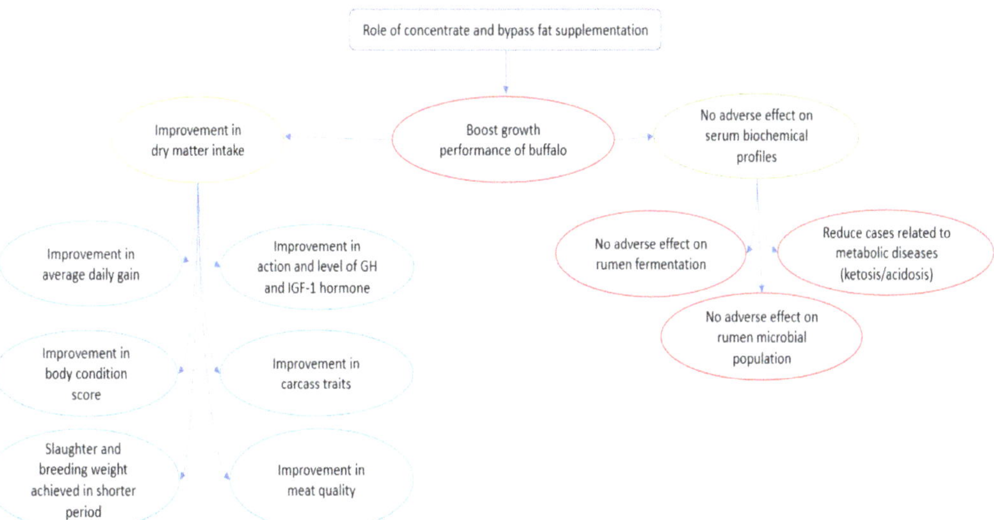

Figure 3. An overview of the role of concentrate and bypass fat supplementation in buffalo nutrition.

5.1. Growth Performance

The body weight gain and the body condition score (BCS) of buffalo calves after weaning represent the growth vigor of the animal. This is a substantial feature in the selection of animals [88]. The body condition score of a buffalo is classified as a subjective scoring method to assess the outer appearance of the animal, which interacts with its body fat to provide a better understanding of the biological relationships between body fat, reproductive performance and production of milk. These could assist in the implementation of optimal management practices to achieve maximum production and to preserve better health status [87]. Indeed, the BCS may also provide an immediate evaluation of the body condition of the animal and can be readily integrated into operational decision-making [89]. The use of BCS started in studies of ewes, beef cattle and Holstein dairy cows in 1961, 1976 and 1989, respectively. It used a 0 to 5 scale with a chart developed for references of body condition scoring [87]. In Pakistan, there was a study on Nilli Ravi buffalo that assessed the BCS by using a linear scale of 1 to 9 (1–3 for thin, 4–6 for average and 7–9 for fat), where the scoring was performed visually by assessing the covering of fat over the tail head, rump, sacral bone and loin and withers area [90]. However, the BCS of buffaloes was improved and established in India [87] using a 1 to 5 scale for assessing the animals (Figure 4). India has the highest buffalo population in the world and shows dramatic increases of population numbers yearly [91]. In addition, it is the native tract for the best buffalo breeds of the world [92]. The improved chart for BCS of buffalo with a 1 to 5 scale using 0.25 increments has been widely used in Asian countries [92]. The BCS score assessment was carried out by taking into consideration the anatomical features and amounts of fat reserves at various skeletal checkpoints. Validation of the precision of the BCS score had been carried out via ultrasonic measurements of subcutaneous fat [92]. Thus, the use of the BCS 1 to 5 score is suitable for assessing the reproduction and production status of buffaloes. Furthermore, the most widely used criterion is also to determine the growth of animals by assessing their body weights. Body weight is significantly associated with the types of feed and ration offered.

According to Vahora et al. [79], improvements in the average daily weight gain, body length, height and heart girt in buffaloes fed with a basal diet were significantly associated with the incorporation of protein and fat supplements. Furthermore, offering supplementation at 0.6 kg/animal/day to grazing buffaloes after weaning for a period of two years enabled the calves to reach an average of 578.38 kg body weight, with improved body condition scores [36]. Similarly, supplementing beef heifers with dietary energy supplements increased the average daily gain and utilization of energy from native forage which contained low-quality nutrients [93]. Sawyer et al. [94] also showed that supplementing with a low ratio of energy at 40 g/day of crude protein might potentially replace greater quantities (160 g/day of crude protein) while still maintaining rumen function. However, animals grazing on low-quality dormant range and fed with a supplementation had no change in body weight during pregnancy, breeding and longevity compared to those feds grazing without supplement or a lower rumen undegradable plant-based protein supplement [95]. However, adding supplemental fat in the ration at a rate of 5–7% dry matter (DM) resulted in an improved lamb weight at 15% to 20% [96]. Ngidi et al. [97] reported that the apparent digestibility of fat increased, whereas true digestibility lessened when fat was added at up to 8% of diet DM. Meanwhile, Kumar and Thakur [98] concluded that supplementation of bypass fat at 2.5% to 4% of dry matter intake increased average daily gain and feed conversion ratio in buffalo calves and improved the growth performance without an adverse effect on nutrient utilization. It was concluded that addition of 2% to 4% of fat potentially stimulated feed intake and increased ruminant's digestibility energy intake [99].

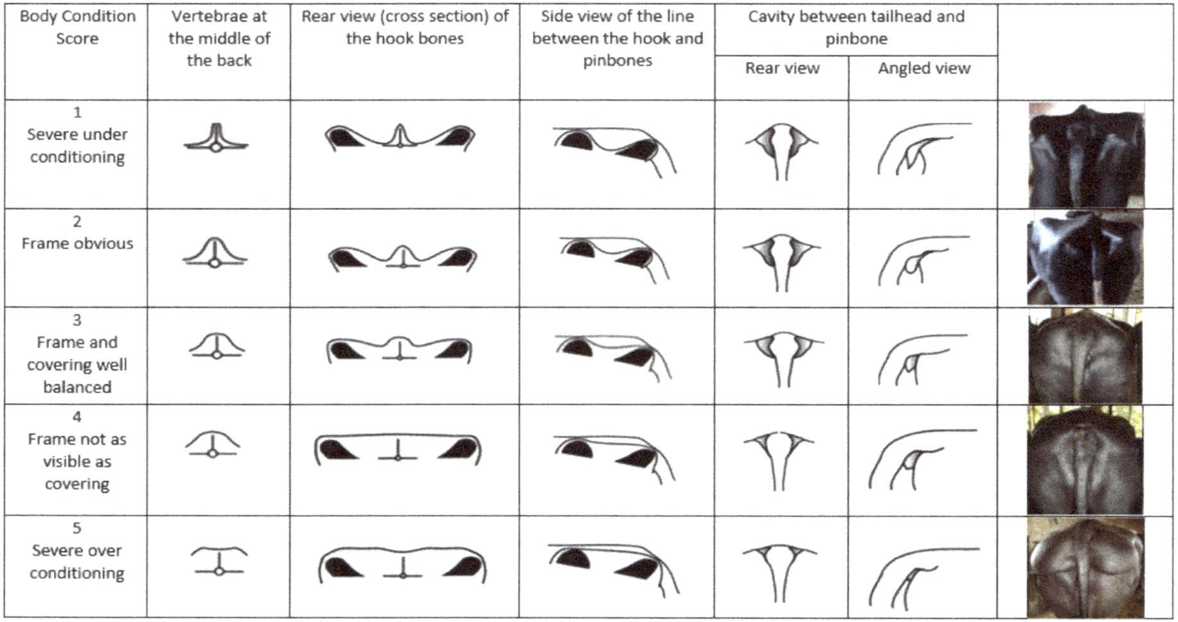

Figure 4. Body condition scoring chart for buffalo in a scale using 1.0 increments. Reprinted with permission from ref. [87]. Copyright 2017 Singh et al.

5.2. Serum Biochemistry and Hormone Profiles

The main function of blood in the body is to maintain the physiological equilibrium, but many physiological conditions may alter this equilibrium [100]. Blood contains a myriad of constituents that provide a valuable medium for clinical investigations and nutritional evaluations of an organism [101]. Serum biochemicals can be affected by age, nutrition, physiological status, sex, genetics, environmental factors and stresses, such as diseases and transportation [102]. The consequences of these variables can be measured by evaluating the physiological responses, since it is known that environmental and nutritional factors predispose animals to the occurrence of disease and decrease animal productivity. Such deviations from normal alter animal body constituents, especially the body fluids, thus health risk conditions can be well-understood by evaluating blood components. Disease incidence and malnutrition could be diagnosed by analyzing the deviations of the various hematological and serum biochemical parameters from the normal reference values [103]. In fact, these values are used to compare and interpret the metabolic state or condition of animals as a reference point [104].

In the ruminant industry, analysis of blood metabolic profiles for assessing the nutritional and health status of goats, cows and buffaloes are not widely used [105], although the health and nutritional status of animals are important. However, the conventional and common practices used to evaluate the nutritional status of animals include the body condition scoring and the gain of body weights. Therefore, the use of blood metabolites is less common in assessing nutritional status of ruminants, especially among smallholder farmers in developing countries [106,107]. Needless to say, there are some blood metabolites that are related to the nutritional status of ruminants, and they reflect the animal's response to nutrition. They include blood total protein, cholesterol, triglyceride, glucose and urea.

Blood metabolites could be used regularly, objectively and reliably to assess the nutritional status of animals. However, the use of blood metabolite analysis in field buffalo and cattle is rare, particularly in developing countries such as Southern Africa and India due to the high cost of analyzing the samples, lack of equipment and expertise [106,107]. Similarly, the use of blood metabolites is quite uncommon in the field of buffalo management. This is unfortunate since several factors, namely the animal physiological status, nutrition, breed, age and season, are found to affect the blood metabolites, and in combination with data from body condition scores and body weights, blood metabolite analysis increases the accuracy of assessing the nutritional and welfare states of the ruminant population [107]. The success of the blood metabolite profile test alone is limited because several non-dietary factors also influence the concentrations of blood metabolites, such as herd origin, lactation stage, milk yield and season of the year [108].

According to Campanile et al. [109], the buffalo heifers fed *Brachiaria* hay with concentrate (high-energy diet) showed similarities in non-esterified fatty acid and triglyceride levels when compared to the group fed a low-energy diet (without concentrate) (0.53 vs. 0.47 mmol/L, 17.1 vs. 18.7 mg/dL, respectively), but there were significant increments in glucose, total cholesterol and high-density lipoprotein (90.5 vs. 73.6 mg/mL, 80.4 vs. 58.7 mg/dL, 64.0 vs. 45.4 mg/dL, respectively). Bertoni et al. [110] have revealed that buffaloes fed isonitrogenous diets with different energy contents showed constant blood urea content regardless of the different diets, while cattle with similar treatment showed a significant decrease in blood urea with increasing dietary energy. This indicated the decline in ammonia content in the rumen of cattle as a result of the limited ability to recycle blood urea into the rumen. On the other hand, blood urea nitrogen of buffaloes would be at optimum range following feeding with different energy content diets that ranged between 7.00 and 8.50 g/dL [111]. Tiwari et al. [112] reported that the normal glucose levels in growing buffaloes and Holstein cattle were 51 to 64 and 74 to 76 mg/dL respectively, when provided concentrate and roughages at an equal ratio. Other studies reported that Murrah buffalo fed a basal diet with concentrate supplementation and a mixture of concentrate with bypass fat supplementation had no effect on the blood urea nitrogen (BUN)

(49.30 vs. 50.41, mg/L), total protein (7.57 vs. 7.75, g/dL), albumin (2.71 vs. 2.84, g/dL), globulin (4.86 vs. 4.91, g/dL) and cholesterol (106.27 vs. 106.36, mg/dL) levels [113]. However, the buffaloes that were fed with supplement bypass fat showed slightly increased high-density lipoprotein and calcium levels as compared to animals fed without the bypass fat (65.55 vs. 57.22 mg/dL and 7.03 vs. 5.70 mg/dL, respectively) [113,114]. Nevertheless, the buffaloes that were supplemented with bypass fat showed a slight decrement of blood glucose, non-esterified fatty acid and low-density lipoprotein levels as compared to animals fed without the bypass fat (63.50 vs. 65.98 mg/dL, 0.66 vs. 0.93 mmol/L, 33.66 vs. 37.49 mg/dL, respectively) [113].

Similarly, hormonal profiles can also be used to determine the health and nutritional status of ruminants. Animal growth is influenced by many hormones, blood metabolites and growth factors acting both in an endocrine and an autocrine manner and requires the coordinated actions of several hormones, such as growth hormone (GH) and insulin-like growth factor-I (IGF-I) [115]. The somatropic axis is the important hormonal system for growth development of animals. It consists of GH, IGF-I, carrier proteins and receptors [116]. The hormones of GH and IGF-I have both independent characteristic and combined impacts on ruminant metabolism and production. The growth hormone is synthesized in the pituitary gland and acts directly on the liver and adipose tissue to regulate gluconeogenesis, protein synthesis, lipogenesis, lipolysis and insulin secretion by binding to the growth hormone receptor (GHR) [117–119]. On the other hand, IGF-I is a critical somatomedin that is synthesized in the liver. It plays an important role in some physiological processes, contributes to improved feed conversion rate and increases protein synthesis [120]. The IGF-I binds to insulin-like growth factor binding protein-3 (IGFBP-3) to influence the growth, development and reproduction in animals [121]. Meanwhile, the existence of the axis between GH and IGF-I has played a vital role in the regulation of metabolism. GHR combines with GH to stimulate a series of metabolic activities by producing IGF-I in the target tissues, especially in the liver [122,123].

In some cases, nutritional factors are critical regulators of IGFs [124]. Deficiency in either energy or protein intake significantly decreases the IGF-I levels [124]. Similarly, Clemmons et al. [125] reported that low energy in the diet caused the IGF-I level to decrease 4-fold. In fact, limiting the energy in the diet increases the GH levels and reduces the IGF-I secretion [126,127]. Meanwhile, over-consumption increases the IGF-I, although excess calories are not nearly as strong a stimulus as nutritional restriction [128]. A short-term feeding study on protein deprivation revealed a potent and independent role of protein on the IGF-I levels. Deficiency of essential amino acids has a severe depressing effect on the IGF-I levels [124]. However, a high-carbohydrate diet increases the IGF-I levels relative to a high-fat diet, possibly by maintaining hepatic sensitivity to GH [129]. Several researchers have recorded that yaks and buffaloes have evolved and developed complex strategies to respond to the deficiencies of nutrition and hypoxia stress [130–132].

According to Campanile et al. [109], buffalo heifers fed with *Brachiaria* hay and supplemented with concentrate (high-energy diet) showed increases in GH and IGF-1 levels compared to buffalo fed with a basal diet without concentrate (low-energy diet) (6.3 vs. 5.6 pg/mL, 95.5 vs. 79.1 ng/mL, respectively). Other studies also revealed that Murrah buffalo increased IGF-1 and remain constant in GH levels when fed a diet supplemented with a mixture of concentrate with bypass fat when compared to being fed with a diet supplemented solely by concentrate (119.10 vs. 116.24 ng/mL, 4.91 vs. 4.93 ng/mL, respectively) [113,133]. Even though a few studies have shown the effects of dietary fat and protein supplements on the serum biochemistry profiles as well as the hormonal profiles in the blood, unfortunately, limited work has been performed on the comparison of blood and hormonal profiles between buffalo breeds.

5.3. Rumen Fermentation Pattern

Rumen digestion is fundamentally a fermentation process within the rumen by various types of bacteria, protozoa and fungi [134]. These microbes are responsible for the breakdown of feed and water intake into volatile fatty acids (VFA), gases (i.e., methane, carbon dioxide) and microbial proteins that are useful for the animal. For the rumen microbes to function properly, the rumen environment parameters including pH, temperature and moisture must be maintained [134]. Therefore, the outcome of rumen fermentation depends on adequate nutrition with respect to composition and quality of feedstuffs, which is reflected in the voluntary intake and digestibility of ruminants [135].

Buffaloes are like cattle which utilize micro-organisms in the rumen to digest the feed. It has been reported that many farms until today fed their buffaloes using cow requirements as a reference point [136]. This might be due to the similarities in rumen fermentation characteristics between buffalo and cow. According to Smith et al. [137], both animals have similarities in heat tolerance, milk composition and ability to utilize highly fibrous feed [138,139]. Nevertheless, studies also revealed that buffaloes indeed have higher forage digestibility due to higher populations of cellulolytic bacteria, total fungi and lower protozoa numbers in their rumen, as compared to cattle rumen [140,141]. Therefore, it is important to understand the rumen fermentation and the ruminal microbial differences between buffaloes and cows when formulating a feeding regime for them [138].

Changes in diet, from forage to high-protein diet, can affect the fermentation process of ruminants such as cattle and buffalo. The increase in dietary crude protein (CP) leads to increased concentration of total rumen volatile fatty acid (VFA), which is consistent with the improved degradability following increased bacterial populations and microbial enzyme activities [142,143]. A study by Kang et al. [144] reported that Swamp buffalo fed with a high concentrate (160 g/kg DM) supplement ratio had constant ruminal pH (average 6.59 to 6.61) and temperature (average 39.1 to 39.3 °C), but had increased values of ammonia, total VFA, propionic acid, butyric acid of 9.8 vs. 13.8 mg/dL, 20.1 vs. 26.2 mol/100 mol and 10.7 vs. 12.3 mol/100 respectively, when compared to animals fed with low concentrate (120 g/kg DM).

Carbohydrate in the diet is mainly characterized by the proportion of non-structural (NSC) and structural carbohydrate [145,146]. In the rumen, the NSC are initially utilized by the rumen microbes and are degraded quickly to produce volatile fatty acids [147]. Sutton et al. [148] confirmed that high NSC in the concentrate diet resulted in higher propionate concentration, while McCarthy et al. [149] also concluded that increasing the content of ruminal fermentable starch enhanced the total volatile fatty acid (TVFA) concentration. In fact, decreased ruminal pH was also reported following high intake levels of dietary crude protein, and this was attributed to the increased ruminal total VFA output [150,151]. Ruminal ammonia-nitrogen (NH_3-N) is primarily derived from ruminal degradable proteins and is used for the synthesis of microbial protein [152,153]. An increase in ruminal NH_3-N content occurs following an increase in dietary CP levels [152] and is largely attributed to the increased ruminal CP degradability [154]. However, an increase in dietary concentrate or carbohydrate is not a successful strategy to mitigate either the enteric methane production or ammonia emission from the manure. Therefore, incorporating supplemented concentrate with bypass fat has the potential of reducing the methane and ammonia productions.

Unfortunately, a study by Budi et al. [155] found that the increased level of bypass fat significantly reduced the in vitro dry matter degradability, but the levels of TVFA and ammonia nitrogen remained constant, while Naik et al. [156] revealed that following in vitro fermentation, the levels of TVFA and ammonia nitrogen were much higher in animals fed with a diet with concentrate and bypass fat compared to those fed without the supplement. Saijpaul et al. [157] recommended that the high level of bypass fat supplementation could substitute up to 40% of natural fat in a concentrate mixture (6% natural fat) contained in total mixed rations (50:50 roughage to concentrate) with no changes in in vitro fermentation parameters of TVFA, total nitrogen and ammonia. Other studies also reported that supplementation of bypass fat between 5% and 15% in buffalo diet had no adverse effect on the in vivo rumen fermentation characteristics such as pH, NH_3-N and VFA levels [158–160]. The specific ingredient of dietary buffer in calcium salt bypass fat functioned to maintain ruminal pH and to minimize the rate of dissociation of calcium salts in the rumen [82], thus the rate of biohydrogenation was potentially reduced. In summary, supplementation of solely concentrate in a basal diet could affect the buffalo rumen fermentation characteristics but no adverse effects were found on the reported parameters when buffaloes fed given the mixture of concentrate and bypass fat in the diet.

5.4. Rumen Microbial Populations

It is also known that the feed and feeding regime in ruminants may also significantly affect the ruminal microbial community. Indeed, it has been demonstrated that the composition of the rumen bacterial population significantly correlated with feed efficiency [161,162]. A study showed that changes in the structure of the ruminal microbial population potentially promoted feed efficiency, intake and the average daily gain of ruminants [163]. The microorganisms also often provided the host ruminant with nutrients to produce energy [164]. Thus, increased ruminant production could be achieved by using proper feeding formulation and management that were able to manipulate the ruminal microbes and ecosystem. For example, a proper ratio of supplemented feed in the diet that provided adequate energy and protein, allowed for an optimal ruminal fermentation process that maximized production efficiency while decreasing energy loss such as methane that polluted the environment [165]. Indeed, there were also reports on the effects of dietary changes involving protein and energy on rumen microbial population and the rumen environment in ruminants [166].

The effects of dietary energy and protein supplementation on the rumen microbial population have been studied extensively in sheep, goat and cow, but less so in buffaloes. Indeed, a study by Dahllöf et al. [167] has reported that dietary modifications in cattle feeding have major effects on the communities of rumen bacteria. Faniyi et al. [168] also stated that a shift in the ruminant diet from forage-based to a high-concentrate diet resulted in significant alterations of the ruminal environment and rumen bacterial population structure. Indeed, various studies also showed that there were clear changes in the structure of the rumen bacterial population as the dietary forage to concentrate ratio gradually increased from 80:20 to 60:40 or even to 20:80, with increases in Proteobacteria and decreases in Firmicutes [168–172]. The increased abundance of Proteobacteria during high-concentrate diets was suggestive of an increased need for bacterial species that could metabolize the newly available fermentable carbohydrates [169], thus favoring the growth of amylolytic and other starch-digesting bacterial species and reducing the number of cellulolytic bacteria. This suggested that when animals were shifted from a forage diet to a high-concentrate diet, the microbial diversity in terms of the different species numbers remained but the composition or the species makeup changed drastically in order to adapt to the new rumen environment.

Fibrobacter succinogenes, a fibrolytic bacterium that digests fiber, was reported to be gradually decreased as animals were adapted to a high-concentrate diet, and their numbers were 40-fold lower than in those animals fed on hay [169]. A study by Tajima et al. [173] reported a 20-fold decrease in the *Fibrobacter* population size by day 3 and a 57-fold decrease by day 28 in animals on high-concentrate diets. The population of *Butyrivibrio fibrisolvens*, another fibrolytic bacterium with high affinity toward maltose and sucrose, also declined 20-fold during adaptation to a high-concentrate diet [172]. Due to the ability of this species to use both cellulose and starch, the *Butyrivibrio fibrisolvens* population showed a slight decrease in the rumen environment [169]. However, the drop in the population of *Butyrivibrio fibrisolvens* during a diet with a high proportion of concentrate (30:80) might be because of pH changes due to the increased number of fermentable substrates present within the rumen. This was consistent with the findings of a recent study which showed that the population of *Butyrivibrio fibrisolvens* increased in high-fiber diets and decreased in high-energy diets [174,175].

Changes in bacterial populations with increasing dietary crude protein were attributed to the increased ruminal amino acids, peptides, branched chain VFA and NH_3-N [176,177].

Moreover, it was reported by Wang et al. [178] that increased dietary crude protein would enhance ruminal microbial growth, particularly of the bacterial populations in the rumen fluid that consisted of *Ruminococcus albus, Ruminococcus flavefaciens, Butyrivibrio fibrisolvens, Prevotella ruminicola, Fibrobacter succinogenes* and *Ruminobacter amylophilus*. In general, the improved microbial enzyme activity has a significant relationship with the increase of the bacterial populations in the rumen [143]. In particular the activities of cellobiase, xylanase, pectinase, carboxymethyl-cellulase, pectinase, protease and α-amylase increased following an increase in dietary protein. However, another researcher showed that lambs fed with a diet containing crude protein levels from 111.7 to 143.6 g/kg DM did not show any effect on the rumen microbial population of *R. albus, R. flavefaciens* and *F. succinogenes* [179]. These contradictory findings might be due to the differences in the animals studied and the dietary composition [151,179].

5.5. Qualities of Carcass and Buffalo Meat and Their Implications on Human Health

Nutrition is a dominant component of livestock production systems that influences several aspects of meat quality. Consumer demands on meat quality have motivated the meat producers to focus on the nutritional aspects of livestock rearing since carcass and meat qualities are affected by the amount and type of nutrient intake. These include dressing yield, meat to bone ratio, protein to fat ratio, fatty acid composition, caloric value, color, physicochemical and processing properties, shelf life and sensory attributes [180].

Dressing percentage is one of the important parameters that reflects the potential meat yield from an animal. Usually, the weight of hot carcass is used to compute the dressing percentage. When the weight of cold carcass is used, the dressing percentage is less due to the chilling shrinkage that ranges between 3% and 4.5% of the initial weight of the hot carcass for buffaloes at the age of 6 months up to 4 years [181]. Other than chilling, a feeding diet with supplementation might inconsistently influence the carcass quality. Anjaneyulu et al. [182] reported that supplemented dietary protein did not affect the carcass composition of male buffalo. However, the dressing percentages and yields of the lean meat were higher when buffaloes were fed a high-energy diet compared to those fed a low-concentrate diet [112]. Nevertheless, based on the current information, it was concluded that feed supplementation had little effect on the carcass quality of buffaloes.

Meat quality is a major factor that is used for marketing the product [183]. The meat quality is evaluated through physical, biochemical, histological and sensory analyses. The nutritional composition and pH of meat are parts of the biochemical analyses used in assessing the meat quality, which contribute to the edibility or the desirability of the product [184]. Three crucial factors can affect the quality and composition of the meat produced [185]. They are: (1) the feedstuffs and proportion of the diet used for feeding the animal, (2) types of diet, supplement, breed or genetic cross of the animal and (3) the age at which the animal is slaughtered. Meanwhile, diet has been shown to be one of the most important environmental factors that influences the carcass yield, meat cutability and quality [186,187]. The effects of feeding on meat quality are generally studied in terms of the content and composition of the lean and fat tissues, and the subsequent effects on the nutrient content of protein, fat, energy and moisture content.

In particular, the total fatty acid content and the types of fatty acid found in meat have major influences on the meat quality and acceptability of the meat by consumers [187]. Feeding a high-energy diet potentially affects the rate of conditioning and consequently, the carcass composition, conformation, meat yield and meat and fat quality [188].

Carabeef, also known as buffalo meat, is considered to be a highly nutritious and valuable food. It is a source of high biological value of protein, omega 3 and omega 6 fatty acids and low in fat and cholesterol levels [189]. However, different nutrient contents of buffalo meat have been reported due to different feeding regimes [190]. Lambertz et al. [191] proposed a concentrate supplementation at the rate of approximately 1.5% of body weight to enhance the carcass characteristics of Swamp buffaloes via expressing superior dressing percentage, better muscling and redder meat, with higher contents of protein and fat.

Furthermore, addition of a fat supplement enables facilitated absorption of liposoluble nutrients, making it possible to modify the meat fat composition according to consumers' demand [192]. Moreover, the high energy levels allow the increase of pulp proportion in the diet of fattening animals, which act as precursors of the fatty acid responsible for the lack of consistency of fat from carcasses [193].

6. Conclusions

From the available literature, it can be summarized that supplementation of concentrate and bypass fat in a potential buffalo diet is very important to overcome the problem of negative energy balance without adversely affecting the dry matter intake, rumen fermentation, blood metabolites and rumen microbial populations. Supplementation of concentrate and bypass fat provides additional benefits due to improved ruminant body weight, body condition score and the economics of the ruminant industry. Further research is necessary to find out the effects of supplementation with concentrate and bypass fat on growing buffaloes fed with various types of basal rations at different productive levels.

Author Contributions: H.A.H. conceived the manuscript purpose and design and critically revised the manuscript. A.F.M.A. wrote the initial version of the manuscript and revised according to H.A.H.'s suggestions. All authors (H.A., N.M.N., Y.-M.G., M.Z.-S., M.Z.A.B., A.S., P.A. and A.J.) read and approved the final manuscript. All authors have read and agreed to the published version of the manuscript.

Funding: The whole research work and manuscript writing were fully funded by the Ministry of Science, Technology, and Innovation (MOSTI), Malaysia, under the FLAGSHIP Research Grant Scheme (Project Title: Enhancing the Productivity of Farm Buffalo (*Bubalus bubalis*) Through Modification of Rearing Practices; Reference number: FP05514B0020-2(DSTIN)/6300858). Amirul Faiz M. A. was a recipient of a Graduate Research Fellowship from the Universiti Putra Malaysia.

Institutional Review Board Statement: Not applicable.

Data Availability Statement: Not applicable.

Acknowledgments: The authors would like to thank the staff members of Jabatan Penternakan Haiwan dan Perusahaan Ternak (JPHPT) and the Buffalo Breeding and Research Centre, Telupid Sabah, as well as all postgraduate students and staff from Nutrition Lab, Faculty of Veterinary Medicine, Universiti Putra Malaysia, for their assistance in this study. This study was financially supported by the Flagship Research grant ABI/FS/2016/01 of the Ministry of Science, Technology, and Innovation, Malaysia, and Ministry of Higher Education Malaysia through Higher Institution Centres of Excellence (HiCoE) and Universiti Putra Malaysia.

Conflicts of Interest: The authors declare that they have no competing interests.

Abbreviations

CP	Crude protein
RUP	Rumen undegradable protein
kg	Kilogram
g	Gram
mg	milligram
ADG	Average daily gain
TVFA	Total volatile fatty acid
DM	Dry matter
BW	Body weight
BCS	Body condition score
NEFA	Non-esterified fatty acid
dl	Deciliter
GH	Growth hormone
IGF-1	Insulin-like growth factor- 1
GHR	Growth hormone receptor
BUN	Blood urea nitrogen

References

1. Işik, M.; Gül, M. Economic and social structures of water buffalo farming in Muş province of Turkey. *Rev. Bras. Zootec.* **2016**, *45*, 400–408. [CrossRef]
2. Cockrill, W.R. The working buffalo. In *The Husbandry and Health of the Domestic Buffalo*; Food and Agricultural Organization of the United Nations: Rome, Italy, 1974.
3. Scherf, B.D. *World Watch List for Domestic Animal Diversity*, 3rd ed.; Food and Agriculture Organization (FAO): Rome, Italy, 2000.
4. Kumar, S.; Nagarajan, M.; Sandhu, J.S.; Kumar, N.; Behl, V.; Nishanth, G. Mitochondrial DNA analyses of Indian water buffalo support a distinct genetic origin of river and swamp buffalo. *Anim. Genet.* **2007**, *38*, 227–232. [CrossRef] [PubMed]
5. Andrabi, S.M.H. Applied andrology in water buffalo. *Anim. Androl. Theor. Appl.* **2014**, 380–403. [CrossRef]
6. Shaari, N.A.L.; Jaoi-Edward, M.; Loo, S.S.; Salisi, M.S.; Yusoff, R.; Ab Ghani, N.I.; Saad, M.Z.; Ahmad, H. Karyotypic and mtDNA based characterization of Malaysian water buffalo. *BMC Genet.* **2019**, *20*, 1–6. [CrossRef]
7. McGrath, J.; Duval, S.M.; Tamassia, L.F.; Kindermann, M.; Stemmler, R.; de Gouvea, V.N.; Acedo, T.S.; Immig, I.; Williams, S.N.; Celi, P. Nutritional strategies in ruminants: A lifetime approach. *Res. Veter. Sci.* **2018**, *116*, 28–39. [CrossRef]
8. Henry, B.; Charmley, E.; Eckard, R.; Gaughan, J.B.; Hegarty, R. Livestock production in a changing climate: Adaptation and mitigation research in Australia. *Crop. Pasture Sci.* **2012**, *63*, 191–202. [CrossRef]
9. Martin, G.B.; Kadokawa, H. "Clean, green and ethical" animal production. Case study: Reproductive efficiency in small ruminants. *J. Reprod. Develop.* **2006**, *52*, 145–152. [CrossRef]
10. Triwitayakorn, K.; Moolmuang, B.; Sraphet, S.; Panyim, S.; Na-Chiangmai, A.; Smith, D.R. Analysis of genetic diversity of the Thai swamp buffalo (Bubalus bubalis) using cattle microsatellite DNA markers. *Asian Australas. J. Anim. Sci.* **2006**, *19*, 617–621. [CrossRef]
11. Zhang, Y.; Colli, L.; Barker, J.S.F. Asian water buffalo: Domestication, history and genetics. *Anim. Genet.* **2020**, *51*, 177–191. [CrossRef]
12. Tuyen, D.K.; Ly, N.V. The role of swamp buffalo in agricultural production of small farm holder. In Proceedings of the National Workshop on Swamp Buffalo Development, Hanoi, Vietnam, 17–18 December 2001; p. 17.
13. Iannuzzi, L. A genetic physical map in river buffalo (Bubalus bubalis, 2n = 50). *Caryologia* **1998**, *51*, 311–318. [CrossRef]
14. Colli, L.; Milanesi, M.; Vajana, E.; Iamartino, D.; Bomba, L.; Puglisi, F.; Cruz, L. New insights on water buffalo genomic diversity and post-domestication migration routes from medium density SNP chip data. *Front. Genet.* **2018**, *9*, 53. [CrossRef]
15. Barker, J.S.F.; Moore, S.S.; Hetzel, D.J.S.; Evans, D.; Byrne, K. Genetic diversity of Asian water buffalo (Bubalus bubalis): Microsatellite variation and a comparison with protein-coding loci. *Anim. Genet.* **1997**, *28*, 103–115. [CrossRef]
16. Thu, N.V. *Exploring Approaches to Research in Animal Sciences in Vietnam*; Australian Centre for International Agricultural Research (ACIAR): Bruce, Australia, 1995; Volume 68, p. 216.
17. Nanda, A.S.; Nakao, T. Role of buffalo in the socioeconomic development of rural Asia: Current status and future prospectus. *Anim. Sci. J.* **2003**, *74*, 443–455. [CrossRef]
18. Benjamin, B.R. Cross-breeding among buffaloes, an unexploited natural resource. In *Buffalo Newsletter, Bulletin of the FAO Inter-Regional Cooperative Research Network on Buffalo for Europe—Near East*; The International Buffalo Federation (IBF): Rome, Italy, 1996.
19. Lemcke, B. Water buffalo. In *The New Rural Industries—A Handbook for Farmers and Investors*; Hyde, K., Ed.; Rural Industries Research and Development Corporation: Sydney, Australia, 1997.
20. Borghese, A. *Buffalo Production and Research. REU Technical Series*; Food and Agriculture Organization of the United Nations: Rome, Italy, 2005; p. 67.
21. Mohd Azmi, A.M.; Abu Hassim, H.; Nor, N.M.; Ahmad, H.; Meng, G.; Abdullah, P.; Bakar, A.; Vera, J.; Deli, N.M.; Salleh, A.; et al. Comparative growth and economic performances between indigenous swamp and Murrah crossbred buffaloes in Malaysia. *Animals* **2021**, *11*, 957. [CrossRef]
22. Hamid, M.; Ahmed, S.; Rahman, M.; Hossain, K. Status of buffalo production in Bangladesh compared to SAARC countries. *Asian J. Anim. Sci.* **2016**, *10*, 313–329. [CrossRef]
23. Hamid, M.; Siddiky, M.; Rahman, M.; Hossain, K. Scopes and opportunities of buffalo farming in Bangladesh: A review. *SAARC J. Agric.* **2017**, *14*, 63–77. [CrossRef]
24. Prasad, R.M.V.; Laxmi, P.J. Studies on the temperament of Murrah buffaloes with various udder and teat shapes and its effect on milk yield. *Buffalo Bull.* **2014**, *33*, 170–176.
25. Moioli, B.; Georgoudis, A.; Napolitano, F.; Catillo, G.; Giubilei, E.; Ligda, C.; Hassanane, M. Genetic diversity between Italian, Greek and Egyptian buffalo populations. *Livest. Prod. Sci.* **2001**, *70*, 203–211. [CrossRef]
26. Borghese, A.; Mazzi, M. Buffalo population and strategies in the world. *Buffalo Prod. Res.* **2005**, *67*, 1–39.
27. Borghese, A. Development and perspective of buffalo and buffalo market in Europe and near East. Proceedings of 9th World Buffalo Congress, Buenos Aires, Argentina, 25–28 April 2010; pp. 20–31.
28. Wanapat, M. Potential uses of local feed resources for ruminants. *Trop. Anim. Heal. Prod.* **2008**, *41*, 1035–1049. [CrossRef]
29. Ly, L.V. Overview on buffalo development in Vietnam. In Proceedings of the Regional Workshop on Water Buffalo Development, Bangkok, Thailand, 8–10 February 2001.
30. Ariff, O.M.; Sharifah, N.Y.; Hafidz, A.W. Status of beef industry of Malaysia. *Malays. J. Anim. Sci.* **2015**, *18*, 1–21.
31. Saadullah, M. Buffalo production and constants in Bangladesh. *J. Anim. Plant Sci.* **2012**, *22*, 221–224.

32. Nahar, T.N.; Alam, M.K.; Akhtar, S. Study the assessment of nutritional composition and bacterial load in buffalo milk in some selected areas of Bangladesh. In Proceeding of the Annual Research Review Workshop, Umuahia, Nigeria, 17–19 April 2012; p. 13.
33. Momin, M.M.; Khan, M.K.I.; Miazi, O.F. Performance traits of buffalo under extensive and semi-intensive Bathan system. *Iran. J. Appl. Anim. Sci.* **2016**, *6*, 823–831.
34. Afzal, M.; Anwar, M.; Mirza, M.; Andrabi, S. Comparison of growth rate of male buffalo calves under open grazing and stall feeding system. *Pak. J. Nutr.* **2009**, *8*, 187–188. [CrossRef]
35. Zamri-Saad, M.; Azhar, K.; Zuki, A.B.; Punimin, A.; Hassim, H.A. Enhancement of performance of farmed buffaloes pasture management and feed supplementation in Sabah, Malaysia. *Pertanika J. Trop. Agric. Sci.* **2017**, *40*, 553–564.
36. Alves, T.C.; Franzolin, R. Growth curve of buffalo grazing on a grass pasture. *Rev. Bras. Zootec.* **2015**, *44*, 321–326. [CrossRef]
37. Shamsuddin, M.; Bhuiyan, M.M.U.; Sikder, T.K.; Sugulle, A.H.; Chanda, P.K.; Alam, M.G.S.; Galloway, D. *Constraints Limiting the Efficiency of Artificial Insemination of Cattle in Bangladesh*; International Atomic Energy Agency: Vienna, Austria, 2001.
38. Sarkar, S.; Hossain, M.M.; Amin, M.R. Socio-economic status of buffalo farmers and the management practices of buffaloes in selected areas of Bagerhat district of Bangladesh. *Bangladesh J. Anim. Sci.* **2013**, *42*, 158–164. [CrossRef]
39. Amin, M.R.; Siddiki, M.A.; Kabir, A.K.M.A.; Faruque, M.O.; Khandaker, Z.H. Status of buffalo farmer and buffaloes at Subornochar upozila of Noakhali District. *Progress. Agric.* **2015**, *26*, 71–78. [CrossRef]
40. Bartocci, S.; Di Lella, T. Capacità di utilizzazione digestiva degli alimenti. *Agric. Ricerca* **1994**, *153*, 49–56.
41. Puppo, S.; Bartocci, S.; Terramoccia, S.; Grandoni, F.; Amici, A. Rumen microbial counts and in vivo digestibility in buffaloes and cattle given different diets. *Anim. Sci.* **2002**, *75*, 323–329. [CrossRef]
42. Terramoccia, S.; Bartocci, S.; Amici, A.; Martillotti, F. Protein and protein-free dry matter rumen degradability in buffalo, cattle and sheep fed diets with different forage to concentrate ratios. *Livest. Prod. Sci.* **2000**, *65*, 185–195. [CrossRef]
43. Manish, P. Nutritional Requirement of Lactating Buffaloes. Benision Media Article. 2018. Available online: http://benisonmedia.com/nutritional-requirement-of-lactating-buffaloes (accessed on 18 January 2021).
44. Sarwar, M.; Khan, M.A.; Nisa, M.; Bhatti, S.A.; Shahzad, M.A. Nutritional management for buffalo production. *Asian Australas. J. Anim. Sci.* **2009**, *22*, 1060–1068. [CrossRef]
45. Iqbal, Z.M.; Abdullah, M.; Javed, K.; Jabbar, M.A.; Ahmed, N.; Ditta, Y.A.; Mustafa, H.; Shahzad, F. Effect of varying levels of concentrate on growth performance and feed economics in Nili-Ravi buffalo heifer calves. *Turk. J. Veter. Anim. Sci.* **2017**, *41*, 775–780. [CrossRef]
46. Terramoccia, S.; Bartocci, S.; Borghese, A. Nutritional requirements in buffalo cows and heifers. In *Buffalo Production and Research, FAO Inter-Regional Cooperative Research Network on Buffalo, REU Technical Series*; Food and Agriculture Organization Of the United Nations: Rome, Italy, 2005; Volume 67, pp. 145–160.
47. Zicarelli, L. Can we consider buffalo a non precocious and hypofertile species? *Ital. J. Anim. Sci.* **2007**, *6*, 143–154. [CrossRef]
48. Paul, S.S.; Mandal, A.B.; Pathak, N.N. Feeding standards for lactating riverine buffaloes in tropical conditions. *J. Dairy Res.* **2002**, *69*, 173–180. [CrossRef]
49. Campanile, G. Nutrition and milk production in dairy buffalo. In Proceedings of the III Simposio de Búfalos de las Américas and 2nd Buffalo Symposium of the Europe and Americans, Medelin, Colombia, 6–8 September 2006; pp. 132–141.
50. Pathak, N.N. Comparison of feed intake, digestibility of nutrients and performance of cattle (B. indicus and B. indicus × B. taurus crosses) and buffaloes (swamp and Indian). In *Coping with Feed Scarcity in Smallholder Livestock Systems in Developing Countries*; Ayantunde, A.A., Fernandez-Rivera, S., McCrabb, G., Eds.; Animal Science UR: Wageningen, The Netherlands; University of Reading: Reading, UK; Science Institute of Technology: Zurich, Switzerland; International Livestock Research Institute: Nairobi, Kenya, 2005; pp. 209–233.
51. Hayashi, Y.; Shah, S.; Shah, S.K.; Kumagai, H. Dairy production and nutritional status of lactating buffalo and cattle in small-scale farms in Terai, Nepal. *Livest. Res. Rural Dev.* **2005**, *17*, 65–74.
52. Paul, S.S.; Lal, D. *Nutrient Requirements of Buffaloes*; Satish Serial Publishing House: Delhi, India, 2010; pp. 5–17.
53. National Research Council—NRC. *Nutrient Requirements of Domestic Animals: Nutrient Requirements of Dairy Cattle*, 7th ed.; National Academy of Sciences: Washington, DC, USA, 2001; p. 333.
54. Agricultural and Food Research Council—AFRC. *Technical Committee Reports*; Agricultural and Food Research Council—AFRC: London, UK, 1991.
55. Bülbül, T. Energy and nutrient requirements of buffaloes. *Kocatepe Vet. Dergisi* **2010**, *3*, 55–64.
56. Fiest, M. *Basic Nutrition of Bison*; Saskatchewan Agriculture and Food: Moose Jaw, SK, Canada, 1999.
57. Tripaldi, C. Buffalo milk quality. In *Buffalo Production and Research, FAO Inter-Regional Cooperative Research Network on Buffalo*; REU Technical Series; Food and Agriculture Organization Of the United Nations: Rome, Italy, 2005; Volume 67, pp. 173–183.
58. Perry, T.W.; Cullison, A.E.; Lowrey, R.S. *Feeds and Feeding*, 6th ed.; Pearson Education, Inc.: Upper Saddle River, NJ, USA, 2004.
59. Bartocci, S.; Terramoccia, S.; Puppo, S. New acquisitions on the digestive physiology of the Mediterranean buffalo. In *Buffalo Production and Research*; Reu Technical Series; Borghese, A., Ed.; Food and Agriculture Organization of the United Nations: Rome, Italy, 2005; Chapter VIII; Volume 67, pp. 161–172.
60. Paulino, M.F.; Detmann, E.; Valente, E.E.L.; Barros, L.V. Nutrition of grazing cattle. In Proceedings of the 4th Symposium on Strategic Management of Pasture; Universidade Federal de Viçosa: Viçosa, Brazil, 2008; pp. 131–169.

61. Figueiras, J.F.; Detmann, E.; Franco, M.O.; Batista, E.D.; Reis, W.L.S.; Paulino, M.F.; Filho, S.C.V. Effects of supplements with different protein contents on nutritional performance of grazing cattle during the rainy season. *Asian Australas. J. Anim. Sci.* **2016**, *29*, 1710–1718. [CrossRef]
62. Sabia, E.; Napolitano, F.; Claps, S.; Braghieri, A.; Piazzolla, N.; Pacelli, C. Feeding, nutrition and sustainability in dairy en-terprises: The case of Mediterranean buffaloes (Bubalus bubalis). In *The Sustainability of Agro-Food and Natural Resource Systems in the Mediterranean Basin*; Springer: Cham, Switzerland, 2015; pp. 57–64.
63. Detmann, E.; Valente, É.E.; Batista, E.D.; Huhtanen, P. An evaluation of the performance and efficiency of nitrogen utilization in cattle fed tropical grass pastures with supplementation. *Livest. Sci.* **2014**, *162*, 141–153. [CrossRef]
64. Sahoo, A.; Elangovan, A.V.; Mehra, U.R.; Singh, U.B. Catalytic supplementation of urea-molasses on nutritional performance of male buffalo (Bubalus bubalis) calves. *Asian Australas. J. Anim. Sci.* **2004**, *17*, 621–628. [CrossRef]
65. Wynn, P.C.; Warriach, H.M.; Morgan, A.; McGill, D.M.; Hanif, S.; Sarwar, M.; Iqbal, A.; Sheehy, P.A.; Bush, R.D. Perinatal nutrition of the calf and its consequences for lifelong productivity. *Asian Australas. J. Anim. Sci.* **2009**, *22*, 756–764. [CrossRef]
66. Lazzarini, Í.; Detmann, E.; Valadares Filho, S.D.C.; Paulino, M.F.; Batista, E.D.; de Almeida Rufino, L.M.; Dos Reis, W.L.S.; de Oliveira Franco, M. Nutritional performance of cattle grazing during rainy season with nitrogen and starch supplementation. *Asian Australas. J. Anim. Sci.* **2015**, *29*, 1120–1128. [CrossRef] [PubMed]
67. Costa, V.A.C.; Detmann, E.; Paulino, M.F.; Valadares Filho, S.D.C.; Henriques, L.T.; Carvalho, I.P.C.D. Total and partial digestibility and nitrogen balance in grazing cattle supplemented with non-protein and, or true protein nitrogen during the rainy season. *Rev. Bras. Zootec.* **2011**, *40*, 2815–2826. [CrossRef]
68. Raleigh, R.J.; Lesperance, A.L. Range cattle nutrition. In *Digestive Physiology and Nutrition of Ruminants*; Church, D.C., Ed.; Oregon State University Bookstores: Corvallis, OR, USA, 1972.
69. Loerch, S.C. Effects of feeding growing cattle high-concentrate diets at a restricted intake on feedlot performance. *J. Anim. Sci.* **1990**, *68*, 3086–3095. [CrossRef]
70. Del Curto, T.R.C.; Cochran, D.L.; Harmon, A.A.; Beharka, K.A.; Jacques, G.T.; Vanzant, E.S. Supplementation of dormant tallgrass-prairie for age: I. Influence of varying supplemental protein and (or) energy levels on forage utilization characteristics of beef steers in confinement. *J. Anim. Sci.* **1990**, *68*, 515–522. [CrossRef]
71. Ba, N.X.; Van, N.H.; Ngoan, L.D.; Leddin, C.M.; Doyle, P.T. Effects of amount of concentrate supplement on forage intake, diet digestibility and live weight gain in yellow cattle in Vietnam. *Asian Australas. J. Anim. Sci.* **2008**, *21*, 1736–1744. [CrossRef]
72. Agle, M.; Hristov, A.N.; Zaman, S.; Schneider, C.; Ndegwa, P.M.; Vaddella, V.K. Effect of dietary concentrate on rumen fermentation, digestibility, and nitrogen losses in dairy cows. *J. Dairy Sci.* **2010**, *93*, 4211–4222. [CrossRef]
73. Liu, H.; Xu, T.; Xu, S.; Ma, L.; Han, X.; Wang, X.; Zhang, X.; Hu, L.; Zhao, N.; Chen, Y.; et al. Effect of dietary concentrate to forage ratio on growth performance, rumen fermentation and bacterial diversity of Tibetan sheep under barn feeding on the Qinghai-Tibetan plateau. *Peer J.* **2019**, *7*, e7462. [CrossRef]
74. Shahudin, M.S.; Ghani, A.A.A.; Zamri-Saad, M.; Zuki, A.B.; Abdullah, F.F.J.; Wahid, H.; Hassim, H.A. The necessity of a herd health management programme for dairy goat farms in Malaysia. *Pertanika J. Trop. Agric. Sci.* **2018**, *41*, 1–18.
75. Suharti, S.; Astuti, D.A.; Wina, E.; Toharmat, T. Rumen microbial population in the in vitro fermentation of different ratios of forage and concentrate in the presence of whole lerak (Sapindus rarak) fruit extract. *Asian Australas. J. Anim. Sci.* **2011**, *24*, 1086–1091. [CrossRef]
76. Van Dung, D.; Shang, W.; Yao, W. Effect of crude protein levels in concentrate and concentrate levels in diet on in vitro fermentation. *Asian Australas. J. Anim. Sci.* **2014**, *27*, 797. [CrossRef]
77. Krause, K.M.; Oetzel, G. Understanding and preventing subacute ruminal acidosis in dairy herds: A review. *Anim. Feed. Sci. Technol.* **2006**, *126*, 215–236. [CrossRef]
78. Tahir, M.; Hetta, M.; Larsen, M.; Lund, P.; Huhtanen, P. In vitro estimations of the rate and extent of ruminal digestion of starch-rich feed fractions compared to in vivo data. *Anim. Feed. Sci. Technol.* **2013**, *179*, 36–45. [CrossRef]
79. Vahora, S.; Parnerkar, S.; Kore, K. Effect of feeding bypass nutrients to growing buffalo heifers under field conditions. *Livest. Res. Rural Dev.* **2012**, *24*, 39.
80. Yadav, C.M.; Chaudhary, J.L. Effect of feeding protected protein on growth performance and physiological reaction in crossbred heifers. *Indian J. Anim. Nutr.* **2010**, *27*, 401–407.
81. Behan, A.A.; Loh, T.C.; Fakurazi, S.; Kaka, A.; Samsudin, A.A.; Kaka, U. Effects of supplementation of rumen protected fats on rumen ecology and digestibility of nutrients in sheep. *Animals* **2019**, *9*, 400. [CrossRef]
82. Naik, P.K.; Saijpaul, S.; Kaur, K. Effect of supplementation of indigenously prepared rumen protected fat on rumen fermentation in buffaloes. *Indian J. Anim. Sci.* **2010**, *80*, 902.
83. Naik, P.K. Bypass fat in dairy ration—A review. *Anim. Nutr. Feed Technol.* **2013**, *13*, 147–163.
84. Bharadwaj, A.; Sengupta, B.P.; Sethi, R.K. Effect of feeding protected and unprotected protein on nutrients intake and reproductive performance of lactating buffaloes. *Indian J. Anim. Sci.* **2000**, *70*, 428–430.
85. Chatterjee, A.; Walli, T.K. Effect of feeding formaldehyde treated mustard cake as bypass protein on milk yield and milk composition of Murrah buffaloes. *Indian J. Dairy Sci.* **2003**, *56*, 299–306.
86. Park, B.K.; Choi, N.-J.; Kim, H.C.; Kim, T.I.; Cho, Y.M.; Oh, Y.K.; Im, S.K.; Kim, Y.J.; Chang, J.S.; Hwang, I.H.; et al. Effects of amino acid-enriched ruminally protected fatty acids on plasma metabolites, growth performance and carcass characteristics of Hanwoo steers. *Asian Australas. J. Anim. Sci.* **2010**, 23, 1013–1021. [CrossRef]

87. Singh, R.; Randhawa, S.S.; Randhawa, C.S. Body condition scoring by visual and digital methods and its correlation with ultrasonographic back fat thickness in transition buffaloes. *Buffalo Bull.* **2017**, *36*, 169–182.
88. Moran, J.B. Growth and development of buffaloes. In *Buffalo Production*; Tulloh, N.M., Holmes, J.H.G., Eds.; Elsevier: Amsterdam, NY, USA, 1992; pp. 191–221.
89. Garnsworthy, P.C. The effect of energy reserves at calving on performance of dairy cows. In *Nutrition and Lactation in the Dairy Cow*; Garnsworthy, P.C., Ed.; Butterwooths: London, UK, 1988.
90. Mirza, R.H.; Javed, K.; Abdullah, M.; Pasha, T.N. Genetic and non-genetic factors affecting body condition score in Nili Ravi buffaloes and its correlation with milk yield. *J. Anim. Plant Sci.* **2013**, *23*, 1486–1490.
91. FAO, Food and Agriculture Organization of the United Nation. *FAO Food and Nutrition Paper*; FAO: Rome, Italy, 2004; pp. 22661–23667.
92. Anitha, A.; Rao, K.S.; Suresh, J.; Moorthy, P.S.; Reddy, Y.K. A body condition score (BCS) system in Murrah buffaloes. *Buffalo Bull.* **2011**, *30*, 79–96.
93. Lalman, D.L.; Petersen, M.K.; Ansotegui, R.P.; Tess, M.W.; Clark, C.K.; Wiley, J.S. The effects of ruminally undegradable protein, propionic acid, and monensin on puberty and pregnancy in beef heifers. *J. Anim. Sci.* **1993**, *71*, 2843–2852. [CrossRef] [PubMed]
94. Sawyer, J.E.; Mulliniks, J.T.; Waterman, R.C.; Petersen, M.K. Influence of protein type and level on nitrogen and forage use in cows consuming low-quality forage. *J. Anim. Sci.* **2012**, *90*, 2324–2330. [CrossRef] [PubMed]
95. Mulliniks, J.T.; Hawkins, D.E.; Kane, K.K.; Cox, S.H.; Torell, L.A.; Scholljegerdes, E.J.; Petersen, M.K. Metabolizable protein supply while grazing dormant winter forage during heifer development alters pregnancy and subsequent in-herd retention rate. *J. Anim. Sci.* **2013**, *91*, 1409–1416. [CrossRef]
96. Delmotte, C.; Rondia, P.; Raes, K.; Dehareng, F.; Decruyenaere, V. Omega 3 and CLA naturally enhanced levels of animal products: Effects of grass and linseed supplementation on the fatty acid composition of lamb meat and sheep milk. In Proceedings of the 11th Seminar of the FAO-CIHEAM Sub-Network on Sheep and Goat Nutrition Advanced Nutrition and Feeding Strategies to Improve Sheep and Goat Production, Catania, Italy, 8–10 September 2005; Volume 24.
97. Ngidi, M.E.; Loerch, S.C.; Fluharty, F.L.; Palmquist, D.L. Effects of calcium soaps of long-chain fatty acids on feedlot performance, carcass characteristics and ruminal metabolism of steers. *J. Anim. Sci.* **1990**, *68*, 2555–2565. [CrossRef]
98. Kumar, B.; Thakur, S.S. Effect of supplementing bypass fat on the performance of buffalo calves. *Indian J. Anim. Nutr.* **2007**, *24*, 233–236.
99. Mierlita, D.; Padeanu, I.; Daraban, S.; Lup, F.; Maerescu, C.; Chereji, I. Effect of dietary supplementation with different types of protected fat on bioproductive performances and quality of carcass in sheeps. In *Analele Universității din Oradea, Fascicula: Ecotoxicologie, Zootehnie și Tehnologii de Industrie Alimentară*; Editura Universităţii din Oradea: Oradea, Romania, 2010; pp. 607–614.
100. Ahmad, I.; Gohar, A.; Ahmad, N.; Ahmad, M. Haematological profile in cyclic, non-cyclic and endometritic cross-bred cattle. *Int. J. Agric. Biol.* **2003**, *5*, 332–334.
101. Aderemi, F.A. Effect of replacement of wheat bran with cassava supplement or un-supplemented with enzyme on the haematology and serum biochemistry of pullet chick. *Trop. J. Anim. Sci.* **2004**, *7*, 147–153.
102. Opara, M.N.; Ike, K.A.; Okoli, I.C. Haematology and plasma biochemistry of the wild adult African grasscutter (Thryonomis swinderianus, Temminck). *J. Am. Sci.* **2006**, *2*, 17–22.
103. Otto, F.; Vilela, F.; Harun, M.; Taylor, G.; Baggasse, P.; Bogin, E. Biochemical blood profile of Angoni cattle in Mozambique. *Israel J. Vet. Med.* **2000**, *55*, 95–102.
104. Babatunde, G.M.; Fajimi, A.O.; Oyejide, A.O. Rubber seed oil versus palm oil in broiler chicken diets. Effects on performance, nutrient digestibility, haematology and carcass characteristics. *Anim. Feed Sci. Technol.* **1992**, *35*, 133–146.
105. Grunwaldt, E.G.; Guevara, J.C.; Estevez, O.R.; Vicente, A.; Rousselle, H.; Alcuten, N.; Stasi, C.R. Biochemical and haemato-logical measurements in beef cattle in Mendoza plain rangelands (Argentina). *Trop. Anim. Health Prod.* **2005**, *37*, 527–540. [CrossRef]
106. Ndlovu, T.; Chimonyo, M.; Okoh, A.I.; Muchenje, V.; Dzama, K.; Raats, J.G. Assessing the nutritional status of beef cattle: Current practices and future prospects. *Afr. J. Biotechnol.* **2007**, *6*, 2727–2734. [CrossRef]
107. Maurya, S.K.; Singh, O.P. Assessment of blood biochemical profile and nutritional status of buffaloes under field conditions. *Buffalo Bull.* **2015**, *34*, 161–167.
108. Lee, A.; Twardock, A.; Bubar, R.; Hall, J.; Davis, C. Blood metabolic profiles: Their use and relation to nutritional status of dairy cows. *J. Dairy Sci.* **1978**, *61*, 1652–1670. [CrossRef]
109. Campanile, G.; Baruselli, P.; Vecchio, D.; Prandi, A.; Neglia, G.; Carvalho, N.; Sales, J.; Gasparrini, B.; Occhio, M. Growth, metabolic status and ovarian function in buffalo (Bubalus bubalis) heifers fed a low energy or high energy diet. *Anim. Reprod. Sci.* **2010**, *122*, 74–81. [CrossRef]
110. Bertoni, G.; Amici, A.; Lombardelli, R.; Bartocci, S. Variations of metabolic profile and hormones in blood of buffaloes, cattle and sheep males fed the same diets. In *Publication-European Association for Animal Production*; Butterworths: London, UK, 1993; Volume 62, p. 345.
111. Divers, T.J.; Peek, S.F. *Rebhun's Diseases of Dairy Cattle*; Elsevier Health Sciences: Amsterdam, The Netherlands, 2007.
112. Tiwari, C.; Chandramoni, A.; Jadhao, S.; Gowda, S.; Khan, M. Studies on blood biochemical constituents and rumen fermentation in growing buffalo calves fed ammoniated straw-based rations supplemented with different protein sources. *Anim. Feed. Sci. Technol.* **2001**, *89*, 119–130. [CrossRef]

113. Katiyar, S.; Mudgal, V.; Sharma, R.K.; Jerome, A.; Phulia, S.K.; Balhara, A.K.; Singh, I. Alterations in haemato-biochemical profile following by-pass nutrients supplementation in early lactating Murrah buffaloes. *Buffalo Bull.* **2019**, *38*, 203–215.
114. Ranjan, A.; Sahoo, B.; Singh, V.K.; Srivastava, S.; Singh, S.P.; Pattanaik, A.K. Effect of bypass fat supplementation on productive performance and blood biochemical profile in lactating Murrah (Bubalus bubalis) buffaloes. *Trop. Anim. Heal. Prod.* **2012**, *44*, 1615–1621. [CrossRef]
115. Kitagawa, H.; Kitoh, K.; Ito, T.; Ohba, Y.; Nishii, N.; Katoh, K.; Obara, Y.; Motoi, Y.; Sasaki, Y. Serum growth hormone and insulin-like growth factor-1 concentrations in Japanese black cattle with growth retardation. *J. Vet. Med. Sci.* **2001**, *63*, 167–170. [CrossRef]
116. Agudelo-Gómez, D.; Hurtado-Lugo, N.; Cerón-Muñoz, M.F. Growth curves and genetic parameters in colombian buffaloes (Bubalus bubalis Artiodactyla, Bovidae). *Rev. Colomb. Cienc. Pec.* **2009**, *22*, 178–188.
117. Sørensen, M.; Chaudhuri, S.; Louveau, I.; Coleman, M.; Etherton, T. Growth hormone binding proteins in pig adipose tissue: Number, size and effects of pGH treatment on pGH and bGH binding. *Domest. Anim. Endocrinol.* **1992**, *9*, 13–24. [CrossRef]
118. Baumann, G. Growth hormone-binding proteins: State of the art. *J. Endocrinol.* **1994**, *141*, 1–6. [CrossRef]
119. Renaville, R.; Hammadi, M.; Portetelle, D. Role of the somatotropic axis in the mammalian metabolism. *Domest. Anim. Endocrinol.* **2002**, *23*, 351–360. [CrossRef]
120. Gao, X.; Xu, X.-R.; Ren, H.-Y.; Zhang, Y.-H.; Xu, S.-Z. The effects of the GH, IGF-I and IGF-IBP3 gene on growth and development traits of Nanyang cattle in different growth period. *Yi Chuan Hered.* **2006**, *28*, 927–932.
121. Othman, O.E.M.; Alam, S.S.; El-Aziem, S.H.A. Single nucleotide polymorphism in Egyptian cattle insulin-like growth factor binding protein-3 gene. *J. Genet. Eng. Biotechnol.* **2014**, *12*, 143–147. [CrossRef]
122. Ayuk, J.; Sheppard, M.C. Growth hormone and its disorders. *Postgrad. Med. J.* **2006**, *82*, 24–30. [CrossRef] [PubMed]
123. Akis, I.; Oztabak, K.; Gonulalp, I.; Mengi, A.; Un, C. IGF-1 and IGF-1R gene polymorphisms in East Anatolian Red and South Anatolian Red cattle breeds. *Генетика* **2010**, *46*, 439–442. [CrossRef]
124. Giovannucci, E.; Pollak, M.; Liu, Y.; Platz, E.A.; Majeed, N.; Rimm, E.B.; Willett, W.C. Nutritional predictors of insulin-like growth factor I and their relationships to cancer in men. *Cancer Epidemiol. Biomark. Prev.* **2003**, *12*, 84–89.
125. Clemmons, D.R.; Klibanski, A.; Underwood, L.E.; McArthur, J.W.; Ridgway, E.C.; Beitins, I.Z.; Van Wyk, J.J. Reduction of plasma immunoreactive somatomedin C during fasting in humans. *J. Clin. Endocrinol. Metab.* **1981**, *53*, 1247–1250. [CrossRef]
126. Merimee, T.J.; Zapf, J.; Froesch, E.R. Insulin-like growth factors in the fed and fasted states. *J. Clin. Endocrinol. Metab.* **1982**, *55*, 999–1002. [CrossRef]
127. Ho, K.Y.; Veldhuis, J.D.; Johnson, M.L.; Furlanetto, R.; Evans, W.S.; Alberti, K.G.; Thorner, M.O. Fasting enhances growth hormone secretion and amplifies the complex rhythms of growth hormone secretion in man. *J. Clin. Investig.* **1988**, *81*, 968–975. [CrossRef]
128. Thissen, J.-P.; Ketelslegers, J.-M.; Underwood, L.E. Nutritional regulation of the insulin-like growth factors. *Endocr. Rev.* **1994**, *15*, 80–101. [CrossRef]
129. Clemmons, D.R.; Seek, M.M.; Underwood, L.E. Supplemental essential amino acids augment the somatomedin-C/insulin-like growth factor I response to refeeding after fasting. *Metabolism* **1985**, *34*, 391–395. [CrossRef]
130. Shao, B.; Long, R.; Ding, Y.; Wang, J.; Ding, L.; Wang, H. Morphological adaptations of yak (Bos grunniens) tongue to the foraging environment of the Qinghai-Tibetan Plateau1. *J. Anim. Sci.* **2010**, *88*, 2594–2603. [CrossRef]
131. Qiu, Q.; Zhang, G.; Ma, T.; Qian, W.; Wang, J.; Ye, Z.; Cao, C.; Hu, Q.; Kim, J.; Larkin, D.M.; et al. The yak genome and adaptation to life at high altitude. *Nat. Genet.* **2012**, *44*, 946–949. [CrossRef]
132. Zhang, Z.; Xu, D.; Wang, L.; Hao, J.; Wang, J.; Zhou, X.; Wang, W.; Qiu, Q.; Huang, X.; Zhou, J.; et al. Convergent evolution of rumen microbiomes in high-altitude mammals. *Curr. Biol.* **2016**, *26*, 1873–1879. [CrossRef]
133. Khan, S.; Singh, M.; Mehla, R.K.; Thakur, S.; Meena, B. Plasma hormones and milk production performances in early lactation buffaloes supplemented with a mixture of prilled fat, sweetener and toxin binder. *Biotehnol. Stocarstvu* **2016**, *32*, 15–26. [CrossRef]
134. Church, D.C. *Ruminant Animal: Digestive Physiology and Nutrition*; Prentice-Hall: Eaglewood Cliffs, NJ, USA, 1993; p. 564.
135. Getachew, G.; DePeters, E.J.; Robinson, P.H.; Fadel, J.G. Use of an in vitro rumen gas production technique to evaluate mi-crobial fermentation of ruminant feeds and its impact on fermentation products. *Anim. Feed Sci. Technol.* **2005**, *123*, 547–559. [CrossRef]
136. Calabro, S.; Piccolo, V.; Infascelli, F. Evaluation of diet for buffalo dairy cows using the Cornell Net Carbohydrate and Protein System. *Asian Australas. J. Anim. Sci.* **2003**, *16*, 1475–1481. [CrossRef]
137. Smith, D.; Smith, T.; Rude, B.; Ward, S. Short communication: Comparison of the effects of heat stress on milk and component yields and somatic cell score in Holstein and Jersey cows. *J. Dairy Sci.* **2013**, *96*, 3028–3033. [CrossRef]
138. Iqbal, M.W.; Zhang, Q.; Yang, Y.; Li, L.; Zou, C.; Huang, C.; Lin, B. Comparative study of rumen fermentation and microbial community differences between water buffalo and Jersey cows under similar feeding conditions. *J. Appl. Anim. Res.* **2017**, *46*, 740–748. [CrossRef]
139. Chanthakhoun, V.; Wanapat, M.; Kongmun, P.; Cherdthong, A. Comparison of ruminal fermentation characteristics and microbial population in swamp buffalo and cattle. *Livest. Sci.* **2012**, *143*, 172–176. [CrossRef]
140. Wanapat, M.; Puramongkon, T.; Siphuak, W. Feeding of cassava hay for lactating dairy cows. *Asian Australas. J. Anim. Sci.* **2000**, *13*, 478–482. [CrossRef]
141. Gurbuz, Y.; Davies, D.R. Organic matter digestibility and condensed tannin content of some hybrid sorghum. *Anim. Nutr. Feed Technol.* **2010**, *10*, 19–28.

142. Broderick, G. Effects of varying dietary protein and energy levels on the production of lactating dairy cows. *J. Dairy Sci.* **2003**, *86*, 1370–1381. [CrossRef]
143. Wang, C.; Liu, Q.; Guo, G.; Huo, W.; Ma, L.; Zhang, Y.; Pei, C.; Zhang, S.; Wang, H. Effects of dietary supplementation of rumen-protected folic acid on rumen fermentation, degradability and excretion of urinary purine derivatives in growing steers. *Arch. Anim. Nutr.* **2016**, *70*, 441–454. [CrossRef]
144. Kang, S.; Wanapat, M.; Phesatcha, K.; Norrapoke, T. Effect of protein level and urea in concentrate mixture on feed intake and rumen fermentation in swamp buffaloes fed rice straw-based diet. *Trop. Anim. Heal. Prod.* **2015**, *47*, 671–679. [CrossRef] [PubMed]
145. Bi, Y.; Zeng, S.; Zhang, R.; Diao, Q.; Tu, Y. Effects of dietary energy levels on rumen bacterial community composition in Holstein heifers under the same forage to concentrate ratio condition. *BMC Microbiol.* **2018**, *18*, 1–11. [CrossRef]
146. Van Houtert, M. Challenging the rational for altering VFA ratios in growing ruminants. *Feed Mix* **1996**, *4*, 8–11.
147. Chesson, A.; Forsberg, C.W. Polysaccharide degradation by rumen microorganisms. In *The Rumen Microbial Ecosystem*; Springer Science and Business Media LLC: Berlin, Germany, 1997; pp. 329–381.
148. Sutton, J.D.; Morant, S.V.; Bines, J.A.; Napper, D.J.; Givens, D.I. Effect of altering the starch: Fibre ratio in the concentrates on hay intake and milk production by Friesian cows. *J. Agric. Sci.* **1993**, *120*, 379–390. [CrossRef]
149. McCarthy, R.D., Jr.; Klusmeyer, T.H.; Vicini, J.L.; Clark, J.H.; Nelson, D.R. Effects of source of protein and carbohydrate on ruminal fermentation and passage of nutrients to the small intestine of lactating cows. *J. Dairy Sci.* **1989**, *72*, 2002–2016. [CrossRef]
150. Kolver, E.; De Veth, M. Prediction of ruminal pH from pasture-based diets. *J. Dairy Sci.* **2002**, *85*, 1255–1266. [CrossRef]
151. Oh, Y.-K.; Kim, J.-H.; Kim, K.-H.; Choi, C.-W.; Kang, S.-W.; Nam, I.-S.; Kim, D.-H.; Song, M.-K.; Kim, C.-W.; Park, K.-K. Effects of level and degradability of dietary protein on ruminal fermentation and concentrations of soluble non-ammonia nitrogen in ruminal and omasal digesta of Hanwoo steers. *Asian Australas. J. Anim. Sci.* **2008**, *21*, 392–403. [CrossRef]
152. Da Silva, L.; Pereira, O.; Da Silva, T.; Filho, S.V.; Ribeiro, K. Effects of silage crop and dietary crude protein levels on digestibility, ruminal fermentation, nitrogen use efficiency, and performance of finishing beef cattle. *Anim. Feed. Sci. Technol.* **2016**, *220*, 22–33. [CrossRef]
153. Reynolds, C.K.; Kristensen, N.B. Nitrogen recycling through the gut and the nitrogen economy of ruminants: An asynchronous symbiosis. *J. Anim. Sci.* **2008**, *86*, 293–305. [CrossRef]
154. Reynal, S.; Broderick, G. Effect of dietary level of rumen-degraded protein on production and nitrogen metabolism in lactating dairy cows. *J. Dairy Sci.* **2005**, *88*, 4045–4064. [CrossRef]
155. Budi, T.; Budi, S.; Elizabeth, W. Protected fat: Preparation and digestibility. In Proceedings of the Workshop on Advances in Small Ruminant Research, Ciawi, Indonesia, 3–4 August 1993; pp. 3–4.
156. Naik, P.K.; Saijpaul, S.; Rani, N. Effect of ruminally protected fat on in vitro fermentation and apparent nutrient digestibility in buffaloes (Bubalus bubalis). *Anim. Feed. Sci. Technol.* **2009**, *153*, 68–76. [CrossRef]
157. Saijpaul, S.; Naik, P.K.; Rani, N. Effects of rumen protected fat on in vitro dry matter degradability of dairy rations. *Indian J. Anim. Sci.* **2010**, *80*, 993.
158. Chalupa, W.; Vecchiarelli, B.; Elser, A.E.; Kronfeld, D.; Sklan, D.; Palmquist, D.L. Ruminal fermentation in vivo as influenced by long-chain fatty acids. *J. Dairy Sci.* **1986**, *69*, 1293–1301. [CrossRef]
159. Ohajuruka, O.A.; Zhigou, W.U.; Palmquist, D.L. Ruminal metabolism, fiber and protein digestion by lactating dairy cows fed calcium soap or animal vegetable fat. *J. Dairy Sci.* **1991**, *74*, 2601–2609. [CrossRef]
160. Schauff, D.; Clark, J. Effects of feeding diets containing calcium salts of long-chain fatty acids to lactating dairy cows. *J. Dairy Sci.* **1992**, *75*, 2990–3002. [CrossRef]
161. Guan, L.L.; Nkrumah, J.D.; Basarab, J.A.; Moore, S.S. Linkage of microbial ecology to phenotype: Correlation of rumen mi-crobial ecology to cattle's feed efficiency. *FEMS Microbiol. Lett.* **2008**, *288*, 85–91. [CrossRef]
162. Carberry, C.A.; Kenny, D.A.; Han, S.; McCabe, M.S.; Waters, S.M. Effect of phenotypic residual feed intake and dietary forage content on the rumen microbial community of beef cattle. *Appl. Environ. Microbiol.* **2012**, *78*, 4949–4958. [CrossRef]
163. Bevans, D.W.; Beauchemin, K.A.; Schwartzkopf-Genswein, K.S.; McKinnon, J.J.; McAllister, T.A. Effect of rapid or gradual grain adaptation on subacute acidosis and feed intake by feedlot cattle1,2. *J. Anim. Sci.* **2005**, *83*, 1116–1132. [CrossRef]
164. Russell, J.B.; Rychlik, J.L. Factors that alter rumen microbial ecology. *Science* **2001**, *292*, 1119–1122. [CrossRef]
165. Makkar, H.P.; Beever, D. Optimization of feed use efficiency in ruminant production systems. In Proceedings of the FAO Symposium, Bangkok, Thailand, 27 November 2012; FAO Animal Production and Health Division: Rome, Italy, 2013.
166. Cui, K.; Qi, M.; Wang, S.; Diao, Q.; Zhang, N. Dietary energy and protein levels influenced the growth performance, ruminal morphology and fermentation and microbial diversity of lambs. *Sci. Rep.* **2019**, *9*, 16612. [CrossRef]
167. Dahllöf, I.; Baillie, H.; Kjelleberg, S. rpoB-based microbial community analysis avoids limitations inherent in 16s rRna gene intraspecies heterogeneity. *Appl. Environ. Microbiol.* **2000**, *66*, 3376–3380. [CrossRef]
168. Faniyi, T.O.; Adegbeye, M.J.; Elghandour, M.; Pilego, A.B.; Salem, A.Z.; Olaniyi, T.A.; Adediran, O.; Adewumi, M.K. Role of diverse fermentative factors towards microbial community shift in ruminants. *J. Appl. Microbiol.* **2019**, *127*, 2–11. [CrossRef]
169. Fernando, S.C.; Purvis, H.T.; Najar, F.Z.; Sukharnikov, L.O.; Krehbiel, C.R.; Nagaraja, T.G.; Roe, B.A.; DeSilva, U. Rumen microbial population dynamics during adaptation to a high-grain diet. *Appl. Environ. Microbiol.* **2010**, *76*, 7482–7490. [CrossRef]
170. De Menezes, A.B.; Lewis, E.; O'Donovan, M.; O'Neill, B.F.; Clipson, N.; Doyle, E.M. Microbiome analysis of dairy cows fed pasture or total mixed ration diets. *FEMS Microbiol. Ecol.* **2011**, *78*, 256–265. [CrossRef]

171. Petri, R.M.; Schwaiger, T.; Penner, G.B.; Beauchemin, K.A.; Forster, R.J.; McKinnon, J.J.; McAllister, T.A. Characterization of the core rumen microbiome in cattle during transition from forage to concentrate as well as during and after an acidotic challenge. *PLoS ONE* **2013**, *8*, e83424. [CrossRef]
172. Goad, D.W.; Goad, C.L.; Nagaraja, T.G. Ruminal microbial and fermentative changes associated with experimentally induced subacute acidosis in steers. *J. Anim. Sci.* **1998**, *76*, 234–241. [CrossRef]
173. Tajima, K.; Aminov, R.I.; Nagamine, T.; Matsui, H.; Nakamura, M.; Benno, Y. Diet-dependent shifts in the bacterial population of the rumen revealed with real-time PCR. *Appl. Environ. Microbiol.* **2001**, *67*, 2766–2774. [CrossRef]
174. Koike, S.; Kobayashi, Y. Development and use of competitive PCR assays for the rumen cellulolytic bacteria: Fibrobacter succinogenes, Ruminococcus albus and Ruminococcus flavefaciens. *FEMS Microbiol. Lett.* **2001**, *204*, 361–366. [CrossRef]
175. Mrazek, J.; Tepšič, K.; Avguštin, G.; Kopečný, J. Diet-dependent shifts in ruminal butyrate-producing bacteria. *Folia Microbiol.* **2006**, *51*, 294–298. [CrossRef]
176. Vinh, N.; Wanapat, M.; Khejornsar, P.; Kongmun, P. Studies of diversity of rumen microorganisms and fermentation in swamp buffalo fed different diets. *J. Anim. Veter. Adv.* **2011**, *10*, 406–414. [CrossRef]
177. Liu, Q.; Wang, C.; Pei, C.; Li, H.; Wang, Y.; Zhang, S.; He, J.; Wang, H.; Yang, W.; Bai, Y.; et al. Effects of isovalerate supplementation on microbial status and rumen enzyme profile in steers fed on corn stover based diet. *Livest. Sci.* **2014**, *161*, 60–68. [CrossRef]
178. Wang, C.; Liu, Q.; Guo, G.; Huo, W.J.; Liang, Y.; Pei, C.X.; Wang, H. Effects of different dietary protein levels and rumen-protected folic acid on ruminal fermentation, degradability, bacterial populations and urinary excretion of purine derivatives in beef steers. *J. Agric. Sci.* **2017**, *155*, 1477–1486. [CrossRef]
179. Yang, C.-T.; Si, B.-W.; Diao, Q.-Y.; Jin, H.; Zeng, S.-Q.; Tu, Y. Rumen fermentation and bacterial communities in weaned Chahaer lambs on diets with different protein levels. *J. Integr. Agric.* **2016**, *15*, 1564–1574. [CrossRef]
180. Nardone, A.; Valfrè, F. Effects of changing production methods on quality of meat, milk and eggs. *Livest. Prod. Sci.* **1999**, *59*, 165–182. [CrossRef]
181. Rao, V.A.; Thulasi, G.; Ruban, S.W.; Thangaraju, P. Optimum age of slaughter of non-descript buffalo: Carcass and yield characteristics. *J. Agric. Sci.* **2009**, *42*, 133–138.
182. Anjaneyulu, A.S.R.; Sengar, S.S.; Lakshmanan, V.; Joshi, D.C. Meat quality of male buffalo calves maintained on different levels of protein. *Buffalo Bull.* **1985**, *4*, 45–47.
183. De La Torre, A.; Debiton, E.; Juanéda, P.; Durand, D.; Chardigny, J.-M.; Barthomeuf, C.; Bauchart, D.; Gruffat, D. Beef conjugated linoleic acid isomers reduce human cancer cell growth even when associated with other beef fatty acids. *Br. J. Nutr.* **2006**, *95*, 346–352. [CrossRef]
184. Issanchou, S. Consumer expectations and perceptions of meat and meat product quality. *Meat Sci.* **1996**, *43*, 5–19. [CrossRef]
185. Miller, R.K. Factors affecting the quality of raw meat. In *Meat Processing: Improving Quality*; Woodhead Publishing: Cambridge, UK, 2002; pp. 27–63.
186. Webb, E.C. Carcass fat quality and composition. In *Consistency of Quality: Proceedings of the 11th International Meat Symposium, Centurion, South Africa, 29–30 January 2003*; Agricultural Research Council: Irene, South Africa, 2003; pp. 48–55.
187. Wood, J.D.; Enser, M.; Fisher, A.V.; Nute, G.R.; Sheard, P.R.; Richardson, R.I.; Hughes, S.I.; Whittington, F.M. Fat deposition, fatty acid composition and meat quality: A review. *Meat Sci.* **2008**, *78*, 343–358. [CrossRef]
188. Webb, E.C. Manipulating beef quality through feeding. *South Afr. J. Food Sci. Nutr.* **2006**, *7*, 1–24.
189. Infascelli, F.; Roscia, M.; Buffardi, F. *La Carne di Bufalo*; Edizioni Pubblicità Italia: Rome, Italy, 2009.
190. Al-Jammas, M.; Agabriel, J.; Vernet, J.; Ortigues-Marty, I. The chemical composition of carcasses can be predicted from proxy traits in finishing male beef cattle: A meta-analysis. *Meat Sci.* **2016**, *119*, 174–184. [CrossRef]
191. Lambertz, C.; Panprasert, P.; Holtz, W.; Moors, E.; Jaturasitha, S.; Wicke, M.; Gauly, M. Carcass characteristics and meat quality of swamp buffaloes (Bubalus bubalis) fattened at different feeding intensities. *Asian Australas. J. Anim. Sci.* **2014**, *27*, 551–560. [CrossRef]
192. Manso, T.; Castro, T.; Mantecón, A.; Jimeno, V. Effects of palm oil and calcium soaps of palm oil fatty acids in fattening diets on digestibility, performance and chemical body composition of lambs. *Anim. Feed. Sci. Technol.* **2006**, *127*, 175–186. [CrossRef]
193. Velasco, S.; Cañeque, V.; Pérez, C.; Lauzurica, S.; Díaz, M.; Huidobro, F.; Manzanares, C.; González, J. Fatty acid composition of adipose depots of suckling lambs raised under different production systems. *Meat Sci.* **2001**, *59*, 325–333. [CrossRef]

MDPI

Article

Effects of Concentrate and Bypass Fat Supplementations on Growth Performance, Blood Profile, and Rearing Cost of Feedlot Buffaloes

Amirul Faiz Mohd Azmi [1], Hafandi Ahmad [1], Norhariani Mohd Nor [1], Goh Yong Meng [1], Mohd Zamri Saad [2], Md Zuki Abu Bakar [1], Punimin Abdullah [3], Anuraga Jayanegara [4] and Hasliza Abu Hassim [1,4,5,*]

[1] Department of Veterinary Preclinical Sciences, Faculty of Veterinary Medicine, Universiti Putra Malaysia, UPM, Serdang 43400, Malaysia; amirulfaizazmi@gmail.com (A.F.M.A.); hafandi@upm.edu.my (H.A.); norhariani@upm.edu.my (N.M.N.); ymgoh@upm.edu.my (G.Y.M.); zuki@upm.edu.my (M.Z.A.B.)

[2] Department of Veterinary Laboratory Diagnosis, Faculty of Veterinary Medicine, Universiti Putra Malaysia, UPM, Serdang 43400, Malaysia; mzamri@upm.edu.my

[3] Faculty of Science and Natural Resources, Universiti Malaysia Sabah, Jalan UMS, Kota Kinabalu 88400, Malaysia; puniminabdullah@ums.edu.my

[4] Animal Feed and Nutrition Modelling (AFENUE) Research Group, Department of Nutrition and Feed Technology, Faculty of Animal Science, IPB University, Bogor 16680, Indonesia; anuraga.jayanegara@gmail.com

[5] Laboratory of Sustainable Animal Production and Biodiversity, Institute of Tropical Agriculture and Food Security, Universiti Putra Malaysia, UPM, Serdang 43400, Malaysia

* Correspondence: haslizaabu@upm.edu.my; Tel.: +60-39769-3417

Citation: Mohd Azmi, A.F.; Ahmad, H.; Mohd Nor, N.; Meng, G.Y.; Saad, M.Z.; Abu Bakar, M.Z.; Abdullah, P.; Jayanegara, A.; Abu Hassim, H. Effects of Concentrate and Bypass Fat Supplementations on Growth Performance, Blood Profile, and Rearing Cost of Feedlot Buffaloes. *Animals* **2021**, *11*, 2105. https://doi.org/10.3390/ani11072105

Academic Editors: María-Luz García and María-José Argente

Received: 23 May 2021
Accepted: 17 June 2021
Published: 15 July 2021

Simple Summary: Studies have shown that providing concentrate and bypass fat as feed supplements resulted in better performance of large ruminants. However, there is limited information about the effects of these supplements on the performance of buffaloes. This study evaluates the effects of concentrate and bypass fat supplementations on the growth performance, blood metabolites, and feeding cost of Murrah cross and Swamp buffaloes. Following diet supplementation, the feed intake, body weight, and body condition score were significantly improved without any side effects on the blood metabolites of both buffalo breeds. Although the mixture of concentrate and bypass fat supplement (26:4) used in this study was found to increase the cost of feed, overall, it resulted in a greater return.

Abstract: This study investigates the effects of supplementation of the basal diet with concentrate and rumen bypass fat on the dry matter intake (DMI), growth performance, blood metabolites and hormonal changes, and the feeding cost of feedlot water buffaloes. Thirty-six healthy, three- to four-month-old male Murrah crossbred (n = 18) and Swamp (n = 18) buffaloes with a similar average initial body weight of 98.64 ± 1.93 kg were each randomly allocated into three dietary experimental groups. Buffaloes were fed with Diet A, which consisted of 100% *Brachiaria decumbens*, Diet B, consisting of 70% *Brachiaria decumbens* and 30% concentrate, and Diet C, consisting of 70% *Brachiaria decumbens*, 26% concentrate, and 4% rumen bypass fat for a period of 730 days. Feed intake was measured daily, while blood samples were collected for every eight months. Furthermore, body scores were noted prior to and at the end of the experimental period. The results showed that the average daily gain for buffaloes fed with Diet C was the highest. The DMI, BCS, FI, and FCR for the three groups showed significant ($p < 0.05$) differences, in the following order: Diet C > Diet B > Diet A. At the end of the two-year feeding trial, buffaloes fed with Diet B had significantly ($p < 0.05$) higher cholesterol levels than Diet A and Diet C. In addition, buffaloes fed with Diet C had significantly ($p < 0.05$) higher levels of serum total protein, growth hormone, and insulin-like growth factor-I hormone compared to Diet A and Diet B. On the other hand, buffaloes fed with Diet B and Diet C showed significant ($p < 0.05$) decrease in glucose levels. Supplemented diet improved the buffalos' weight gain to achieve the market weight in a shorter period of time, thus, giving farmers a greater return. In conclusion, concentrate and bypass fat supplementations in the diet of water buffaloes improved the growth performance without adverse effect on the blood metabolites, which enabled better farmer profitability.

Keywords: blood biochemical; buffalo; cost analysis; growth performance; supplementation

1. Introduction

Energy and protein are important constituents of animal diets. They play vital roles in the production and reproduction of animals. Therefore, nutrient requirements that are recommended by the National Research Council [1] are widely referred to when formulating diets for ruminants. However, the nutrient requirement equations presented by the NRC are mostly based on the requirements of cattle (*Bos taurus*). In general, the nutrient requirements of buffaloes are different from cattle, mainly due to the differences in climatic adaptability, nutrient utilization [2] and the digestibility of each nutrient in the feed [3]. Therefore, it is challenging to formulate diets for buffaloes due to limited references on the nutrient requirements and diet formulation for buffaloes (*Bubalus bubalis*).

Buffaloes are important to beef and dairy animals in several parts of the world, particularly in the tropical and subtropical regions of the globe. Cost-effective productions of quality buffalo beef and milk depend on accurate information on the buffaloes' energy and protein requirements. In many developing countries, buffaloes are largely fed on natural pastures for survival, consisting of poor-quality roughages with low energy and high fibre [4]. This eventually resulted in poor growth [5], delayed age at puberty, low calving rate, and poor reproductive performance [6]. Nevertheless, the slow growth rate could be enhanced cost-effectively by proper feed and feeding, which include diet supplementation. Indeed, previous studies have reported that improving feed formulation by including concentrate supplementation for Murrah cross and Swamp buffaloes resulted in better growth and reproductive performances [7,8] that eventually resulted in greater returns to the farmer [8].

In livestock farming, the cost of feeding accounted for between 63% and 84% of the total cost of production and determined the economic viability of the livestock production system [9]. Furthermore, the late age-at-puberty and the delay in reaching market weight due to improper feeding increased farm operation costs [10]. Therefore, feeding buffaloes according to their protein and energy requirements is key to enhancing profitability [11]. The use of supplements in diet either as an individual concentrate or as part of a balanced concentrate mixture with bypass fat supplement is a widely observed practice, particularly in cattle farming. It has been shown that supplementation of concentrate and bypass fat with fresh grasses significantly improves feed intake and livestock performance [12,13]. The supplementation increases energy, proteins, minerals, and vitamins intakes, and with good quality forage, helps to overcome the problem of low palatability [10], leading to better production [14,15].

Nevertheless, long-term feeding of high concentrate diet decreases rumen pH [16], leading to a chronic disorder known as subacute ruminal acidosis [17]. However, the growth potential of calves could be fully exploited by incorporating bypass fat supplements in the ration. According to previous studies [18], the recommended inclusion of fat is between 2 and 3% for lactating animals and 10 and 15% of dry matter intake (DMI) (800 to 1000 g/day) for growing animals without any adverse effect on nutrient utilization [16,19]. Indeed, determining the appropriate level of supplementation is one of the important factors that ensures the growth and health of buffaloes. The improper ratio of supplementation might be detrimental to the health and productivity of the animals due to the increase in levels of cholesterol and triglyceride and eventually might create adverse effects on the health of human consumers [19]. Furthermore, scientific reports on the effect of feeding concentrate and bypass fat, especially on blood profile and feed cost analysis, in Murrah cross and Swamp buffaloes are scanty. Therefore, this study was conducted to assess the effects of dietary supplementation of concentrate and bypass fat on growth performance, serum biochemical, and hormonal profiles in Murrah cross and Swamp buffaloes, as well as on the cost of feeding.

2. Materials and Methods

2.1. Statement of Animals Rights

The study was performed and managed according to the Animal Utilization Protocol (AUP), Institutional Animal Care and Use Committee (IACUC), Universiti Putra Malaysia (Approval No. UPM/IACUC/AUP-017/2018, on 8 January 2018). Samplings from the experimental animals were strictly conducted under veterinary supervision.

2.2. Study Area

This study was conducted at the Buffalo Breeding and Research Centre, Sabah, Malaysia (Coordinate 5°30′ N, 117°7′ E). The farm consisted of 749 acres of land with a total of 405 heads of buffaloes. Two types of buffalo breeds were available on this farm, the Swamp and the Murrah × Swamp crossbred, representing 54% and 46% of the total buffalo population, respectively. The crossbred animals were the products of breeding pure Murrah males with Swamp females.

The 398.5 acres of pastureland were planted with establishing pasture, the *Brachiaria decumbens*. The Swamp and Murrah crossbred buffaloes in this farm were kept separated in different paddocks. Wallowing areas were available in each paddock, and drinking water was available ad libitum. The buffaloes were kept extensively and were free to graze all day within the paddocks that were enclosed by barbed wire. The farm practiced an extensive one-month rotational grazing system without feed supplementation, and the intervals were determined according to the size of each paddock to prevent over-grazing. This farm practiced natural breeding with a male to female ratio of 1:20. Breeding season was between November and January each year, and pregnancy diagnosis was carried out every three months following breeding.

2.3. Experimental Animals

A total of thirty-six male buffaloes consisting of Swamp (n = 18) and Murrah cross (n = 18) buffaloes of approximately 3 months old and with an average body weight of 98.64 ± 1.93 kg were randomly divided into three treatment groups with 6 animals per group. Prior to the onset of the experiment, a proper physical examination was conducted for each buffalo. Then, all buffaloes were weighed and treated against ecto- and endoparasites.

2.4. Experimental Design

The study was a completely randomized 2 × 3 factorial arrangement with three treatment diets, two breeds, and six replicates per treatment. Daily feed supply was calculated at 3% body weight (based on dry matter of total mixed ration), given in two equal portions at 07:00 h and 17:00 h [20]. The buffaloes were allowed a 14-day adjustment period to the respective diet before the start of the experiment. Three total mixed rations (TMR) were prepared that contained three components, namely, *Brachiaria decumbens* grass (G), commercial concentrate (C) that composed of corn grain (25.0%), palm kernel cake (32.0%), rice bran (18.0%), soya bean meal (19.7%), calcium carbonate (1.0%), molasses (2.8%), vitamin-mineral premix (0.3%), sodium chloride (0.6%), and dicalcium phosphate (0.6%), and bypass fat (B), which was the OPTI-FAT F8016RXP-rumen bypass supplement from fractionated palm fat without trans-fat. The nutritional composition of grass, concentrate, and bypass fat are presented in Table 1.

Table 1. Nutritional composition of feedstuff (% DM basis).

Nutrient Composition	Diet Components		
	Grass (G)	Concentrate (C)	Bypass Fat (B)
DM [1)] (%)	90.34	90.36	99.75
Ash (% DM)	5.09	5.44	-
CF [2)] (% DM)	26.03	7.49	-
EE [3)] (% DM)	2.03	5.46	100
CP [4)] (% DM)	6.09	18.15	-
NDF [5)] (% DM)	64.27	56.87	-
ADF [6)] (% DM)	33.86	17.38	-
ADL [7)] (% DM)	3.55	2.96	-
NFC [8)] (% DM)	22.05	13.85	-
GE [9)] (MJ/kg)	11.79	15.74	37.65

Note: Data shown are the mean of triplicate analyses of each component. Three components of feedstuffs namely *Brachiaria decumbens* grass (G), commercial concentrate (C) (composition: corn grain (25.0%), palm kernel cake (32.0%), rice bran (18.0%), soya bean meal (19.7%), calcium carbonate (1.0%), molasses (2.8%), vitamin-mineral premix (0.3%), sodium chloride (0.6%), dicalcium phosphate (0.6%)), and bypass fat (B) (sources from calcium salt fractionated palm fat without trans-fat). Abbreviations: [1)] DM, dry matter; [2)] CF, crude fiber, [3)] EE, ether extract; [4)] CP, crude protein; [5)] NDF, neutral detergent fibre; [6)] ADF, acid detergent fibre; [7)] ADL, acid detergent lignin; [8)] NFC, non-fiber carbohydrate; [9)] GE, gross energy.

At the start of the experiment, buffaloes of group Diet A (control) were fed with 100% *Brachiaria decumbens* grass without supplementation. For Diet B, buffaloes were fed 70% grass with 30% concentrate, while for Diet C they were fed 70% grass, 26% concentrate, and 4% bypass fat as summarized in Table 2 [1,21,22]. Diets were supplied twice daily in the form of a total mix ration in which the grasses were offered as cut and carry and the concentrate and bypass fat were mixed according to the ration. The buffaloes were housed in individual pens with free access to clear drinking water and mineral blocks. The feeding trial lasted for 2 years. The area was 30 m^2 per animal, mostly with a compacted dirt floor, while the area close to the feeder was covered with concrete. The feeders were vinyl type and were placed transversely on the upper part of the pens, while the drinkers were located at the divider between two pens.

The total mixed ration that consists of grass, concentrate, and bypass fat was analysed for nutritional composition according to the method of the Association of Official Analytical Chemists (AOAC) [23,24] and the results are summarized in Tables 1 and 2.

2.5. Data Collection

Diet leftovers were weighed daily prior to the morning feeding to determine the average dry matter intake of each treatment. The intake was the difference between the amount of feed offered and the feed refused. The body weight was recorded before the start of the experiment and at three-monthly intervals, prior to morning feeding. The average daily gain was determined by dividing the increase in body weight over the experimental period by the length of the experimental period. The feed conversion ratio (FCR) was determined by measuring the amount of feed intake (kg DM) per kg of body weight gain during the experimental period. The body condition score of each buffalo was determined using a scale of 1 to 5 and was scored at the start and at the end of the experiment by evaluating the eight locations of the animal's body as described by Roche et al. and Anitha et al. [25,26].

Table 2. Nutritional composition of experimental diets, based on a dry matter basis.

Ingredient	% Composition in TMR				
	Diet A	Diet B	Diet C		
Brachiaria decumbens (G)	100	70	70		
Concentrate (C)	-	30	26		
Bypass fat (B)	-	-	4		
Nutrient		**Estimated Content**		**SEM [1)]**	**p-Value**
DM [2)] (%)	90.34	90.31	91.60	0.24	0.374
Ash (% DM)	5.09	5.69	5.93	0.33	0.916
CF [3)] (% DM)	26.03 ^{a}	23.73 ^{b}	21.65 ^{c}	0.66	<0.001
EE [4)] (% DM)	2.03 ^{a}	2.92 ^{b}	16.66 ^{c}	2.87	<0.001
CP [5)] (% DM)	6.09 ^{a}	8.08 ^{b}	6.56 ^{a}	0.36	0.012
NDF [6)] (% DM)	64.27 ^{a}	57.96 ^{b}	49.63 ^{c}	2.17	<0.001
ADF [7)] (% DM)	33.86 ^{a}	28.7 ^{b}	26.65 ^{c}	1.10	<0.001
ADL [8)] (% DM)	3.55	3.32	2.96	0.18	0.442
NFC [9)] (% DM)	22.05	24.84	20.81	2.29	0.074
GE (MJ/kg) [10)]	11.07 ^{a}	12.1 ^{a}	14.59 ^{b}	0.56	<0.001
Hemicellulose (% DM)	30.41 ^{a}	29.25 ^{a}	22.98 ^{b}	1.24	0.002
Cellulose (% DM)	30.32 ^{a}	25.38 ^{b}	23.69 ^{b}	1.04	0.001

Note: Diet A (control): 100% *Brachiaria decumbens*; Diet B: 70 *Brachiaria decumbens* + 30% concentrate; Diet C: 70% *Brachiaria decumbens* + 26% concentrate + 4% bypass fat. Data are the mean of triplicate analyses of each diet. ^{a,b,c} Means in the same row with different superscript are significantly different ($p < 0.05$). Abbreviations: [1)] SEM, standard error of the mean; [2)] DM, dry matter; [3)] CF, crude fiber; [4)] EE, ether extract; [5)] CP, crude protein; [6)] NDF, neutral detergent fiber; [7)] ADF, acid detergent fiber; [8)] ADL, acid detergent lignin; [9)] NFC, non-fibrous carbohydrate; [10)] GE, gross energy.

2.6. Blood and Serum Collection

Blood samples were collected from each buffalo prior to the start of feeding trial and every eight months until end of the feeding trial, approximately 1 h before the morning feeding. A total of 5 mL blood samples were collected in plain and EDTA Vacutainer tubes. The samples were cooled on ice and centrifuged within 24 h at 1000× g for 20 min. The serum and plasma were stored at −20 °C until analysis. The serum samples were used for serum biochemical analysis, including glucose, cholesterol, total protein, urea, and triglyceride, using a blood chemistry analyser (Siemens Dimension Xpand Plus, USA). Blood samples in EDTA tubes were used to determine the levels of growth hormone (GH) and insulin-like growth factor-I (IGF-I) using ELISA kits (Cloud-Clone Corporation, Wuhan, China) according to the manufacturer's recommendation and run using the microwell method. The kit had a sensitivity and inter- and intra-run precision coefficient of variations for IGF-I of 1.56 ng/mL, <12%, and <10%, respectively, and GH of 0.312 ng/mL, <12%, and <10%, respectively. All plates were read using a computerized automated microplate ELISA reader (Infinite 200 series, TECAN). All measurements were made in one run with triplicate for each sample.

2.7. Cost of Feeding

The economic aspect of feedlot buffalo rearing was calculated based on feeding cost/kg live weight gain [27,28] at an exchange rate of 1 USD = 4.07 MYR. To calculate the total operational cost, it was assumed that the cost of feeding represented 87% of the total cost of the activity [29], and the cost of feeding comprised of the costs of basal diet and supplementations (concentrate and bypass fat) [30]. At the time of the study, the values of the feedstuffs (MYR/kg) were 0.23 MYR (0.06 USD) for *Brachiaria decumbens*, 1.11 MYR (0.27 USD) for concentrate, and 3.82 MYR (0.94 USD) for bypass fat. The 2-year management cost of 158.50 MYR (38.94 USD) per animal was added, which included the 0.50 MYR (0.12 USD) cost of deworming, 2.00 MYR (0.49 USD) for an ID tag, and 156.00 MYR (38.33 USD) for fertilizer. The average price of live weight buffalo in Malaysia was 14.60 MYR/kg (3.59 USD). The income through daily live weight gain, the cost of feed per day, and the net income through live weight in the 2-year study was calculated as below [27,28,30]:

Income live weight gain = Average daily gain × Current price of live weight (RM 14.60/kg)

Cost of feed per day = Current price of feed × Dry matter intake (based on 3% of body weight)

Net profit live weight gain = Income live weight gain in 2 years − Cost of feeding in 2 years − Management cost per animal in 2 years

Net profit live weight = Income live weight in 2 years − Cost of feeding in 2 years − Management cost per animal in 2 years

2.8. Statistical Analysis

All data were collected and recorded using Microsoft Excel and analyzed using the software package SPSS (Statistical Package for the Social Science 25.0, Inc., Chicago, IL, USA). Comparisons between breed, diet, and the interactions between breed and diet were performed using general linear model (GLM) procedures according to a 2 × 3 factorial arrangement in a completely randomized design (CRD). The resulting *p*-values were corrected using Tukey's test to identify significant differences between the treatments. Linear mixed effects models were utilized with both 'Diets' and 'Breed' as fixed effects to capture the appropriate structure for GLM, while the feed intake, body weight pattern and gain, FCR, ADG, body condition score, serum biochemical, hormonal profiles, and cost of feeding were considered as a random effect. Furthermore, all data were also analysed using a GLM model for repeated measure procedure with treatment as the between subjects' main effect and period of sampling (months) as the within subject factor. For all the statistical tests used, results were considered significant at $p \leq 0.05$; differences between means were tested using the least significant difference. All procedures were carried out as per Snedecor and Cochran [31].

3. Results

3.1. Dry Matter Intake

The concentrate and bypass fat supplementations in the basal diet resulted in a significant ($p < 0.05$) effect and gradual changes in both buffalo breeds (Figures 1 and 2). Total dry matter intake per day ranged between 4.70 and 12.64 kg/day for Murrah cross and between 3.72 and 11.30 kg/day for Swamp buffaloes, which was significantly ($p < 0.05$) influenced by the dietary treatment and level of supplement. Similarly, higher daily intake was observed in Murrah cross that were fed with Diet B and C. Both breeds fed with supplemented diets showed between 16.32 and 19.58 kcal ME/kg higher dry matter intake than the control group (Diet A) with 15.89 kcal ME/kg. There was no significant ($p > 0.05$) difference in the dry matter intake between the breeds, but a significant ($p < 0.05$) correlation was observed between months and dry matter intake for each breed during this two-year feeding trial.

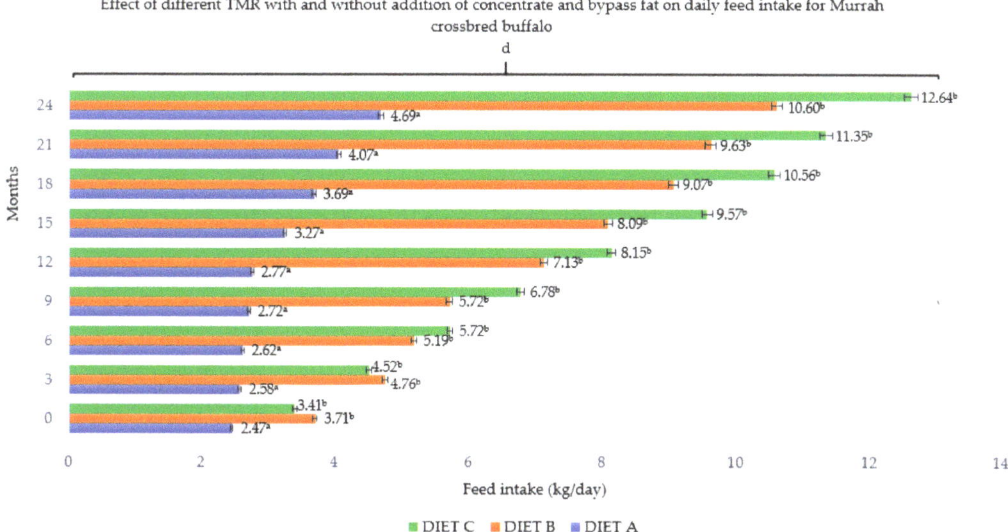

Figure 1. Effect of different TMR with and without addition of concentrate and bypass fat on daily feed intake for Murrah cross buffalo. Diet A (control)—(100% *Brachiaria decumbens*); Diet B—(70% *Brachiaria decumbens* + 30% concentrate); Diet C—(70% *Brachiaria decumbens* + 26% concentrate + 4% bypass fat); the values are presented as mean; ^{a,b} different superscripts indicate significant difference on daily feed intake for Murrah crossbred buffalo at $p < 0.05$; ^{d} indicating significant ($p < 0.05$) difference comparing between months and feed intake.

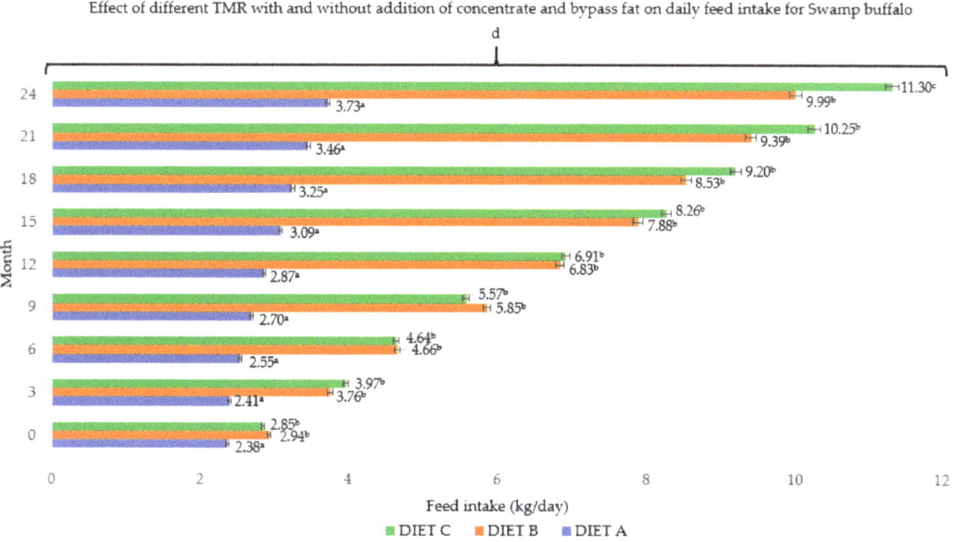

Figure 2. Effect of different TMR with and without addition of concentrate and bypass fat on daily feed intake for Swamp buffalo. Diet A (control)—(100% *Brachiaria decumbens*); Diet B—(70% *Brachiaria decumbens* + 30% concentrate); Diet C—(70% *Brachiaria decumbens* + 26% concentrate + 4% bypass fat); the values are presented as mean; ^{a,b,c} different superscripts indicate significant difference on daily feed intake for Swamp buffalo at $p < 0.05$; ^{d} indicating significant ($p < 0.05$) difference comparing between months and feed intake.

3.2. Body Weight Pattern

The bodyweight patterns for Murrah cross and Swamp buffaloes are presented in Figures 3 and 4. The effects of different diet on body weight of both buffalo breeds were highly significant ($p < 0.05$) throughout the experimental period. Higher body weight patterns were observed in both Murrah cross and Swamp buffaloes fed with Diet C containing 26% concentrate and 4% bypass fat, followed by Diet B and A. In general, the crossbreed showed significantly ($p < 0.05$) higher body weights than the Swamp buffaloes. The targeted 250 kg market weight was achieved in 12 months for Murrah cross and in 15 months for Swamp buffaloes that were fed with diet containing supplementations. In addition, there was a significant ($p < 0.05$) correlation between diet and month for the body weight pattern.

3.3. Average Daily Gain (ADG)

The average daily gain of the buffaloes fed with the three dietary treatments are summarized in Figures 5 and 6. Buffaloes that were fed with Diet B and C showed significantly higher ($p < 0.05$) ADG in each of the three monthly intervals than buffaloes fed with Diet A (control). The mean daily gain for crossbred buffaloes fed with Diet A, B, and C throughout the two-year study were 0.10, 0.32, and 0.42 kg/day, respectively. On the other hand, the mean ADG for Swamp buffaloes were 0.06, 0.32, 0.39 kg/day, respectively. There was no correlation ($p > 0.05$) between the diet and breed for average daily weight gain.

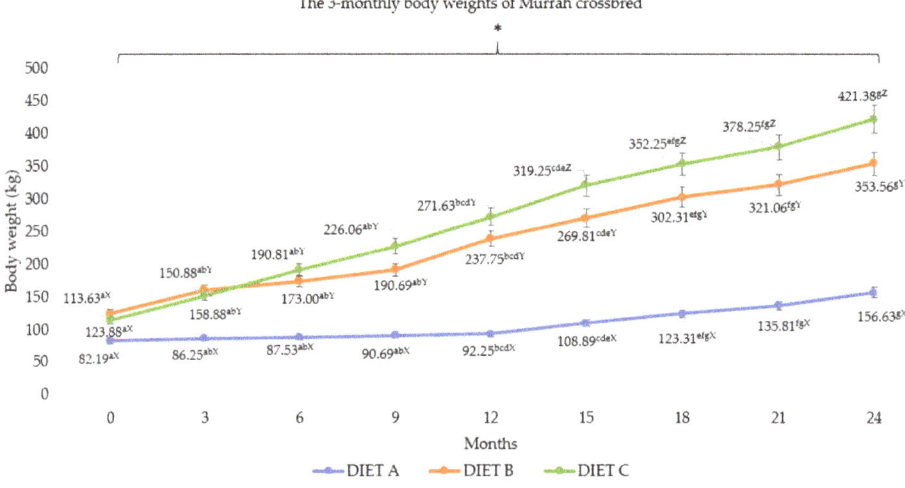

Figure 3. The three-monthly body weight (Mean ± SEM) of Murrah cross. Diet A (control)—(100% *Brachiaria decumbens)*; Diet B—(70% *Brachiaria decumbens* + 30% concentrate); Diet C—(70% *Brachiaria decumbens* + 26% concentrate + 4% bypass fat); the values are presented as mean; ^{a,b,c,d,e,f,g} different superscripts indicate significant difference of three-monthly body weight of Murrah crossbred buffalo for each Diet A, B, and C at $p < 0.05$; ^{X,Y,Z} different superscripts indicate significant difference between Diet A, B, and C for each three month interval of feeding at $p < 0.05$; * indicating significant ($p < 0.05$) difference comparing body weight and month.

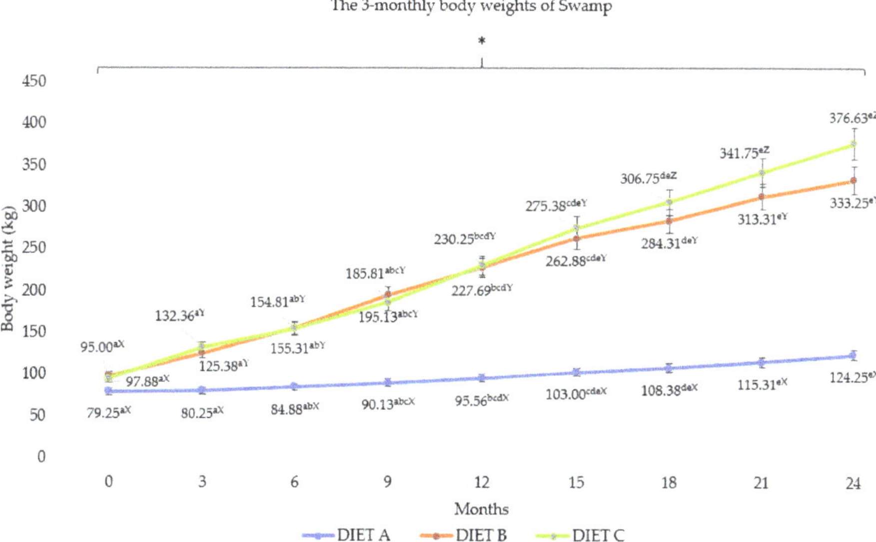

Figure 4. The three-monthly body weight (Mean ± SEM) of Swamp. Diet A (control)—(100% *Brachiaria decumbens*); Diet B—(70% *Brachiaria decumbens* + 30% concentrate); Diet C—(70% *Brachiaria decumbens* + 26% concentrate + 4% bypass fat); the values are presented as mean; ^{a,b,c,d,e} different superscripts indicate significant difference of three-monthly bodyweight of Swamp buffalo for each Diet A, B, and C at $p < 0.05$; ^{X,Y,Z} different superscripts indicate significant difference between Diet A, B, and C for each three month interval of feeding at $p < 0.05$; * indicating significant ($p < 0.05$) difference comparing body weight and month.

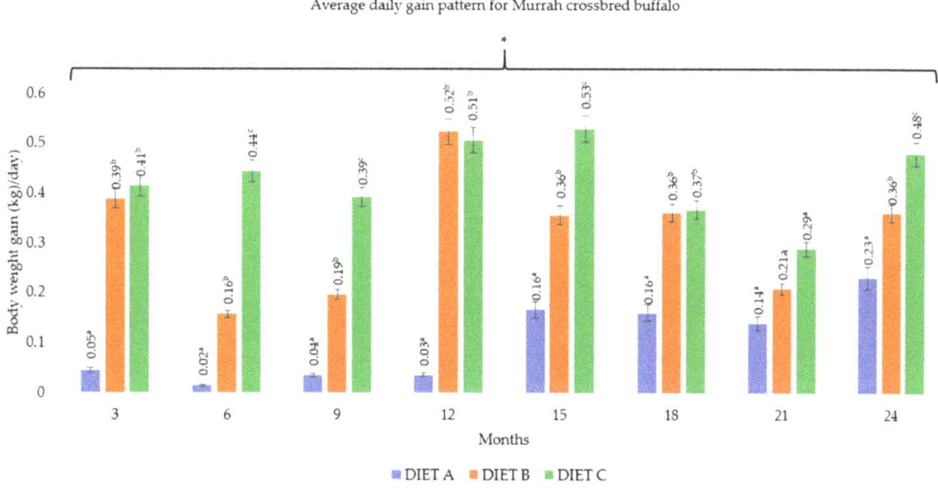

Figure 5. The average daily gain pattern at three-month intervals for Murrah crossbred buffaloes. Diet A (control)—(100% *Brachiaria decumbens*); Diet B—(70% *Brachiaria decumbens* + 30% concentrate); Diet C—(70% *Brachiaria decumbens* + 26% concentrate + 4% bypass fat); the values are presented as mean; ^{a,b,c} different superscripts indicate significant difference of average daily gain pattern for Murrah crossbred buffalo at $p < 0.05$; * indicating significant ($p < 0.05$) difference comparing body weight gain and month.

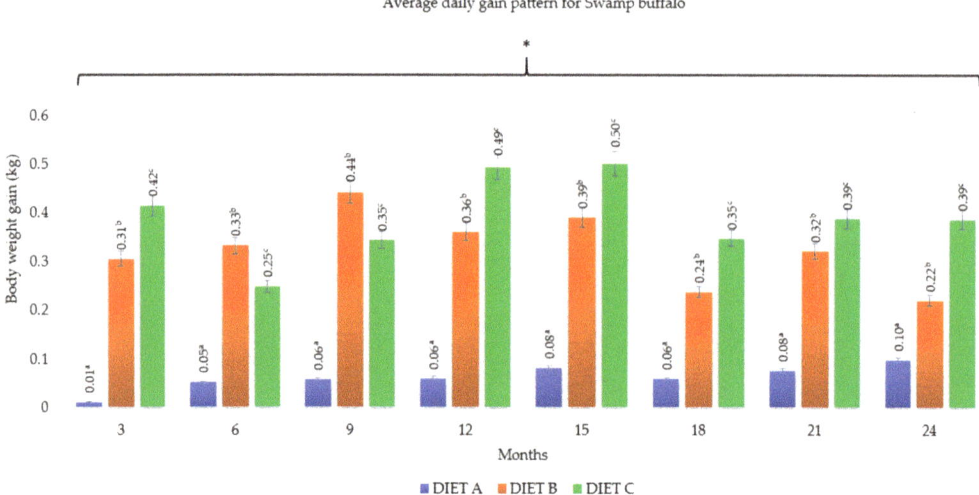

Figure 6. Effect of different diet on average daily gain at three-month intervals for Swamp buffalo. Diet A (control)—(100% *Brachiaria decumbens)*; Diet B—(70% *Brachiaria decumbens* + 30% concentrate); Diet C—(70% *Brachiaria decumbens* + 26% concentrate + 4% bypass fat); the values are presented as mean; ^{a,b,c} different superscripts indicate significant difference of average daily gain pattern for Swamp buffalo at $p < 0.05$; * indicating significant ($p < 0.05$) difference comparing body weight gain and month.

The highest average daily weight gain was observed during the first 12-month period of the study, and the lowest was between 18 and 21 months of the study. The ADG range of buffaloes fed with supplement between 12 and 15 months was 0.51 kg/day to 0.53 kg/day for Murrah cross and 0.49 kg/day to 0.50 kg/day for Swamp buffaloes ($p < 0.05$), respectively. Furthermore, between 18 and 21 months, the ADG was significantly ($p < 0.05$) lower with 0.21 kg/day to 0.37 kg/day for Murrah cross and 0.24 kg/day to 0.39 kg/day for Swamp buffaloes, respectively. However, there was no correlation ($p > 0.05$) between body weight gain and month of the year.

3.4. Overall Intake and Growth Performance

The body weight gain (BWG), average daily gain (ADG), and feed conversion ratio (FCR) were significantly ($p < 0.05$) different between the different dietary treatment groups (Table 3). However, only the final body weight recorded a significant ($p < 0.05$) difference between the breeds. Average feed intake (kg/day) for Murrah cross and Swamp buffaloes fed with Diet A, B, and C during the two-year trial was 2.99 vs. 2.80, 6.57 vs. 6.15, and 7.41 vs. 6.37, respectively, which were statistically ($p < 0.05$) significant for diet but not breed. Murrah cross significantly ($p < 0.05$) had a 10.46% increase in final body weight when compared to Swamp breed. Furthermore, total BWG and ADG of buffaloes fed with supplement were significantly ($p < 0.05$) higher by three- to four-folds for Murrah cross and five- to six-folds for Swamp buffaloes as compared to the control diet. In general, the total feed intake, daily feed intake, total BWG, and ADG for Murrah cross were higher than Swamp buffaloes, and supplemented diets (Diet C and B) resulted in better growth performance.

Table 3. Feed intake, body weight gain (BWG), average daily gain (ADG), and feed conversion ratio (FCR) of buffaloes fed with and without supplement over two-year experiment.

Attribute	Diet						Breed		*p*-Value			
	Murrah Cross			Swamp			Murrah Cross	Swamp				
Parameter	Diet A	Diet B	Diet C	Diet A	Diet B	Diet C			SEM	Diet	Breed	Interaction
Total intake (kg)	2183.12 [a]	4798.91 [b]	5407.43 [c]	2043.23 [a]	4487.06 [b]	4649.74 [b]	4129.82	3726.68	206.76	0.001	0.332	0.014
Initial BW (kg)	82.19	123.88	113.63	79.25	97.88	95.00	106.57	90.71	5.07	0.081	0.064	0.088
Final BW (kg)	156.63 [a]	353.56 [b]	421.38 [c]	124.25 [a]	333.25 [b]	376.63 [c]	310.52 [y]	278.04 [z]	34.54	<0.001	0.047	0.210
BWG (kg)	74.44 [a]	229.69 [b]	307.75 [c]	45.00 [a]	235.38 [b]	281.63 [c]	203.96	187.34	30.68	<0.001	0.582	0.054
ADG (kg/day)	0.10 [a]	0.32 [b]	0.42 [c]	0.06 [a]	0.32 [b]	0.39 [c]	0.28	0.26	0.04	<0.001	0.582	0.054
Feed intake (DM kg/day)												
Brachiaria decumbens	2.99	4.60	5.19	2.80	4.31	4.46	4.26	3.86	0.27	0.079	0.101	0.127
Concentrate	-	1.97	1.93	-	2.06	1.66	1.95	1.86	-	-	-	-
Bypass fat	-	-	0.29	-	-	0.25	0.29	0.25	-	-	-	-
Total feed intake per day (kg/day)	2.99 [a]	6.57 [b]	7.41 [c]	2.80 [a]	6.15 [b]	6.37 [b]	5.66	5.11	0.55	<0.001	0.309	0.014
FCR	30.57 [a]	21.07 [b]	17.65 [b]	52.24 [a]	19.16 [b]	16.63 [b]	23.10	29.34	3.87	<0.001	0.141	<0.001

Note: Diet A (control): 100% *Brachiaria decumbens*; Diet B: 70% *Brachiaria decumbens* + 30% concentrate); Diet C: 70% *Brachiaria decumbens* + 26% concentrate + 4% bypass fat; [a,b,c,y,z] means with different superscript letters in the same column are significantly different at $p < 0.05$. Abbreviations: SEM: standard error of means, BW: body weight, BWG: body weight gained, ADG: average daily gained, FI: feed intake, FCR: feed conversion ratio.

The higher BWG among Murrah cross and Swamp buffaloes resulted in different FCR for the different diets ($p < 0.05$). The feed conversion ratio for buffaloes was reduced by between 31.08% and 42.26% for Murrah cross and between 63.32% and 68.16% for Swamp buffaloes when fed Diet B and C. The buffaloes were able to reach the targeted market weight of 250 kg between 9 and 12 months for Murrah cross and between 12 and 15 months for Swamp buffaloes when fed Diet C, but took between 12 and 15 months for both Murrah cross and Swamp buffaloes fed with Diet B. Moreover, the buffaloes were able to reach the targeted breeding weight of 350 kg in between 15 and 18 months for Murrah cross and between 21 and 24 months for Swamp buffaloes fed with Diet C and between 21 and 24 months and 24 to 26 months for Murrah cross and Swamp buffaloes fed with Diet B, respectively. However, Murrah cross and Swamp buffaloes fed with Diet A reached neither market weight nor breeding weight within the 24-month feeding trial. Moreover, there were correlations ($p < 0.05$) between diet and breed for total intake, initial body weight, FI, and FCR.

3.5. *Body Condition Score*

The body condition score of the buffaloes fed a diet with and without supplementations are shown in Figures 7 and 8. There were no significant ($p < 0.05$) differences in the BCS between buffalo breeds and no correlation ($p < 0.05$) between diet and breed. When fed Diet A, both breeds showed the lowest average BCS at 2.25 ± 0.19 and 2.06 ± 0.23 for crossbred and Swamp buffaloes, respectively. They showed visible backbone, hips, and shoulder, the ribs were faintly visible, and the tail head area was slightly recessed, significantly ($p < 0.05$) different from buffaloes fed with Diet B and C. Furthermore, there was a significant ($p < 0.05$) correlation between weaning and diets. On the other hand, buffaloes fed with Diets B and C showed a gradual but significant ($p < 0.05$) increase in the BCS. At the end of the study, the BCS ranged between 2.94 ± 0.19 and 3.50 ± 0.20 for Murrah cross and 2.81 ± 0.16 and 3.31 ± 0.22 for Swamp buffaloes. Compared with Diet A, both breeds fed with Diet B showed improvement in BCS: 30.7% for Murrah cross and 36.4% for Swamp buffaloes. Buffaloes that were fed with Diet C showed 55.6% and 60.7% improvement in BCS for Murrah cross and Swamp buffaloes, respectively. Buffaloes fed with supplemented diets (Diet B and C) were considered moderately lean where the hip bones were visible faintly, ribs were not visible, tail head area was not recessed, and body outline appeared smooth.

3.6. *Serum Biochemical and Hormonal Profiles*

The concentration of plasma glucose, cholesterol, total protein, urea, triglyceride, growth hormone (GH), and insulin like growth factor-I (IGF-I) following a two-year study are tabulated in Tables 4 and 5. The blood glucose, cholesterol, GH, and IGF-I were significantly ($p < 0.05$) influenced by the supplemented diet, while other blood metabolites were not. All blood parameters were also shown to be significantly influenced by the period ($p < 0.05$). The blood glucose level in buffaloes fed with Diet B and C showed a significant ($p < 0.05$) decrease of between 15% and 16% compared with Diet A (control). Significant ($p < 0.05$) increments of cholesterol and triglyceride were observed in both breeds fed with Diet B compared with Diet C and A. Similarly, buffaloes fed with Diet C did not show any significant ($p < 0.05$) impact on blood profiles.

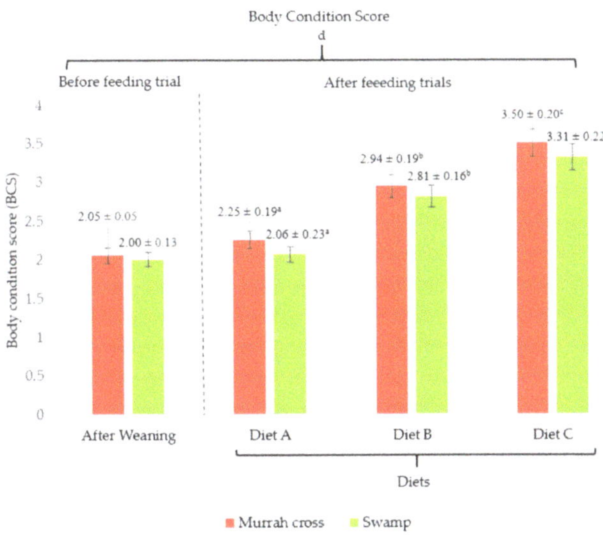

Figure 7. Effect of different diet on body condition score (BCS) at the end of the two-year feeding trial for Murrah cross and Swamp buffalo. Diet A (control)—(100% *Brachiaria decumbens);* Diet B—(70% *Brachiaria decumbens* + 30% concentrate); Diet C—(70% *Brachiaria decumbens* + 26% concentrate + 4% bypass fat); the values are presented as mean; a,b,c different superscripts indicate significant difference of average daily gain pattern for Swamp buffalo at $p < 0.05$; d indicating significant ($p < 0.05$) difference comparing body condition score between before and after feeding trials.

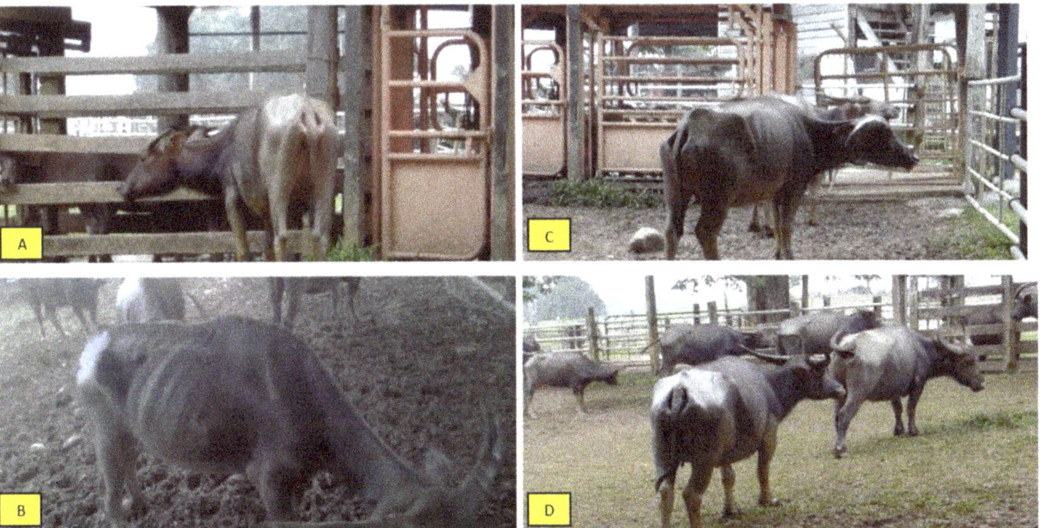

Figure 8. The body condition score of buffalo fed with and without supplemented diet. Notes: (**A**): after weaning calf (before feeding trials); (**B**): buffalo fed with Diet A; (**C**): buffalo fed with Diet B; (**D**): buffalo fed with Diet C.

Table 4. Effect of different diet on serum biochemical and hormonal profiles at the end of the two-year feeding trial for both buffalo.

Attribute	Diet						Breed		SEM	Ref. Interval	Ref.
	Murrah Cross			Swamp			Murrah Cross	Swamp			
Diets	Diet A	Diet B	Diet C	Diet A	Diet B	Diet C					
Glucose (mmol/L)											
0 month	4.80	5.10	5.30	5.40	5.20	5.10	5.07	5.23	0.09		
8 months	4.55 [a]	4.00 [b]	4.10 [b]	5.00 [a]	4.97 [a]	4.86 [b]	4.22	4.94	0.18		
16 months	4.29 [a]	3.82 [b]	3.85 [b]	4.11 [a]	4.14 [a]	3.92 [b]	3.99	4.39	0.08	1.97–5.13	
24 months	4.23 [a]	3.53 [b]	3.58 [c]	5.43 [a]	4.53 [b]	3.58 [c]	3.78	4.51	0.17		
Overall mean	4.47	4.11	4.21	4.76	4.47	4.40	4.26	4.54	0.09		
Cholesterol (mmol/L)											
0 month	3.87	3.95	3.89	3.52	2.99	3.14	3.90	3.22	0.17		
8 months	3.14 [a]	2.39 [b]	2.32 [b]	2.90 [a]	2.43 [b]	2.55[b]	2.62	2.63	0.13		
16 months	2.56 [a]	1.86 [b]	1.79 [b]	2.10 [a]	2.44 [b]	2.35[b]	2.07	2.30	0.13	0.75–2.67	
24 months	1.68 [a]	2.00 [b]	1.98 [c]	2.38 [a]	2.50 [b]	2.13[c]	1.89	2.34	0.13		
Overall mean	2.81 [a]	2.55 [b]	2.45 [b]	2.73 [a]	2.59 [b]	2.54[b]	2.60	2.62	0.06		
Total protein (g/L)											Abd Ellah et al. [32]
0 month	85.23	83.15	86.76	78.37	81.64	80.93	85.05	80.31	1.24		
8 months	72.80	77.31	79.34	80.15	81.05	81.99	76.48	81.06	1.36		
16 months	75.46	78.95	83.03	81.64	80.57	85.34	79.15	82.52	1.40	56.30–98.30	
24 months	79.65[a]	79.53 [a]	89.14 [b]	79.70 [a]	78.00 [a]	89.08 [b]	82.77	82.26	2.10		
Overall mean	78.29[a]	79.74 [a]	84.57 [b]	79.97 [a]	80.32 [a]	84.34 [b]	80.87	81.54	1.07		
Urea (mmol/L)											
0 month	6.3	6.77	6.61	6.97	6.52	6.89	6.56	6.79	0.10		
8 months	6.01	6.62	6.50	6.33	6.44	6.48	6.38	6.42	0.09		
16 months	5.90	6.39	6.43	5.55	5.89	5.74	6.24	5.73	0.15	5.40–21.24	
24 months	5.80	5.98	6.38	5.05	5.20	5.38	6.05	5.21	0.21		
Overall mean	6.00	6.44	6.48	5.98	6.01	6.12	6.31	6.04	0.09		
Triglyceride (mmol/L)											
0 month	0.33	0.31	0.33	0.18	0.21	0.19	0.32	0.19	0.03		
8 months	0.27	0.24	0.24	0.16	0.24	0.19	0.25	0.20	0.02		
16 months	0.21	0.18	0.20	0.15	0.25	0.17	0.20	0.19	0.01	0.05–0.65	
24 months	0.17	0.13	0.18	0.20	0.28	0.14	0.16	0.21	0.02		
Overall mean	0.26	0.22	0.24	0.17	0.25	0.17	0.24	0.20	0.01		

Table 4. *Cont.*

Attribute	Diet						Breed		SEM	Ref. Interval	Ref.
	Murrah Cross			Swamp			Murrah Cross	Swamp			
Diets	Diet A	Diet B	Diet C	Diet A	Diet B	Diet C					
IGF-I (ng/mL)											
0 month	114.01	116.74	112.49	108.89	107.08	108.44	114.41	108.14	1.53		
8 months	116.43 ^{a}	128.35 ^{b}	133.09 ^{b}	110.72 ^{a}	124.34 ^{b}	124.06 ^{b}	125.96	119.71	3.30		
16 months	119.25 ^{a}	147.52 ^{b}	152.11 ^{b}	116.49 ^{a}	144.42 ^{b}	149.38 ^{b}	139.63	136.76	6.52	117–300	Ashmawy [33]
24 months	122.80 ^{a}	158.30 ^{b}	171.61 ^{c}	119.44 ^{a}	153.93 ^{b}	159.57 ^{c}	150.90	144.31	8.72		
Overall mean	118.12 ^{a}	137.73 ^{b}	142.33 ^{b}	113.89 ^{a}	132.45 ^{b}	135.36 ^{b}	132.73	127.23	4.65		
GH (ng/mL)											
0 month	1.92	1.93	1.91	1.6	1.66	1.59	1.92	1.62	0.07		
8 months	1.91 ^{a}	2.11 ^{b}	2.30 ^{b}	1.72 ^{a}	1.89 ^{b}	1.91 ^{b}	2.11	1.84	0.08		
16 months	1.86 ^{a}	2.37 ^{b}	2.54 ^{c}	1.78 ^{a}	2.07 ^{b}	2.14 ^{c}	2.26	2.00	0.12	0.05–17.00	Mishra et al. [34]
24 months	1.87 ^{a}	2.58 ^{b}	2.71 ^{c}	1.83 ^{a}	2.15 ^{b}	2.46 ^{c}	2.39	2.15	0.15		
Overall mean	1.89 ^{a}	2.25 ^{b}	2.37 ^{b}	1.73 ^{a}	1.94 ^{b}	2.03 ^{c}	2.17	1.90	0.10		

Note: Diet A (control): 100% *Brachiaria decumbens*; Diet B: 70% *Brachiaria decumbens* + 30% concentrate); Diet C: 70% *Brachiaria decumbens* + 26% concentrate + 4% bypass fat; the values are presented as mean ± SEM (standard error of mean); ^{a,b,c} means with different superscript letters in the same column are significantly different at $p < 0.05$. Abbreviations: SEM: standard error of means, GH: growth hormone, IGF-I: insulin-like growth factor-I.

Table 5. The significant values for the effect of different diet on serum biochemical and hormonal profiles for both buffalo.

	p-Value			Interaction			
Parameters	**Diet**	**Breed**	**Period**	**Diet * Breed**	**Breed * Period**	**Diet * Period**	**Diet * Breed * Period**
Glucose (mmol/L)	<0.001	0.674	0.04	0.963	0.104	0.051	0.682
Cholesterol (mmol/L)	<0.001	0.266	<0.001	0.757	0.095	0.080	0.466
Total Protein (g/L)	0.049	0.469	<0.001	0.983	0.295	0.361	0.301
Urea (mmol/L)	0.341	0.299	<0.001	0.246	0.118	0.873	0.215
Triglyceride (mmol/L)	0.066	0.789	<0.001	0.246	0.316	0.078	0.961
Hormones							0.634
IGF-1 (ng/mL)	0.017	0.592	<0.001	0.752	0.303	0.077	0.462
GH (ng/mL)	<0.001	0.076	<0.001	0.665	0.081	0.056	0.075

Note: * indicates interaction between parameter studied.

The GH and IGF-I were significantly ($p < 0.05$) highest in both breeds that were fed Diet C, followed by Diet B and Diet A (Table 5). The increased IGF-I levels in buffaloes fed with Diet B and C were between 22.4% and 28.4% for Murrah cross and between 22.0% and 25.6% for Swamp buffaloes. Meanwhile, the GH level was recorded between 27.5% and 31.0% for Murrah cross and between 14.9% and 25.6% for Swamp buffaloes, higher than the control diet (Diet A). Furthermore, there were insignificant correlations between diet, breed, and period in these parameters ($p > 0.05$).

3.7. Analysis of the Cost of Feeding

Table 6 shows the total costs of feed for Murrah cross and Swamp buffaloes in the two-year feeding trial. They were found to be comparable. The higher average daily intake of diets with supplementation by Murrah cross resulted in higher feed costs of between 2.17 MYR (0.53 USD) and 2.98 MYR (0.73 USD) per animal/day, while Swamp buffalo was between 2.02 MYR (0.50 USD) and 2.56 MYR (0.63 USD) per animal/day. Buffaloes fed with Diet A showed significantly ($p < 0.05$) fewer costs of feeding, between 0.46 MYR (0.11 USD) and 0.43 MYR (0.10 USD) per animal/day.

The costs of diets with supplementation were roughly five- to seven-fold higher than the diet without supplementation, following higher consumption of supplemented feed. However, the net profit after two years of the feeding trial showed that buffaloes fed with supplemented diets provided significantly ($p < 0.05$) higher returns that ranged between 1,319.24 MYR (324.14 USD) and 1,421.94 MYR (349.37 USD) per animal for Diet B and between 1,708.84 MYR (419.86 USD) and 2,013.12 MYR (494.62 USD) per animal for Diet C compared with control Diet A that ranged between 154.80 MYR (38.03 USD) and 176.25 MYR (43.31 USD) per animal. In addition, Murrah cross showed significantly ($p < 0.05$) greater returns than the Swamp buffaloes.

Table 6. Cost analysis (RM) of buffaloes fed with different supplemented diet.

	Diet						Breed					
	Murrah Cross			Swamp			Murrah Cross	Swamp	SEM	*p*-Value		
	A	B	C	A	B	C				Diet	Breed	Interaction
A. Income from live weight gain (MYR/day/animal)	1.49 ^{a}	4.59 ^{b}	6.16 ^{c}	0.90 ^{a}	4.71 ^{b}	5.63 ^{c}	4.08 ^{y}	3.75 ^{z}	0.61	<0.001	0.012	0.001
B. Cost of feeding (RM/day)												
Brachiaria grass	0.46	0.71	0.80	0.43	0.66	0.68	0.66	0.59				
Concentrate	-	1.46	1.43	-	1.36	1.23	1.45	1.30				
Bypass fat	-	-	0.75	-	-	0.65	0.75	0.65				
Total cost of average daily DMI (MYR/day/animal)	0.46 ^{a}	2.17 ^{b}	2.98 ^{c}	0.43 ^{a}	2.02 ^{b}	2.56 ^{c}	1.87	1.67	0.30	<0.001	0.505	0.044
C. Fixed cost in 2 years												
Deworming	0.50	0.50	0.50	0.50	0.50	0.50	0.50	0.50				
ID tag	2.00	2.00	2.00	2.00	2.00	2.00	2.00	2.00				
Fertilizer	156.00	156.00	156.00	156.00	156.00	156.00	156.00	156.00				
Total (MYR/2year/animal)	158.50	158.50	158.50	158.50	158.50	158.50	158.50	158.50				
D. Gross return over feed cost (RM/day/animal) (A-B)	1.03 ^{a}	2.43 ^{b}	3.18 ^{c}	0.47 ^{a}	2.68 ^{b}	3.08 ^{c}	2.21	2.08	0.32	<0.001	0.289	0.042
E. Net profit from live weight gain for 2 years (MYR/animal)	176.25 ^{a}	1421.94 ^{b}	2013.12 ^{c}	154.80 ^{a}	1319.24 ^{b}	1708.84 ^{c}	1203.77 ^{y}	1060.96 ^{z}	220.33	<0.001	<0.001	<0.001
F. Net profit from live weight for 2 years (MYR/animal)	1793.48 ^{a}	3423.07 ^{b}	3821.95 ^{c}	1342.26 ^{a}	3229.21 ^{b}	3472.89 ^{c}	3012.83 ^{y}	2681.45 ^{z}	286.39	<0.001	<0.001	0.001

Note: 1 USD = 4.07 MYR currency conversion 5 March 2021, *MYR* Malaysian Ringgit. Estimations: income from live weight, RM 14.60/kg/animal; *Brachiaria* grass, RM 0.23/kg dry matter; Concentrate mixture, RM 1.11/kg; bypass fat, RM 3.82/kg. Diet A (control): 100% *Brachiaria decumbens*; Diet B: 70% *Brachiaria decumbens* + 30% concentrate); Diet C: 70% *Brachiaria decumbens* + 26% concentrate + 4% bypass fat; ^{a,b,c,y,z} means with different superscript letters in the same column are significantly different at $p < 0.05$. Abbreviation: SEM: standard error of means.

4. Discussion

Genetic improvements of buffaloes include the choice of breed, crossbreeding, and selection within breeds [35]. These selections aim to improve outputs such as better body weight gain, reproductive performance, and carcass traits [35]. However, the overall aim of large ruminant breeding is to improve profitability [35]: the shorter the period taken for buffalo to achieve the market weight, the more the return gained by the farmers. In this study, the Murrah crossbred showed significantly heavier bodyweight than Swamp buffaloes, similar to buffaloes in Indonesia [36], Thailand [37], and the Philippines [38]. This is because breed influences the bodyweight of buffalo calves [38]. Similarly, proper nutritional management greatly influences the bodyweight of animals [39]. Livestock farmers in most developing tropical countries were forced to maximize the limited feed resources for their livestock, resulting in significant inefficiency in the ruminant metabolic processes [40]. This affects the average daily gain and feed conversion ratio [41], and eventually the body weight. Therefore, the formulation of diets based on energy and protein intake per unit of live weight gain might give a similar performance pattern for growing animals, especially for buffalo calves.

The average body weight of weaned buffalo calves in the present study was 98.64 kg, which was higher than the 86.5 kg reported earlier [42]. A similar study revealed that approximately nine months were needed for the calves to achieve bodyweight of more than 100 kg, irrespective of the breed of the buffaloes [38]. However, both Swamp and crossbreeds showed rapid pre-weaning growth, although the crossbreeds showed significantly better growth, leading to a better overall daily weight gain of 0.89 kg/day, compared to 0.65 kg/day reported by Vaz et al. [42]. To achieve slaughter or breeding weight in a short period requires proper diet formulation. Supplementation in basal diets significantly ($p < 0.05$) affected the growth performance parameters such as ADG, BWG, and BCS. Therefore, the final body weight of the buffaloes in this two-year feeding trial reached between 350 and 400 kg in 12 to 15 months, which is suitable for slaughter [43]. However, Murrah cross was reported to reach slaughter weight much faster than Swamp buffaloes even when fed with the same diet. Furthermore, male buffalo calves were able to reach 350 kg body weight with an initial weight of 200 kg in a short fattening period of about four months [44].

One of the reasons for the better growth performance with the supplemented diets was the feed intake, which was better in the treatment groups [19]. In fact, when grazing cattle were offered concentrate supplement, they showed 0.9 kg/d higher total dry matter intake [45], similar to buffaloes fed with Diet B in this study. However, the improved feed intake by buffaloes fed with Diet C in this study disagreed with other studies that concluded that supplementation of rumen bypass fat at levels higher than 2.5% could reduce the DMI [46,47] due to the release of peptides in the gut as a response to a higher fat diet composition [48]. Furthermore, the addition of long-chain fatty acid capsulated with Ca salts at a high level was capable of depressing animal daily intake [49]. Thus, the inclusion of 5% to 20% of protected fat in diets could significantly decrease the feed intake of dairy cows [50], although no adverse effect was observed on rumen fermentation [51]. Similarly, Duckett et al. [52] reported a decrease in feed intake even with 1% fat supplementation. Nevertheless, Fiorentini et al. [53] reported that higher intake of dry matter and organic matter were observed in animals fed with supplementation consisting of protected fat compared to animals fed with unprotected fat palm oil, linseed oil, and whole soybeans, as observed in this study. A previous study had shown that different percentages of bypass fat that ranged between 30% and 80% would not affect the rumen fermentation characteristics such as total volatile fatty acid, total nitrogen, ammonia, and apparent rumen degradability content [54]. Furthermore, Naik et al. [55] concluded that there was no difference in the apparent digestibility of dry matter, organic matter crude protein, total carbohydrate, NDF, ADF, cellulose except in ether extract, and hemicellulose in buffalo rumen. Indeed, the apparent digestibility of all nutrients except ether extract and hemicellulose did not give any changes at different levels of bypass fat due to non-interference of bypass fat with digestibility of nutrients due to its relatively stable nature and minimum dissociation in the rumen [56].

The pattern of ADG in this study was similar to previous studies that fed male buffalo calves with low, medium, and high energy diets that contained 90%, 100%, and 110% of NRC recommendations [57,58]. The ADG was recorded at 516, 559, and 607 g/day, respectively, and was influenced by the energy levels. However, a higher daily weight gain of 980 g/day for yearling buffalo calves compared to 420 g/day in our study has been reported and this might be due to the use of different buffalo breeds [59]. Furthermore, the variations in average daily gain of animals could also be affected by the differences in initial body weight, genetic resources, age, and nutritional management [60,61].

An important indicator of good weight gain and health status of livestock is the body condition score. The use of BCS to indicate adequate production management has the advantage of being fast, accurate, economical, and non-invasive [62]. In this study, body weight and BCS were found to have a positive correlation and be in accordance with the type of supplement added. Buffaloes that were offered concentrate and bypass supplements had higher BCS due to better body weight gain. Similarly, buffaloes fed with concentrate have been shown to have good BCS [63], while cattle fed bypass fat showed improved feed intake [64]. On the other hand, the addition of 4 kg/d to 8 kg/d of concentrates for five weeks was able to improve the body mass gain without changes in BCS of dairy cows [65], while another study revealed that additional bypass fat did not improve body weight gain and BCS [66]. The variation in results might be due to the feed, energy and protein sources, period of feeding trials, type of animals, breeds, and the age of animals [66].

In this study, blood profiles were within the normal range, no adverse effect was observed, and no clinical signs related to metabolic disorder were detected following diet supplementation in buffaloes. Feeding concentrate and bypass fat to growing buffalo calves had little impact on the blood total protein and cholesterol but lowered the level of blood glucose [28,67,68], which remained within the normal range of 1.97–5.13 mmol/L. The findings were similar to previous studies which reported a slight increase in total protein and lipid profiles in buffaloes supplemented with concentrate and bypass fat and contributed to positive energy balance [19] and suggested that nutrient supply could also influence the lipogenic enzyme activities [69,70]. The decreased blood glucose level was calculated to be between 43% and 52%, much higher than the previous report of 10.24% [71]. On the other hand, glucose levels negatively correlate with age, with a higher value at weaning age and a lower value at an older age. This is in relation with the increased intake of starter diet post-weaning [72], causing high ruminal fermentation that switched the energy reliance to volatile fatty acids leading to lower blood glucose level in advanced age [73,74]. Therefore, supplementation of fat in the diet for a short period of time (less than 100 days) resulted in a positive correlation with the increased blood glucose and cholesterol levels [19,75] following an enhanced uptake of dietary fatty acid [12]. According to Tyagi et al. [76], supplementing with concentrate and 2.5% of bypass fat did not give any change in blood cholesterol level in growing cows. However, in the present study, serum glucose and cholesterol levels showed a decreasing pattern after 24 months of the feeding trial. This may be due to animal studies facing a long-term period of feeding trial; thus, the animal was undergoing normal homeostasis as the pre-requisite for maintaining health [77].

Growth hormone and insulin-like growth factor-I are parts of the somato-trophic axis that have multifunctional roles in the metabolism and physiology of mammals [78]. A study showed that injecting young cattle with exogenous bovine GH increased the ADG levels; thus, ADG and BW had a positive correlation with serum GH and IGF-I [79]. However, there was no difference in the concentrations of GH between Murrah cross and Swamp buffaloes in this study; thus, breed had no significant effect on plasma GH levels [80,81], although a study found an association between GH concentration and breed of animals [82]. Nevertheless, the CP:ME ratio might influence the GH and IGF-I levels in Holstein heifers [32]. Therefore, a higher dietary energy level decreases the GH serum concentration [33]. Our study showed that GH and IGF-I worked well in promoting proteosynthesis and lipolysis since both hormones were significantly high in Murrah cross and Swamp buffaloes compared to the control. We observed that the serum level of IGF-I increased with an increased dietary energy level, in agreement with previous studies in heifers and dairy cattle [34,83]. Furthermore, this study also showed that the experimental trial period significantly affected all blood and hormonal profiles [37].

In general, the crossbreeds had significantly heavier body weight than Swamp buffaloes from birth until 24 months old. A similar observation was made among buffaloes in Indonesia [36], Thailand [37], and the Philippines [43]. Body weights of buffalo calves are influenced by many factors such as feeding management [84], breeds of buffalo [85,86], and environmental factors [87]. Good feeding management improves farm husbandry and increases revenue for both Swamp and crossbred buffaloes [7,88]. In addition, crossbreeds are able to reach market weight much earlier than the Swamp buffaloes. However, early weaning is costlier for the farm due to the longer weaning-to-production period. This leads to a slightly higher additional cost of the crossbreeds. On the other hand, the birth weight could also affect the reproduction maturity of females [89], when the crossbreed tends to reach the age at first calving earlier than the Swamp buffaloes. Thus, the females can reproduce earlier and for a longer period, which brings more economic benefit to the farm. The earlier age of first calving also reduces the cost of rearing heifers. In addition, studies also showed that heterosis might have impacts on the growth and performance of buffaloes. It can come in three different forms, either individual, maternal, or paternal [90]. According to a recent study, individual heterosis is used in crossing between two breeds, which increased the performance of crossbred progeny relative to that of its purebred parents [91]. Furthermore, maternal and paternal heterosis increased the production of a cow above that of the average of her parent breeds, giving advantages in terms of improvement of reproduction, longevity, calf survivability, increased calf birth weight, shorter period of birth age puberty, and improved bull fertility [92]. However, the impact of the heterosis relies on the level of genetic differences between the original breed, whereby further variation between the two basic breeds causes the heterosis impact to be even greater [91]. Crossbreeding of pure Murrah and Swamp breeds is a common practice developed by farmers in Malaysia and other Asian countries, so that the crossbreeding can inherit superior traits possessed by their parents. The finding of this study is in agreement with previous studies that showed that crossbred buffalo have a greater performance of growth compared to purebred [7,8]. According to Shaari et al. [93], the crossbred Murrah in Malaysia was the product of crossbreeding between male Murrah and female Swamp buffaloes (exhibit 2n = 50 and 2n = 48 number of chromosomes, respectively), producing 49 chromosomes and has shared conserved regions of the genes from their parents. Indeed, a previous study also reported that analysis of mtDNA and phylogenetic trees showed Swamp buffaloes were genetically conserved and the crossbreeds were dominantly Swamp according to the maternal lineage using d-loop mtDNA [94]. Nevertheless, the crossbreed's performance was better than Swamp buffaloes, especially regarding growth, meat, and milk production [8,94].

Nutrition is one of the production factors that reflects the total production and profit of a farm [95]. In this study, total feed cost was significantly ($p < 0.05$) increased for the diets with supplementation but resulted in a significant increase in BW; thus, the targeted selling price was significantly ($p < 0.05$) higher, as earlier observed [30]. The high cost was due to the additional cost of supplements added into the basal diet, where Diet C was three folds and Diet B showed two folds higher than the non-supplemented Diet A (control). Nevertheless, the two-year fattening resulted in significantly higher net profit for Diet C, followed by Diet B. Similar results were observed by Naik et al. [96] and Raval et al. [30].

5. Conclusions

Supplementation with concentrate and bypass fat produced positive effects on the performance of feedlot buffaloes of both breeds. In addition, the supplementations have no adverse effect on serum biochemistry and increased the hormones related to growth. Even though supplementation increased the feed cost per day, subsequently animals could reach the standard market and breeding weight faster and at a significantly younger age, resulting in better income for the farm. Subsequently, Murrah crossbred buffaloes showed a significantly better body condition score and body weight gain and thus reached market and breeding weight at a significantly younger age than Swamp buffaloes. Further studies should address the effect of concentrate and bypass fat supplementations on rumen fermentation, rumen microbial population, and the meat quality of buffaloes.

Author Contributions: Conceived and designed the experiment: H.A.H., M.Z.A.B., M.Z.S., G.Y.M., and H.A.; provided supervision for animal health assessment and feed formulation during experiment: P.A., H.A.H., M.Z.A.B., and M.Z.S.; conducted the experiment and analysed data: A.F.M.A.; data interpretation and scientific discussion: A.F.M.A., N.M.N., P.A., and A.J.; contributed materials and reagents: H.A.H. and H.A.; wrote the manuscript: A.F.M.A. and H.A.H. All authors have read and agreed to the published version of the manuscript.

Funding: The whole research work and manuscript writing were fully funded by the Ministry of Science, Technology, and Innovation (MOSTI), Malaysia, under the FLAGSHIP Research Grant Scheme (Project Title: Enhancing the Productivity of Farm Buffalo (*Bubalus bubalis*) Through Modification of Rearing Practices; reference number: FP05514B0020-2(DSTIN)/6300858). Amirul Faiz M. A. was a recipient of a Graduate Research Fellowship from the Universiti Putra Malaysia.

Institutional Review Board Statement: The authors confirm that the ethical policies of the journal, as noted on the journal's author guidelines page, have been adhered to and the appropriate ethical review committee approval has been received. The animals were cared for in accordance with the animal ethics guidelines of the Animal Utilization Protocol approved by the Institution Animal Care and Use Committee (IACUC) (Approval No. UPM/IACUC/AUP-017/2018, on 8 January 2018), Universiti Putra Malaysia. The sampling from the experimented animals were strictly conducted under veterinary supervision.

Informed Consent Statement: Not applicable.

Data Availability Statement: Availability of data and equipment used and analysed during this study is available from the correspondence author on reasonable request.

Acknowledgments: The authors would like to thank staff of the Department of Veterinary Services Sabah and the Buffalo Breeding and Research Centre, Telupid, Sabah, as well as all postgraduate students and staff of the Nutrition Laboratory, Faculty of Veterinary Medicine, Universiti Putra Malaysia, for their assistance in this study. This study was financially supported by the Flagship Research grant ABI/FS/2016/01 of the Ministry of Science, Technology and Innovation, Malaysia, Ministry of Higher Education Malaysia through Higher Institutions Centres of Excellence (HICoE), and Universiti Putra Malaysia.

Conflicts of Interest: The authors declare no conflict of interest.

Abbreviations

CON: control; DMI: dry matter intake; FCR: feed conversion ratio; BCS: body condition score; ADG: average daily gain; GH: growth hormone; IGF-I: insulin-like growth factor-I; BW: body weight; BWG: body weight gain; h: hour; m: meter; MYR: Malaysian Ringgit; USD: United States Dollar; kg: kilogram; kcal: kilocalorie; ME: metabolize energy; LDL: low density lipoprotein; HDL: high density lipoprotein; g: gram; dl: decilitre; ng: nanogram; ml: millilitre; CP: crude protein; GHR: growth hormone receptor; IGFBP: insulin-like growth factor binding protein.

References

1. NRC National Research Council. *Nutrient Requirements of Dairy Cattle*, 7th ed.; National Academy of Sciences: Washington, DC, USA, 2001.
2. Pasha, T.N. Prospect of nutrition and feeding for sustainable buffalo production. *Buffalo Bull.* **2013**, *32*, 91–110.
3. Franzolin, R. Feed efficiency: A comparison between cattle and buffalo. *Buffalo J.* **1994**, *2*, 39–50.
4. Sarwar, M.; Khan, M.A.; Iqbal, Z. Status paper feed resources for livestock in Pakistan. *Int. J. Agric. Biol.* **2002**, *4*, 186–192.
5. Jabbar, M.A.; Anjum, M.I.; Rehman, S.; Shahzad, W. Comparative efficiency of sunflower meal and cottonseed cakes in the feed of crossbred calves for meat production. *Ratio* **2006**, *26*, 126–128.
6. Qureshi, M.S.; Habib, G.; Samad, H.A.; Siddiqui, M.M.; Ahmad, N.; Syed, M. Reproduction-nutrition relationship in dairy buffaloes. I. Effect of intake of protein, energy and blood metabolites levels. *Asian-Australas. J. Anim. Sci.* **2002**, *15*, 330–339. [CrossRef]
7. Othman, R.; Bakar, M.Z.A.; Kasim, A.; Zamri–Saad, M. Improving the reproductive performance of buffaloes in Sabah, Malaysia. *J. Anim. Health Prod.* **2014**, *2*, 1–4. [CrossRef]
8. Mohd Azmi, A.F.; Abu Hassim, H.; Mohd Nor, N.; Ahmad, H.; Meng, G.Y.; Abdullah, P.; Zamri-Saad, M. Comparative Growth and Economic Performances between Indigenous Swamp and Murrah Crossbred Buffaloes in Malaysia. *Animals* **2021**, *11*, 957. [CrossRef]
9. Razzaque, M.A.; Mohammed, S.A.; Al-Mutawa, T. Dairy heifer rearing in hot arid zone: An economic assessment. *Am. J. Appl. Sci.* **2010**, *7*, 466–472. [CrossRef]
10. Fiaz, M.; Abdullah, M.; Pasha, T.N.; Jabbar, M.A.; Babar, M.E.; Bhatti, J.A.; Nasir, M. Evaluating varying dietary energy levels for optimum growth and early puberty in Sahiwal heifers. *Pak. J. Zool.* **2012**, *44*, 625–634.
11. Tauqir, N.A.; Shahzad, M.A.; Nisa, M.; Sarwar, M.; Fayyaz, M.; Tipu, M.A. Response of growing buffalo calves to various energy and protein concentrations. *Livest. Sci.* **2011**, *137*, 66–72. [CrossRef]
12. Kumar, B.; Thakur, S.S. Effect of supplementing bypass fat on the performance of buffalo calves. *Indian J. Anim. Nutr.* **2007**, *24*, 233–236.
13. Kumar, S. Effect of protein supplementation on growth performance of Buffalo calves. *J. Pharmacogn. Phytochem.* **2018**, *7*, 697–700.
14. Antunović, Z.; Šperanda, M.; Amidžić, D.; Šerić, V.; Stainer, Z.; Domačinović, M.; Boli, F. Probiotic application in lambs nutrition. *Krmiva: Časopis O Hranidbi ŽivotinjaProizv. I Tehnol. Krme* **2006**, *48*, 175–180.
15. Whitley, N.C.; Cazac, D.; Rude, B.J.; Jackson-O'Brien, D.; Parveen, S. Use of a commercial probiotic supplement in meat goats. *J. Anim. Sci.* **2009**, *87*, 723–728. [CrossRef]
16. Santra, A.; Karim, S.A. Effect of dietary roughage and concentrate ratio on nutrient utilization and performance of ruminant animals. *Anim. Nutr. Feed Technol.* **2009**, *9*, 113–135.
17. Chen, Y.; Oba, M. Variation of bacterial communities and expression of Toll-like receptor genes in the rumen of steers differing in susceptibility to subacute ruminal acidosis. *Vet. Microbiol.* **2012**, *159*, 451–459. [CrossRef] [PubMed]
18. Palmquist, D.L. The feeding value of fats. In *Feed Science*; Orskov, E.R., Ed.; Elsevier Science: Amsterdam, The Netherlands, 1988; pp. 293–311.
19. Ranjan, A.; Sahoo, B.; Singh, V.K.; Srivastava, S.; Singh, S.P.; Pattanaik, A.K. Effect of bypass fat supplementation on productive performance and blood biochemical profile in lactating Murrah (*Bubalus bubalis*) buffaloes. *Trop. Anim. Health Prod.* **2012**, *44*, 1615–1621. [CrossRef]
20. Ban-Tokuda, T.; Orden, E.A.; Barrio, A.N.; Lapitan, R.M.; Delavaud, C.; Chilliard, Y.; Kanai, Y. Effects of species and sex on plasma hormone and metabolite concentrations in crossbred Brahman cattle and crossbred water buffalo. *Livest. Sci.* **2007**, *107*, 244–252. [CrossRef]
21. Bulbul, T. Energy and nutrient requirements of buffaloes. *Kocatepe Vet. Derg.* **2010**, *3*, 55–64.
22. Basra, M.J.; Nisa, M.; Khan, M.A.; Riaz, M.; Tuqeer, N.A.; Saeed, M.N. Nili-ravi buffalo III. Energy and protein requirements of 12–15 months old calves. *Int. J. Agric. Biol.* **2003**, *5*, 382–383.
23. Association of Official Analytical Chemists AOAC. *Official Methods of Analysis of AOAC International*, 17th ed.; AOAC International: Gaithersburg, MD, USA, 2000.
24. Association of Official Analytical Chemists AOAC. *Official Methods of Analysis*, 15th ed.; AOAC International: Arlington, VA, USA, 1990; p. 1117.
25. Roche, J.R.; Friggens, N.C.; Kay, J.K.; Fisher, M.W.; Stafford, K.J.; Berry, D.P. Invited review: Body condition score and its association with dairy cow productivity, health, and welfare. *J. Dairy Sci.* **2009**, *92*, 5769–5801. [CrossRef]
26. Anitha, A.; Rao, K.S.; Suresh, J.; Moorthy, P.S.; Reddy, Y.K. A body condition score (BCS) system in Murrah buffaloes. *Buffalo Bull.* **2011**, *30*, 79–96.
27. Lapitan, R.M.; Del Barrio, A.N.; Katsube, O.; Tokuda, T.; Orden, E.A.; Robles, A.Y.; Kanai, Y. Comparison of feed intake, digestibility and fattening performance of Brahman grade cattle (*Bos indicus*) and crossbred water buffalo (*Bubalus bubalis*). *Anim. Sci. J.* **2004**, *75*, 549–555. [CrossRef]
28. Iqbal, Z.M.; Abdullah, M.; Javed, K.; Jabbar, M.A.; Ahmad, N.; Ditta, Y.A.; Mustafa, H.; Shahzad, F. Effect of varying levels of concentrate on growth performance and feed economics in Nili-Ravi buffalo heifer calves. *Turk. J. Vet. Anim. Sci.* **2017**, *41*, 775–780. [CrossRef]

29. Lopes, L.S.; Ladeira, M.M.; Machado Neto, O.R.; da Silveira, A.R.M.C.; Reis, R.P.; Campos, F.R. Economical viability of finishing Nellore and Red Norte bulls in feedlot, in Lavras-MG region. *Ciência E Agrotecnologia* **2011**, *35*, 774–780. [CrossRef]
30. Raval, A.J.; Sorathiya, L.; Kharadi, V.B.; Patel, M.D.; Tyagi, K.K.; Patel, V.R. Choubey, M. Effect of calcium salt of palm fatty acid supplementation on production performance, nutrient utilization and blood metabolites in Surti buffaloes (*Bubalus bubalis*). *Indian J. Anim. Sci.* **2017**, *87*, 1124–1129.
31. Snedecor, G.W.; Cochran, W.G. *Statistical Methods*, 8th ed.; Iowa State University Press: Ames, IA, USA, 1994.
32. Abd Ellah, M.R.; Hamed, M.I.; Derar, R.I. Serum biochemical and haematological reference values for midterm pregnant buffaloes. *J. Appl. Anim. Res.* **2013**, *41*, 309–317. [CrossRef]
33. Ashmawy, N.A. Blood metabolic profile and certain hormones concentrations in Egyptian buffalo during different physiological states. *Asian J. Anim. Vet. Adv.* **2015**, *10*, 271–280. [CrossRef]
34. Mishra, A.; Goswami, T.K.; Shukla, D.C. An enzyme-linked immunosorbent assay (ELISA) to measure growth hormone level in serum and milk of buffaloes (*Bubalus bubalis*). *Indian J. Exp. Biol.* **2007**, *45*, 594–598.
35. Archer, J.A.; Richardson, E.C.; Herd, R.M.; Arthur, P.F. Potential for selection to improve efficiency of feed use in beef cattle: A review. *Aust. J. Agric. Res.* **1999**, *50*, 147–162. [CrossRef]
36. Situmorang, P.; Sitepu, P. Comparative growth performance, semen quality and draught capacity of Indonesian swamp buffalo and its crosses. *ACIAR Proc.* **1991**, *34*, 102–112.
37. Kamonpatana, M.; Sophon, S.; Jetana, T.; Sravasi, S.; Tongpan, R.; Thasipoo, K. A Comparative Study of Reproductive performance and Growth of Female Swamp and Murrah × Swamp Buffalo under Village Conditions. In Proceedings of the Buffalo and Goats in Asia, Genetic Diversity and Its Application, Kuala Lumpur, Malaysia, 10–14 February 1991; Volume 76.
38. Salas, R.C.D.; van der Lende, T.; Udo, H.M.; Mamuad, F.V.; Garillo, E.P.; Cruz, L.C. Comparison of growth, milk yield and draughtability of murrah-Philippine crossbred and Philippine native buffaloes. *Asian-Australas. J. Anim. Sci.* **2000**, *13*, 580–586. [CrossRef]
39. Burque, A.R.; Abdullah, M.; Babar, M.E.; Javed, K.; Nawaz, H. Effect of urea feeding on feed intake and performance of male buffalo calves. *J. Anim. Plant. Sci.* **2008**, *18*, 1–6.
40. Mahmoudzadeh, H.; Fazaeli, H. Growth response of yearling buffalo male calves to different dietary energy levels. *Turk. J. Vet. Anim. Sci.* **2009**, *33*, 447–454.
41. Ahmad, M.; Javed, K.; Rehman, A. Environmental factors affecting some growth traits in Nili-Ravi buffalo calves. In Proceedings of the 7th Word Congress Genetics Applied to Livestock Production, Montpellier, France, 18–23 August 2002.
42. Vaz, R.Z.; Lobato, J.F.P.; Restle, J. Influence of weaning age on the reproductive efficiency of primiparous cows. *Rev. Bras. De Zootec.* **2010**, *39*, 299–307. [CrossRef]
43. Serafy, A.M.; Ashry, A.M. The nutrition of Egyptian water buffaloes from birth to milk and meat production. *Rumin. Prod. Dry Subtrop.* **1989**, *38*, 230–243.
44. Punia, B.S.; Singh, S. Buffalo calf feeding and management. *EH Int. Buffalo Inf. Cent.* **2001**, *20*, 3–11.
45. Jiao, H.P.; Dale, A.J.; Carson, A.F.; Murray, S.; Gordon, A.W.; Ferris, C.P. Effect of concentrate feed level on methane emissions from grazing dairy cows. *J. Dairy Sci.* **2014**, *97*, 7043–7053. [CrossRef] [PubMed]
46. Marinova, P.; Banskalieva, V.; Alexandrov, S.; Tzvetkova, V.; Stanchev, H. Carcass composition and meat quality of kids fed sunflower oil supplemented diet. *Small Rumin. Res.* **2001**, *42*, 217–225. [CrossRef]
47. Haddad, S.G.; Younis, H.M. The effect of adding ruminally protected fat in fattening diets on nutrient intake, digestibility and growth performance of Awassi lambs. *Anim. Feed Sci. Technol.* **2004**, *113*, 61–69. [CrossRef]
48. Nawaz, H.; Ali, M. Effect of supplemental fat on dry matter intake, nutrient digestibility, milk yield and milk composition of ruminants. *Pak. J. Agric. Sci.* **2016**, *53*, 271–275.
49. Andrew, S.M.; Tyrrell, H.F.; Reynolds, C.K.; Erdman, R.A. Net energy for lactation of calcium salts of long-chain fatty acids for cows fed silage-based diets. *J. Dairy Sci.* **1991**, *74*, 2588–2600. [CrossRef]
50. Ferguson, J.D.; Sklan, D.; Chalupa, W.V.; Kronfeld, D.S. Effects of hard fats on in vitro and in vivo rumen fermentation, milk production, and reproduction in dairy cows. *J. Dairy Sci.* **1990**, *73*, 2864–2879. [CrossRef]
51. Naik, P.K. Bypass fat in dairy ration-a review. *Anim. Nutr. Feed Technol.* **2013**, *13*, 147–163.
52. Duckett, S.K.; Andrae, J.G.; Owens, F.N. Effect of high-oil corn or added corn oil on ruminal biohydrogenation of fatty acids and conjugated linoleic acid formation in beef steers fed finishing diets. *J. Anim. Sci.* **2002**, *80*, 3353–3360. [CrossRef]
53. Fiorentini, G.; Carvalho, I.P.C.; Messana, J.D.; Castagnino, P.S.; Berndt, A.; Canesin, R.C.; Berchielli, T.T. Effect of lipid sources with different fatty acid profiles on the intake, performance, and methane emissions of feedlot Nellore steers. *J. Anim. Sci.* **2014**, *92*, 1613–1620. [CrossRef]
54. Saijpaul, S.; Naik, P.K.; Rani, N. Effects of rumen protected fat on in vitro dry matter degradability of dairy rations. *Indian J. Anim. Sci.* **2010**, *80*, 993.
55. Naik, P.K.; Saijpaul, S.; Rani, N. Preparation of rumen protected fat and its effect on nutrient utilization in buffaloes. *Indian J. Anim. Nutr.* **2007**, *24*, 212–215.
56. Reddy, Y.R.; Krishna, N.; Rao, E.R.; Reddy, T.J. Influence of dietary protected lipids on intake and digestibility of straw based diets in Deccani sheep. *Anim. Feed Sci. Technol.* **2003**, *106*, 29–38. [CrossRef]
57. NRC National Research Council. *Nutrient Requirements of Beef Cattle*; National Academy of Sciences: Washington, DC, USA, 1996.

58. Baruah, K.K.; Ranjhan, S.K.; Pathak, N.N. Feed intake, nutrient utilization and growth in male buffalo calves fed different levels of protein and energy. *Buffalo J.* **1988**, *22*, 131–138.
59. Yunus, A.W.; Khan, A.G.; Alam, Z.; Sultan, J.I.; Riaz, M. Effects of substituting cottonseed meal with sunflower meal in rations for growing buffalo calves. *Asian-Australas. J. Anim. Sci.* **2004**, *17*, 659–662. [CrossRef]
60. Barman, K.; Mohini, M. Nutrient utilization for growth and cost of feeding in buffalo calves as influence by rumens in supplementation. *Buffalo J.* **2002**, *18*, 71–82.
61. Nair, P.V.; Verma, A.K.; Dass, R.S.; Mehra, U.R. Growth and nutrient utilization in buffalo calves fed ammoniated wheat straw supplemented with sodium sulphate. *Asian-Australas. J. Anim. Sci.* **2004**, *17*, 325–329. [CrossRef]
62. Freitas Júnior, J.E.D.; Rocha Júnior, V.R.; Rennó, F.P.; Mello, M.T.P.D.; Carvalho, A.P.D.; Caldeira, L.A. Effect of body condition score at calving on productive performance of crossbred Holstein-Zebu cows. *Rev. Bras. De Zootec.* **2008**, *37*, 116–121. [CrossRef]
63. Ghani, A.A.A.; Shahudin, M.S.; Zamri-Saad, M.; Zuki, A.B.; Wahid, H.; Kasim, A.; Hassim, H.A. Enhancing the growth performance of replacement female breeder goats through modification of feeding program. *Vet. World* **2017**, *10*, 630. [CrossRef]
64. Katiyar, G.S.; Mudgal, V.; Sharma, R.K.; Bharadwaj, A.; Phulia, S.K.; Jerome, A.; Singh, I. Effect of rumen-protected nutrients on feed intake, body weights, milk yield, and composition in Murrah buffaloes during early lactation. *Trop. Anim. Health Prod.* **2019**, *51*, 2297–2304. [CrossRef] [PubMed]
65. Muñoz, C.; Herrera, D.; Hube, S.; Morales, J.; Ungerfeld, E.M. Effects of dietary concentrate supplementation on enteric methane emissions and performance of late lactation dairy cows. *Chil. J. Agric. Res.* **2018**, *78*, 429–437. [CrossRef]
66. Mohsin, I.; Shahid, M.Q.; Haque, M.N.; Ahmad, N. Effect of bypass fat on growth and body condition score of male Beetal goats during summer. *South. Afr. J. Anim. Sci.* **2019**, *49*, 810–814. [CrossRef]
67. Jabbar, L.; Cheema, A.M.; Jabbar, M.A.; Riffat, S. Effect of different dietary energy levels, season and age on hematological indices and serum electrolytes in growing buffalo heifers. *J. Anim. Plant. Sci.* **2012**, *22*, 279–283.
68. Singh, A.K.; Chaturvedi, V.B.; Gupta, S.; Kumar, M. Effect of feeding TMR with different ratio of concentrate and roughages on blood biochemical changes in crossbred cattle and buffaloes. *Q. J. Sci. Agric. Eng.* **2018**, *8*, 288–291.
69. Cutrignelli, M.I.; Calabrò, S.; Bovera, F.; Tudisco, R.; D'Urso, S.; Marchiello, M.; Infascelli, F. Effects of two protein sources and energy level of diet on the performance of young Marchigiana bulls. 2. Meat quality. *Ital. J. Anim. Sci.* **2008**, *7*, 271–285. [CrossRef]
70. Eguinoa, P.; Brocklehurst, S.; Arana, A.; Mendizabal, J.A.; Vernon, R.G.; Purroy, A. Lipogenic enzyme activities in different adipose depots of Pirenaican and Holstein bulls and heifers taking into account adipocyte size. *J. Anim. Sci.* **2003**, *81*, 432–440. [CrossRef]
71. Fayed, A.K.; El-Sayed, A.; Ashmawy, N.A.; Ashour, G. Blood metabolites and hormone levels as indicators for growth performance in Egyptian buffalo calves. In Proceedings of the 2nd International Conference on Biotechnology Application in Agriculture (ICBAA), Hurghada, Egypt, 8–12 April 2014; Volume 17, p. 17.
72. Rashid, M.A.; Pasha, T.; Jabbar, M.; Ijaz, A. Changes in blood metabolites of early weaned nili-ravi buffalo (*Bubalus bubalis*) calves. *J. Anim. Plant. Sci.* **2013**, *23*, 1067–1071.
73. Hammon, H.M.; Schiessler, G.; Nussbaum, A.; Blum, J.W. Feed intake patterns, growth performance, and metabolic and endocrine traits in calves fed unlimited amounts of colostrum and milk by automate, starting in the neonatal period. *J. Dairy Sci.* **2002**, *85*, 3352–3362. [CrossRef]
74. Khan, M.A.; Lee, H.J.; Lee, W.S.; Kim, H.S.; Kim, S.B.; Ki, K.S.; Choi, Y.J. Pre-and postweaning performance of Holstein female calves fed milk through step-down and conventional methods. *J. Dairy Sci.* **2007**, *90*, 876–885. [CrossRef]
75. Son, J.; Grant, R.J.; Larson, L.L. Effects of tallow and escape protein on lactational and reproductive performance of dairy cows. *J. Dairy Sci.* **1996**, *79*, 822–830. [CrossRef]
76. Tyagi, N.; Thakur, S.S.; Shelke, S.K. Effect of bypass fat supplementation on productive and reproductive performance in crossbred cows. *Trop. Anim. Health Prod.* **2010**, *42*, 1749–1755. [CrossRef] [PubMed]
77. Giridharan, N.V. Glucose & energy homeostasis: Lessons from animal studies. *Indian J. Med. Res.* **2008**, *148*, 659.
78. Curi, R.A.; De Oliveira, H.N.; Silveira, A.C.; Lopes, C.R. Association between IGF-I, IGF-IR and GHRH gene polymorphisms and growth and carcass traits in beef cattle. *Livest. Prod. Sci.* **2005**, *94*, 159–167. [CrossRef]
79. Purchas, R.W.; Macmillan, K.L.; Hafs, H.D. Pituitary and plasma growth hormone levels in bulls from birth to one year of age. *J. Anim. Sci.* **1970**, *31*, 358–363. [CrossRef]
80. Trenkle, A.; Topel, D.G. Relationship of some endocrine measurements to growth and carcass composition of cattle. *J. Anim. Sci.* **1978**, *46*, 1604–1609. [CrossRef]
81. Irvin, R.; Trenkle, A. Influences of age, breed and sex on plasma hormones in cattle. *J. Anim. Sci.* **1971**, *32*, 292–295. [CrossRef] [PubMed]
82. Shingu, H.; Hodate, K.; Kushibiki, S.; Ueda, Y.; Watanabe, A.; Shinoda, M.; Matsumoto, M. Profiles of growth hormone and insulin secretion, and glucose response to insulin in growing Japanese Black heifers (beef type): Comparison with Holstein heifers (dairy type). *Comp. Biochem. Physiol. Part. C* **2001**, *130*, 259–270. [CrossRef]
83. Andersen, J.B.; Friggens, N.C.; Larsen, T.; Vestergaard, M.; Ingvartsen, K.L. Effect of energy density in the diet and milking frequency on plasma metabolites and hormones in early lactation dairy cows. *J. Vet. Med. Ser. A* **2004**, *51*, 52–57. [CrossRef] [PubMed]
84. Tatsapong, P.; Peangkoum, P.; Pimpa, O.; Hare, M.D. Effect of dietary protein on nitrogen metabolism and protein requirements for maintenance of growing Thai swamp buffalo (*Bubalus bubalis*) calves. *J. Anim. Vet. Adv.* **2010**, *9*, 1216–1222.

85. Pandya, G.M.; Joshi, C.G.; Rank, D.N.; Kharadi, V.B.; Bramkshtri, B.P.; Vataliya, P.H.; Solanki, J.V. Genetic analysis of body weight traits of Surti buffalo. *Buffalo Bull.* **2015**, *34*, 189–195.
86. Charlini, B.C.; Sinniah, J. Performance of Murrah, Surti, Nili-Ravi buffaloes and their crosses in the intermediate zone of Sri Lanka. *Development* **2015**, *27*, 1–17.
87. Javed, K.; Mirza, R.H.; Abdullah, M.; Akhtar, M. Environmental factors affecting live weight and morphological traits in Nili Ravi buffaloes of Pakistan. *Buffalo Bull.* **2013**, *32*, 1161–1164.
88. Zamri-Saad, M.; Azhar, K.; Zuki, A.B.; Punimin, A.; Hassim, H.A. Enhancement of Performance of Farmed Buffaloes Pasture Management and Feed Supplementation in Sabah, Malaysia. *Pertanika J. Trop. Agric. Sci.* **2017**, *40*, 553–564.
89. Naqvi, A.; Shami, S.A. Factors affecting birth weight in Nili-Ravi buffalo calves. *Pak. Vet. J.* **1999**, *19*, 119–122.
90. Harowi, M.; Hamdani, M.D.I. Perbandingan koefisien heterosis antara kambing Boerawa dan Saburai jantan pada bobot sapih di Kecamatan Sumberejo Kabupaten Tanggamus. *J. Ilm. Peternak. Terpadu* **2016**, *4*, 67–72.
91. Azis, R.; Ciptadi, G.; Suyadi, S. Heterosis effect and outbreeding analysis of Boer and PE goat crosses based on birth weight in F1 and F1. *J. Develepmont Res.* **2020**, *4*, 18–23. [CrossRef]
92. Weaber, R. Crossbreeding strategies: Including terminal vs. maternal crosses. In Proceedings of the Range Beef Cow Sympossium XXXIV, Loveland, CO, USA, 17–19 November 2015; Digital Commons of the University of Nebraska: Lincoln, LI, USA, 2015; pp. 117–130.
93. Shaari, N.A.L.; Jaoi-Edward, M.; San Loo, S.; Salisi, M.S.; Yusoff, R.; Ab Ghani, N.I.; Ahmad, H. Karyotypic and mtDNA based characterization of Malaysian water buffalo. *BMC Genet.* **2019**, *20*, 37. [CrossRef] [PubMed]
94. Cruz, L.C. Recent developments in the buffalo industry of Asia. *Rev. Vet.* **2010**, *21*, 7–19.
95. Shahudin, M.S.; Ghani, A.A.A.; Zamri-Saad, M.; Zuki, A.B.; Abdullah, F.F.J.; Wahid, H.; Hassim, H.A. The Necessity of a Herd Health Management Programme for Dairy Goat Farms in Malaysia. *Pertanika J. Trop. Agric. Sci.* **2018**, *41*, 1–18.
96. Naik, P.K.; Saijpaul, S.; Sirohi, A.S.; Raquib, M. Lactation response of cross bred dairy cows fed on indigenously prepared rumen protected fat-A field trial. *Indian J. Anim. Sci.* **2009**, *79*, 1045.

Article

Relationships between Body Condition Score (BCS), FAMACHA©-Score and Haematological Parameters in Alpacas (*Vicugna pacos*), and Llamas (*Lama glama*) Presented at the Veterinary Clinic

Matthias Gerhard Wagener [1,*], Saskia Neubert [1], Teresa Maria Punsmann [1], Steffen B. Wiegand [2] and Martin Ganter [1]

[1] Clinic for Swine, Small Ruminants, Forensic Medicine and Ambulatory Service, University of Veterinary Medicine Hannover, Foundation, Bischofsholer Damm 15, 30173 Hannover, Germany; Saskia.neubert@tiho-hannover.de (S.N.); opunsma@gmail.com (T.M.P.); Martin.ganter@tiho-hannover.de (M.G.)

[2] Department of Anesthesiology, University Hospital, LMU Munich, Marchioninistraße 15, 81377 Munich, Germany; Steffen.wiegand@med.uni-muenchen.de

* Correspondence: Matthias.gerhard.wagener@tiho-hannover.de

Simple Summary: Alpacas and llamas are increasingly presented at the veterinary clinic in Germany. Owners often notice too late when their animal is emaciated or anaemic. Emaciation can be detected by checking the so-called body condition score (BCS). An indication of anaemia can be provided directly by the FAMACHA©-score (FS), which has been adopted from small ruminants. There is still little information on both scores for alpacas and llamas, so a retrospective evaluation of data from the veterinary clinic was carried out. More than half of the animals admitted to the clinic were too lean and more than one in ten alpacas or one in five llamas showed clinical signs of anaemia. Both scores were compared with the findings from the animals' blood counts, which showed that poor nutritional status was associated with anaemia and shifts in inflammatory cells. Regular monitoring of BCS and FS is therefore important in alpacas and llamas to detect emaciation and anaemia in time.

Abstract: South American camelids (SAC) are being more and more presented at the veterinary Clinics in Germany. A bad nutritional condition, which can be easily categorized using a body condition score (BCS) of the animals, is often not noticed by the owners. Further anaemia is also often only detected in an advanced stage in SAC. Clinical detection of anaemia can be performed by assessing the FAMACHA©-score (FS), that is adapted from small ruminants. So far, there is only little information available about BCS and FS in SAC. In this study, both clinical scores were assessed in alpacas and llamas presented at the veterinary clinic and compared with the haematological parameters from the animals. The data were extracted retrospectively from the animals' medical records and compared statistically. More than half of the alpacas (60%) and llamas (70%) had a BCS < 3, while 12% of the alpacas and 21% of the llamas had a FS > 2. A decreased BCS was associated with a decrease in haematocrit, haemoglobin, lymphocytes, and eosinophils, as well as an increase in FS and neutrophils. BCS and FS should be assessed regularly in SAC to detect emaciation and anaemia in time.

Keywords: South American camelids; anaemia; nutritional status; emaciation; clinical scores; haematology

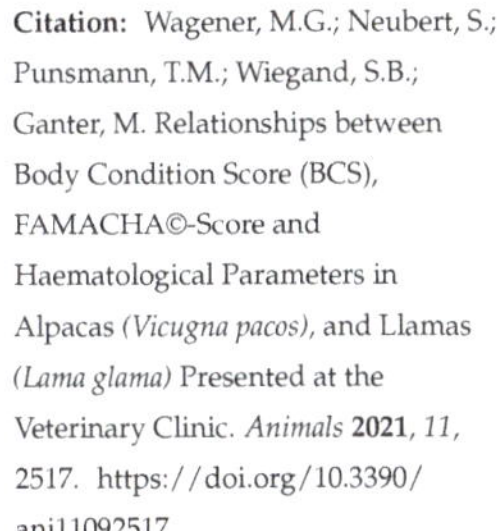

Citation: Wagener, M.G.; Neubert, S.; Punsmann, T.M.; Wiegand, S.B.; Ganter, M. Relationships between Body Condition Score (BCS), FAMACHA©-Score and Haematological Parameters in Alpacas (*Vicugna pacos*), and Llamas (*Lama glama*) Presented at the Veterinary Clinic. *Animals* **2021**, *11*, 2517. https://doi.org/10.3390/ani11092517

Academic Editors: María-Luz García and María-José Argente

Received: 29 July 2021
Accepted: 24 August 2021
Published: 27 August 2021

1. Introduction

Husbandry of alpacas and llamas (SAC: South American camelids) is becoming more and more popular in Europe [1–3]. However, it remains challenging for the owners to detect health problems. Delayed treatment, especially in connection with malnutrition, is common, since SAC are stoic and have a thick fibre coat, [4]. In some cases, emaciated

SAC additionally suffer from life-threatening anaemia that requires immediate blood transfusion [5].

Body condition scoring (BCS) is a common clinical tool to detect emaciation in ruminants and SACs [4,6–9]. For clinical detection of anaemia, the close observation of the colour of the mucous membranes, especially the conjunctives, can be used [10,11].

Body condition scoring was originally developed for sheep and dairy cattle [6–8] but has been adapted for SAC. Different body scoring-systems have been evaluated for alpacas and llamas, which include visual and palpatory examination of different body sites such as the lumbar spine, withers, shoulders, ribs, front and rear legs, or the pelvis [4,9,12–15]. A scale from 1 to 5 (1 = emaciated; 2 = thin; 3 = optimal; 4 = overweight; 5 = obese) is commonly used for determination of the nutritional status [4,15,16]. An optimal BCS (3) can be characterized by a straight line between the dorsal spinous and transverse processes of the lumbar spine [15,16]. A more concave line would be interpreted as a lower BCS, a more convex line as a higher BCS [15,16]. Causes for a bad nutritional status in SAC include management problems such as a restricted animal/feeding place ratio, insufficient nutrient supply, infestation with endoparasites, tooth problems, or any other chronical disease [4,17–19]. To date, there are several descriptions about the assessment of BCS in SAC, and there is also some information available about the relationship between BCS and health in llamas and alpacas [11,19,20]. However, so far, there is a lack of data concerning the BCS of SAC presented in the veterinary clinic.

In addition to malnutrition, anaemia is frequently observed in SACs [5,11], which is often caused by haemonchosis (*Haemonchus contortus*) [11,21]. Other reasons for anaemia in llamas and alpacas include haemotroph mycoplasms (*Candidatus mycoplasma haemolamae*) [22,23], gastric ulcers [24], or undersupply with trace elements that can lead to copper-, cobalt-, or iron deficiency [25–27]. If symptomatic, anaemia can lead to a life-threatening state, requiring immediate treatment [5,28]. Early detection of anaemia is crucial, to monitor the health of a herd and as a marker for anthelmintic treatment response.

The FAMACHA©-score (FS) ("FAMACHA" is an acronym for Dr. Francois "FAffa" MAlan, who created a CHArt with pictures of the conjunctives of sheep with five different red shades) has been established for targeted selective treatment of sheep with haemonchosis [10], which is a major cause of anaemia in sheep. The FS is an easy to use tool, for the clinical detection of anaemia [29]. The different red shades of the chart should be compared with the conjunctives of the animal to be examined and are meant to reflect different packed cell volumes (PCV's). The physiological red colour is expressed as a FS of 1 and the palest, almost white colour as a FS of 5 [10]. FS 1 and 2 are considered as "optimal" and "acceptable", respectively, while FS 3 as "borderline" and FS 4 and 5 as "dangerous" or "fatal" [10]. The FS has been evaluated for small ruminants in many studies worldwide, so far [30–37]. Since the FS targets anaemia and not the potential underlying condition leading to it, scoring of the conjunctives is also suggested as a tool in the diagnosis of other infections that cause anaemia such as fasciolosis [38] or bovine trypanosomosis [39]. Furthermore, there are many other causes for pale conjunctives, for example, acute bleeding.

The FS has also been used as a clinical tool for detecting anaemia in SAC [11,40–43]. Storey et al. investigated the FS in 347 alpacas and 502 llamas on different farms [11]. In 17/21 farms included in their study, *Haemonchus contortus* was the predominant nematode parasite [21]. They found significant associations between the PCV and FS and between the BCS and faecal egg count (FEC), and concluded that the FS is a useful tool to detect anaemia in SAC suffering from haemonchosis [11].

In ruminants, BCS and FS are well established as common clinical examination methods. They can be collected easily and without complex technical equipment directly on site on the animal. However, there are no data yet on the relationships between the BCS and FS and the haematological parameters for SAC presented to the veterinary clinic. In this study, we evaluated those conditions in alpacas and llamas presented to the clinic, to determine the extent to which clinical findings are associated with laboratory diagnostic parameters.

In contrast to previously available data collected from animals in herds in the field [11], this study involves data from sick animals. In addition, further haematological parameters such as the individual leukocyte fractions will be considered as a supplement to the previously known data. The resulting findings could provide important information for veterinarians in the field who do not have the possibility of an immediate haematological examination.

2. Material and Methods

2.1. Data Collection

Relevant data of alpacas and llamas were extracted from the medical files from the archives of the Clinic for Swine, Small Ruminants and Forensic Medicine of the University of Veterinary Medicine Hannover, Germany to an Excel sheet (Microsoft®® Excel®® for Office 365). Patient files were archived as paper files until August 2016 and as digital files in the patient management program "easyVET" (VetZ. EasyVET. Available online at: https://www.vetz.vet/de-de/easyVET, accessed on 29 July 2021) from September 2016 onwards. In this retrospective study, data were assessed between July 2014 and March 2021, since the recording of BCS and FS had been routinely performed during the clinical examination of SAC from July 2014. All the data used in this study were collected during veterinary diagnostic procedures after the owners had given written consent.

2.2. Inclusion Criteria

Only animals with information on species, gender, age, BCS, PCV, haemoglobin (Hb), total leucocytes (white blood count: WBC), and with blood smear results for cell differentiation were included in the analysis. The absolute bodyweight in kg and FS were not recorded for each animal. Therefore, they were only evaluated in a subset of animals.

2.3. Collected Parameters from the Animals

2.3.1. Basic Data about the Animal

Basic data included the animal's clinic-ID, species (alpaca or llama), gender (female or male), day of examination, animal's birthday, animal's age (in days, calculated by subtracting the animal's birthday from the day of examination), and, if available, the animal's bodyweight in kg.

2.3.2. Clinical Scores

Both evaluated clinical scores (BCS and FS) were assessed during the first clinical examination immediately after the animal had been presented to the clinic.

BCS

The BCS was noted as a score from 1 (emaciated) to 5 (obese) with 0.5 steps in between. BCS was assessed by experienced examiners by palpation of the lumbar spine according to the method described recently [15]. An optimal BCS was defined as 3.

FS

The FS was noted as a score for the colour of the mucous membranes of the conjunctives from 1 (physiological red colour) to 5 (pale, almost white). The red shades in between were expressed as scores 2, 3, and 4. FS was assessed by experienced examiners, by presenting the lower palpebral conjunctives. The official FAMACHA©-chard was used for learning the technique, but in most of the cases in the routine the FS was assessed as a subjective impression of the conjunctival colour, without using the colour standard of the card as a direct comparison. A FS > 2 was defined as a hint of anaemia [10]. In animals with reddened conjunctiva due to conjunctivitis, for example, no FS could be assessed.

2.3.3. Haematological Parameters

Routine EDTA-blood samples (EDTA Monovette 9 mL K3E, Sarstedt AG & Co. KG, Nümbrecht, Germany) were taken from the jugular vein from each animal during the general clinical examination [44]. Blood samples were either processed directly or stored at 4 °C when the animals were presented at night or during the weekend.

PCV [L/L]

The packed cell volume (PCV), haematocrit was evaluated by centrifugation of EDTA-blood in a microhematocrit tube for 10 min at 10,000× *g*. The PCV was determined as the ratio of the volume of red blood cells divided by the total blood volume.

Haemoglobin (Hb) [g/L]

Hb was determined photometrically using a cyan solution [45]. A total of 10 μL EDTA blood was added to the 2.5 mL cyan solution (containing 18 mmol/L sodium hydrogen carbonate, 0.768 mmol/L potassium cyanide, 0.608 mmol/L potassium ferricyanide) in a cuvette. The solution was incubated for 5 min and then measured with a photometer (546 nm). Hb in g/L was determined by multiplying the determined extinction with the factor 368.

Total Leucocytes/White Blood Count (WBC) [G/L]

WBC was determined microscopically in a Neubauer counting chamber after 5 min lysis of 100 μL EDTA blood in 900 μL 3% acetic acid solution [45]. If normoblasts were observed in the subsequent differentiation of the blood smear, the total leukocyte count was mathematically corrected according to the following formula: WBC [G/L] = number of nuclei counted in the Neubauer chamber [G/L] × 100/(100 + Number of normoblasts per 100 leukocytes).

Lymphocytes, Segmented Neutrophils, Band Neutrophils, Eosinophils, Basophils, Monocytes, Normoblasts, all [%]

Differentiation of leucocytes was performed microscopically in a blood smear stained according to Pappenheim. In each blood smear, 200 leukocytes were differentiated according to their morphological features and assigned to lymphocytes, segmented neutrophils, band neutrophils, eosinophils, basophils, and monocytes. Normoblasts (nucleated red blood cells) that occurred during differentiation were recorded in addition to the 200 leucocytes.

2.4. Statistical Analysis

Statistical tests were performed using SAS Enterprise Guide 7.1. Descriptive statistics were expressed as median (med), mean, standard deviation (SD), minimum (min), maximum (max), lower quartile (LQ), and upper quartile (UQ) of the investigated parameters in each group. Normal distribution was tested with the Shapiro-Wilk test.

The Wilcoxon-two-sample-test was used to test for statistical differences between alpacas and llamas, as well as differences in gender (male or female) or age (crias [<1 year] or adults [>1 year]) for both species separately. Furthermore, anaemic and non-anaemic animals were compared separately in both species. The reference values of Hengrave-Burri et al. [46] were used to categorize anaemic and non-anaemic animals. Animals whose PCV were below the corresponding lower reference value (alpacas: All crias: 0.29 L/L, adult males: 0.29 L/L, adult females: 0.26 L/L; llamas: All crias: 0.28 L/L, adult males: 0.29 L/L; adult females: 0.27 L/L) were classified as anaemic, while all other animals were classified as non-anaemic.

The Kruskal-Wallis test was used to determine differences for different BCS or FS. Spearman's rank correlation coefficient was used for calculating correlations between the investigated parameters and BCS or FS.

A p-value less than 0.05 was considered significant (* = $p < 0.05$; ** = $p < 0.01$; *** = $p < 0.001$). Significant correlations were interpreted as follows: R = 0.2–0.3: Very weak correlation; R = 0.3–0.5: Weak correlation; R = 0.5–0.8: Moderate correlation; R > 0.8: Strong correlation. Since most of the investigated groups had a value of $p < 0.05$ in the

Shapiro-Wilk test and therefore, failed normal distribution, nonparametric tests were performed. In the descriptive data (Supplementary Material), both the median and mean were given, since the differences in the mean were more obvious.

3. Results

3.1. Population

The total number of all animals (N = 300) was further divided into the species (al-paca (A) or llama (L)), gender (male or female), and the age (crias: < 1 year; adults: > 1 year). A total of 10 different groups were considered:

All alpacas (n = 259)
All llamas (n = 41)
Alpacas, male, cria (n = 26)
Alpacas, male, adult (n = 87)
Alpacas, female, cria (n = 40)
Alpacas, female, adult (n = 106)
Llamas, male, cria (n = 2)
Llamas, male, adult (n = 21)
Llamas, female, cria (n = 2)
Llamas, female, adult (n = 16)

3.2. Clinical Parameters

3.2.1. BCS

About 60% of the alpacas (n = 154/259) and 70% of the llamas (n = 29/41) had a BCS of lower than three. Alpaca crias showed a lower BCS than adult alpacas, with no differences between species or gender (Tables 1 and S1–S4). Alpacas or llamas with anaemia had a lower BCS than alpacas and llamas without anaemia (Tables 1, S5 and S6). When comparing animals with different BCS, there were significant differences in bodyweight, FS, PCV, Hb, and eosinophils in both species (Tables S7 and S8). In llamas, there were additionally differences in lymphocytes, segmented neutrophils, and basophils depending on the BCS of the animal (Tables S7 and S8).

Table 1. Results of the Wilcoxon test to check for differences in species (alpacas vs. llamas), gender (male vs. female), age (juvenile vs. adult), and presence of anaemia (anaemia vs. without anaemia). No data on the influence of age are available for llamas, since the juvenile llamas group contained only two animals. There were no significant differences between male and female animals for any of the parameters. * = $p < 0.05$; ** = $p < 0.01$; *** = $p < 0.001$; n.s. = not significant.

Parameter	Alpaca vs. Llama	Juvenile vs. Adult	Anaemia vs. without Anaemia	
	All	Alpacas	Alpacas	Llamas
Bodyweight (kg)	***	***	***	***
BCS	n.s.	**	***	***
FS	n.s.	*	**	***
PCV (L/L)	n.s.	n.s.	***	***
Hb (g/L)	n.s.	n.s.	***	***
WBC (G/L)	n.s.	n.s.	n.s.	n.s.
Lymphocytes (%)	**	***	n.s.	n.s.
Segmented neutrophils (%)	n.s.	**	n.s.	n.s.
Band neutrophils (%)	n.s.	n.s.	*	n.s.
Eosinophils (%)	n.s.	**	n.s.	**
Basophils (%)	n.s.	n.s.	n.s.	n.s.
Monocytes (%)	n.s.	n.s.	*	n.s.
Normoblasts (%)	n.s.	n.s.	n.s.	*

BCS in alpacas correlated a weak positive correlation with bodyweight and very weak with PCV, Hb, and eosinophils (Tables 2 and S9). A weak negative correlation was observed for BCS and FS (Tables 2 and S9).

Table 2. Results of Spearman's correlation for BCS, FS, and PCV with the investigated clinical and haematological parameters in alpacas (n = 259). * = $p < 0.05$; *** = $p < 0.001$; n.s. = not significant.

Alpacas	BCS		FS		PCV	
	r =		r =		r =	
Bodyweight (kg)	0.41	***	0.03	n.s.	0.16	*
BCS			−0.32	***	0.29	***
FS	−0.32	***			−0.34	***
PCV (L/L)	0.29	***	−0.34	***		
Hb (g/L)	0.28	***	−0.32	***	0.94	***
WBC (G/L)	−0.08	n.s.	−0.08	n.s.	0.00	n.s.
Lymphoytes (%)	0.09	n.s.	−0.08	n.s.	0.12	*
Segmented neutrophils (%)	−0.08	n.s.	0.04	n.s.	−0.13	*
Band neutrophils (%)	−0.12	n.s.	0.13	n.s.	0.08	n.s.
Eosinophils (%)	0.28	***	−0.11	n.s.	−0.10	n.s.
Basophils (%)	0.11	n.s.	−0.12	n.s.	−0.13	*
Monocytes (%)	0.06	n.s.	0.03	n.s.	0.22	***
Normoblasts (%)	−0.09	n.s.	0.26	***	−0.20	***

In llamas, BCS correlated a moderate positive correlation with bodyweight, PCV, Hb, lymphocytes, and eosinophils, whereas FS and segmented neutrophils revealed a negative correlation with BCS in llamas (Tables 3 and S9).

Table 3. Results of Spearman's correlation for BCS, FS, and PCV with the investigated clinical and haematological parameters in llamas (n = 41). * = $p < 0.05$; ** = $p < 0.01$; *** = $p < 0.001$; n.s. = not significant.

Llamas	BCS		FS		PCV	
	r =		r =		r =	
Bodyweight (kg)	0.66	***	−0.36	*	0.56	***
BCS			−0.55	***	0.59	***
FS	−0.55	***			−0.73	***
PCV (L/L)	0.59	***	−0.73	***		
Hb (g/L)	0.60	***	−0.72	***	0.96	***
WBC (G/L)	−0.25	n.s.	−0.06	n.s.	0.18	n.s.
Lymphoytes (%)	0.54	***	−0.17	n.s.	0.07	n.s.
Segmented neutrophils (%)	−0.46	**	0.25	n.s.	−0.16	n.s.
Band neutrophils (%)	−0.13	n.s.	0.22	n.s.	−0.15	n.s.
Eosinophils (%)	0.61	***	−0.54	***	0.50	***
Basophils (%)	0.25	n.s.	−0.04	n.s.	0.09	n.s.
Monocytes (%)	−0.01	n.s.	−0.05	n.s.	0.05	n.s.
Normoblasts (%)	−0.30	n.s.	0.54	***	−0.53	***

3.2.2. FS

About 12% (n = 24/214) of the investigated alpacas and 21% (n = 8/39) of the investigated llamas had an FS > 2. When comparing species or gender separately for each species, there were no significant differences regarding FS (Tables 1, S1, S2 and S4), but there was an effect of age in alpacas: Crias had a lower FS than adult alpacas (Tables 1, S1 and S3). In both species, animals with anaemia had a significantly higher FS than animals without anaemia (Tables 1, S5 and S6). Different FS were associated with differences in BCS, PCV, Hb, MCHC, lymphocytes, eosinophils, and normoblasts in alpacas and with differences in BCS, PCV, Hb, eosinophils, and normoblasts in llamas (Tables S10 and S11).

In alpacas, weak negative correlations were found between FS and BCS, PCV and Hb, and a very weak positive correlation between FS and normoblasts (Tables 2 and S12). In llamas, moderate negative correlations were found between FS and BCS, PCV, Hb, and eosinophils and a weak negative correlation between FS and bodyweight. FS in llamas further correlated moderate positively with normoblasts (Tables 3 and S12).

3.3. Haematological Parameters

3.3.1. PCV

When compared with the reference values of Hengrave Burri (2005) [46], the majority of SAC that was presented to the clinic suffered from anaemia (55% of the alpacas, 49% of the llamas). The lowest PCV in a single alpaca or llama was 0.04 L/L, the maximal PCV was 0.43 L/L in an alpaca and 0.47 L/L in a llama (Table S1). Differences for PCV in species, gender, or age were not statistically significant (Tables 1 and S2–S4). In both alpacas and lamas, a low PCV was found to be associated with a low BCS (Figure 1) and a higher FS (Tables S5 and S6). PCV in alpacas revealed a strong positive correlation with Hb, a weak positive correlation with BCS and monocytes, as well as a weak negative correlation with FS and normoblasts (Tables 2 and S13). PCV in llamas also revealed a strong positive correlation with Hb, a moderate positive correlation with bodyweight, BCS, and eosinophils, as well as a moderate negative correlation with FS and normoblasts (Tables 2 and S13). All the animals with severe anaemia (PCV < 0.10 L/L) had a BCS $\leq$ 3.

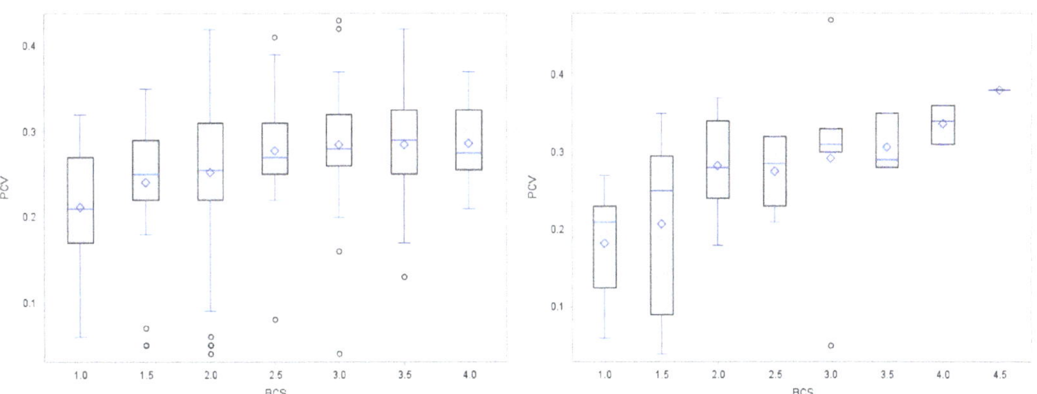

Figure 1. PCV in L/L in alpacas (n = 259, **left**) and llamas (n = 41, **right**) with different BCS. The boxplots display the quartile range and the respective minimum and maximum.

3.3.2. Hb

Hb in alpacas had a range of 20–198 g/L, while in llamas, the range was 15–218 g/L (Table S1). Since PCV and Hb showed a very strong correlation, Hb showed similar associations with the other parameters as PCV (Tables 2 and 3).

3.3.3. WBC

The range of WBC was 0.3–75.0 G/L in alpacas and 2.8–40.4 G/L in llamas (Table S1). Differences in species or gender were not detected, nor was there a difference between alpaca crias and adult alpacas or anaemic and non-anaemic animals (Tables 1 and S2–S5). Influences of different BCS or FS were not evident in the Kruskal-Wallis test (Tables S7 and S10) and correlations with BCS, FS, and PCV did not yield significant results (Tables 2, 3, S9, S12 and S13).

3.3.4. Lymphocytes

Lymphocytes revealed a range of 1–90% in alpacas and 3–65% in llamas (Table S1). Lymphocyte percentages were higher in alpacas than in llamas and higher in alpaca crias than in adult alpacas (Tables 1 and S1–S3). Differences between gender or anaemic and non-anaemic animals were not observed (Tables 1, S4 and S5). The proportion of lymphocytes was different depending on the BCS of the llamas (Tables S7 and S8). BCS and lymphocytes correlated as moderate positive in llamas (Figure 2). However, in alpacas, there was no significant correlation between BCS and lymphocytes (Tables 3 and S9).

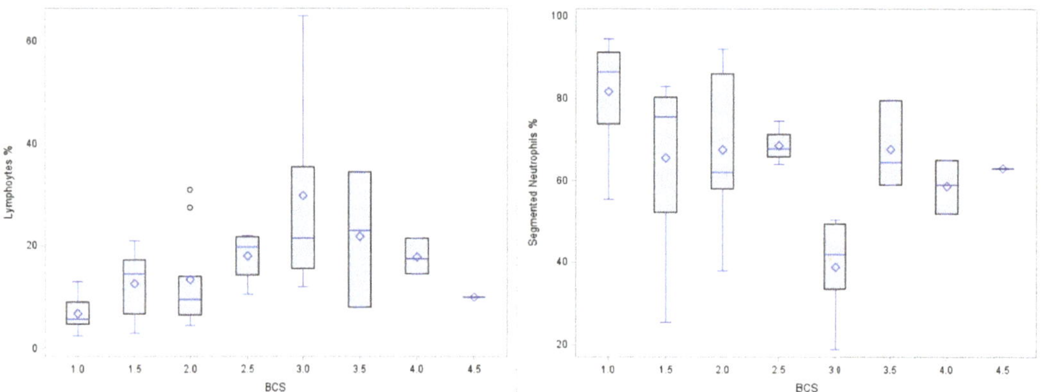

Figure 2. Proportions of lymphocytes (**left**) and segmented neutrophils (**right**) in llamas (n = 41) with different BCS. The boxplots display the quartile range and the respective minimum and maximum.

3.3.5. Neutrophils

The segmented neutrophils' ranges were 3–98% in alpacas and 19–95% in llamas (Table S1). The proportion of segmented neutrophils in alpaca crias was lower than in adult alpacas (Tables 1, S1 and S3), but there were no statistical differences between species, gender, or animals with and without anaemia for segmented neutrophils (Tables 1, S2, S4 and S5). For band neutrophils, there was a statistical difference between anaemic and non-anaemic animals in alpacas. However, the numerical proportion of band neutrophils was lower in anaemic alpacas than in alpacas without anaemia, whereas the opposite was the case in llamas.

In llamas, BCS was associated with segmented neutrophils (Table S7). Segmented neutrophils in llamas with BCS of 1 revealed the highest median and segmented neutrophils in llamas with BCS 3 the lowest (Table S8 and Figure 2). The correlation between BCS and segmented neutrophils in llamas was weak negative (Tables 3 and S9), however, this was seen only in lamas, not in alpacas (Tables 2 and S9). Different segmented neutrophils were not reflected in the FS of alpacas or llamas (Tables S10–S12).

3.3.6. Eosinophils

The percentages of eosinophils were 0–30% in alpacas and 0–39% in llamas (Table S1). There were no differences in eosinophils concerning species or gender (Tables S2 and S4), but adult alpacas had significantly higher amounts of eosinophils than alpaca crias (Tables 1 and S3). In addition, llamas with anaemia had a lower percentage of eosinophils than llamas without anaemia. However, this relationship was not seen in alpacas (Tables 1, S5 and S6). In both species, a relationship between BCS and the proportion of eosinophils existed, with both alpacas and llamas having the lowest median of eosinophils at BCS 1 (Tables S7 and S8). The correlation between BCS and eosinophils was stronger in llamas than in alpacas (Tables 2, 3 and S9). Different FS went in hand with different proportions of eosinophils. Nonetheless, this was only significant for llamas (Tables S10 and S11) and was also reflected in a moderate negative correlation between FS and eosinophils in llamas

(Tables 3 and S12). Further eosinophils in llamas correlated moderately positively with PCV (Tables 3 and S13).

3.3.7. Basophils

Basophils had a range of 0–4% in alpacas and 0–3% in llamas (Table S1). There were no statistical differences concerning species, gender, and age or between anaemic and non-anaemic animals (Tables 1 and S2–S5). However, BCS was associated with the proportion of basophils in llamas (Table S7). Nevertheless, there were no significant correlations between basophils and BCS, FS or PCV in either species (Tables 2, 3, S9, S12 and S13).

3.3.8. Monocytes

The monocyte ranges were 0–21% for alpacas and 0–9% for llamas (Table S1). There were no statistical differences between species, age, or gender (Tables 1 and S2–S4), but alpacas with anaemia revealed lower proportions of monocytes in the differential count than alpacas without anaemia (Tables 1, S5 and S6). However, this was not the case in llamas. Different BCS or FS were not reflected in the proportion of monocytes (Tables S10 and S11) and there were no significant correlations between monocytes and BCS, FS, or PCV in alpacas or llamas (Tables 2, 3, S9, S12 and S13).

3.3.9. Normoblasts

Normoblasts were present in 33% (n = 85/259) of the alpacas and 34% (n = 14/41) of the llamas. Species, gender, and age had no effect on normoblasts (Tables 1 and S2–S4). In anaemic llamas, the amount of normoblasts was higher than in non-anaemic llamas (Tables 1, S5 and S6). However, this was not statistically significant for alpacas. Different BCS had no impact on normoblasts, but FS was connected to normoblasts (Tables S7, S8, S10 and S11). This was also reflected in a positive correlation between FS and normoblasts, that was moderate in llamas but only very weak in alpacas (Tables 2, 3 and S12). Normoblasts further correlated negatively with PCV in both species (Tables 2, 3 and S13).

4. Discussion

We found that most of the llamas and alpacas presented to the clinic had a low BCS and about half of all alpacas and llamas were anaemic. For the clinical parameters, we showed that a low BCS was associated with lower body weight and increased FS in both species. When comparing this with the haematological parameters, a low BCS was also associated with decreased PCV, Hb, and eosinophils in both species. In llamas, a low BCS was additionally associated with a lower percentage of lymphocytes and an increased percentage of segmented neutrophils. In addition, despite the different age of the animals, a good correlation of bodyweight and BCS was found in both species.

A main finding in alpacas and llamas at the clinic is that a poor nutritional status is usually related to a low BCS and in most cases anaemia, as well. The condition of the animals is rarely perceived by the owners themselves. One cause for this is the very dense hair coat of SACs. As the owners do not palpate their animals regularly, the emaciation may remain undetected [4]. In a survey among alpaca and llama owners in Germany [1], fewer than half of all participating 255 farms (38.9%) reported emaciation in their animals from the owners' observation. Anaemia was observed even less frequently: Only 13.3% of the farms reported anaemia in their animals from their own observation [1], which contrasts with the results of the present study. The numbers cannot be easily compared since the animals presented to the clinic are usually sick and most healthy animals from the herds are never presented to the clinic. However, Storey et al. found a much lower proportion of animals with a BCS < 3 in their study on alpaca and lama farms than in the present study. Nonetheless, these animals with lower BCS were overrepresented in the group of anaemic animals [11], which is in line with our data. Storey et al. also found a higher proportion of llamas with anaemia than alpacas. Although this disagrees with our findings, it could be explained by the fact that mainly animals with pathological conditions are presented to the clinic [11].

For detecting anaemia in SAC with a PCV $\leq$ 0.17 L/L with an FS $\geq$ 4, a sensitivity of 50% and a specificity of 94% are given by Storey et al. [11]. FS showed a significantly negative correlation with PCV, Hb, as well as BCS, especially in the llamas.

It should be noted, however, that the PCV determined per FS has a wide range. This range can be explained by the fact that the FS does not specifically assess the PCV as it can also be elevated or depressed due to local inflammatory processes in the conjunctives or circulatory centralisation [34]. In addition, there is also a lack of specific colour scales for SAC. Since the FS was routinely assessed at the clinic by several examiners, this factor of differing expert opinions must also be taken into consideration. No data are available on this for SAC, but the studies by Maia et al. regarding the assessment of different examiners on the FS in sheep and goats indicate that the FS can be collected quite accurately by different persons after adequate training [47]. For everyday clinical practice, a less precise subdivision (for example "physiological red", "pale", and "white") of the FS might be sufficient to gain a first impression of a single animal.

Similar positive correlations of BCS and PCV or Hb as found in our study were also described in sheep or cattle. In pre- and post-partum cattle, Rafia et al. determined r = 0.32 and r = 0.36 for the correlation of BCS and PCV [48]. Torres-Chable et al. investigated the correlation of BCS and PCV in sheep and found r = 0.39 [49]. It is also noteworthy that in this present study, none of the animals with a BCS >3 had severe anaemia (<0.10 L/L).

The association of low BCS and anaemia remains unclear, there could be speculation about atrophy of the bone marrow or chronic inflammation. However, it must also be considered that the animals were usually transported prior to blood sampling and stress reactions can also lead to shifts in leukocyte fractions. Although no statistically relevant associations with WBC and BCS or PCV were found, the shift in the proportions of leukocytes from lymphocytes to segmented neutrophils in animals with low BCS, especially in the llamas, may provide indications for inflammation. A decrease in lymphocytes was also observed in emaciated horses [50].

Reference values play an important role when interpreting laboratory results. The reference values consulted in this study had a lower limit of 0.26–0.29 L/L for PCV, depending on species, gender, or age [46], and were thus in a similar range to reference values for alpacas and llamas from other authors [51–53]. In the animals in our study, no differences in PCV were found with respect to age or gender. However, other authors showed that male SAC usually have higher PCV than females [54,55]. This could be due to the fact that our study did not investigate homogeneous groups, but rather data from animals with different pathological conditions. Moreover, this could possibly be the reason why no significant relationships were found between WBC and the other parameters. However, Rafia et al. found only a very weak negative correlation (r = −0.15) for BCS and WBC in cattle [48]. When interpreting the PCV, it should also be taken into account that the blood samples were usually taken after the animals had been transported. Here, the transport stress could have led to haemoconcentration effects in individual animals.

The positive correlation between BCS and eosinophils remains unclear. In studies on the influence of BCS on haematological parameters in other species, such a relationship was not reported [48,49]. Since anaemia and poor nutritional status are often associated with endoparasites in alpacas [11,21], a negative correlation was expected. It could be speculated here that animals with a lower BCS are in a more immunologically inactive state, which is particularly reflected in the number of eosinophils. However, evidence for this assumption is lacking.

A lower percentage of monocytes in anaemic animals could be due to infection or stress. Nonetheless, since monocytes are generally present in low numbers in a blood smear, a single over- or under-recognized cell during differentiation can account for a large error. The same also applies to the basophils, which are typically found only very sporadically in the blood smear.

Normoblasts or erythroblasts are usually associated with regenerative anaemia, but there are only few exceptions where normoblasts may appear in the peripheral blood in the absence of anaemia [56]. Although there was a clear negative correlation between PCV and normoblasts, it remains questionable why normoblasts were found equally in anaemic and non-anaemic alpacas. Normoblasts can be present in large numbers in alpacas with highly regenerative anaemia, where extramedullary haematoopoiesis seems to play a role [57]. However, there are still too few data on normoblasts in alpacas in general, and more research is needed. Another haematological parameter that provides information about regeneration is the reticulocyte count [58], but this was only determined in some of the animals investigated in our study and was excluded from the evaluation due to the low number.

5. Limitations

The focus of this study was to compare clinical scores (BCS and FS) with haematological findings. Therefore, this does not allow a general statement concerning individual clinical pictures. It is also not possible to draw a conclusion whether a lower BCS was the cause of haematological changes, or whether the BCS was changed as a result of changes that were visible in the blood count. Furthermore, the results of faecal egg counts were not considered. Several animals presented to the clinic had already been dewormed by other veterinarians or by the owners themselves, shortly before presentation at the clinic. This made it impossible to determine the overall contributory of endoparasites to the clinical picture. The clinical diagnoses of the animals were not taken into account in the evaluation. It was not possible to define a main diagnosis for every animal, as some animals had more than one diagnoses. Common diagnoses besides emaciation or anaemia included recumbency or gastric ulcers. Data on pregnancy stage or lactation were also not available for all the animals. Therefore, these parameters were not taken into account.

6. Conclusions

In summary, more than half of the alpacas and llamas presented to the clinic had a BCS < 3. In addition, half of the animals suffered from anaemia. A low BCS was predominantly associated with increased FS and decreased PCV and Hb. There was also evidence that a low BCS was associated with an increase in segmented neutrophils, which may indicate that animals with lower BCS are more likely to be affected by inflammatory diseases.

It is important to educate SAC owners about the fact that a poor nutritional status is closely associated with pathological haematological findings, and to encourage them to check the nutritional status of the animals regularly, so that emaciation can be detected in time. Recording BCS and FS is also particularly useful for identifying potentially anaemic animals. The early detection of pathological conditions in alpacas and llamas can thus make an important contribution to animal welfare.

Supplementary Materials: The following are available online at https://www.mdpi.com/article/10.3390/ani11092517/s1. Table S1: Descriptive statistics of bodyweight, BCS, FS, and haematological parameters in alpacas and llamas; Table S2: Wilcoxon test to compare alpacas and llamas concerning bodyweight, BCS, FS, and haematological parameters; Table S3: Wilcoxon test to compare crias and adults concerning bodyweight, BCS, FS, and haematological parameters in alpacas; Table S4: Wilcoxon test to compare gender concerning bodyweight, BCS, FS, and haematological parameters in alpacas and llamas; Table S5: Wilcoxon test to compare anaemic and non-anaemic animals concerning bodyweight, BCS, FS, and haematological parameters in alpacas and llamas; Table S6: Descriptive statistics of bodyweight, BCS, FS, and haematological parameters in anaemic and non-anaemic alpacas and llamas; Table S7: Kruskal-Wallis test to test the influence of BCS on bodyweight; FS and haematological parameters in alpacas and llamas; Table S8: Descriptive statistics of bodyweight, FS, and haematological parameters in alpacas and llamas with different BCS; Table S9: Spearman's correlation for BCS with bodyweight, FS, and haematological parameters in alpacas and llamas; Table S10: Kruskal-Wallis test to test the influence of FS on bodyweight, BCS, and haematological parameters in alpacas and llamas; Table S11: Descriptive statistics of bodyweight, BCS, and haematological parameters in alpacas and llamas with different FS; Table S12: Spearman's correlation for FS with bodyweight, BCS, and haematological parameters in alpacas and llamas; Table S13: Spearman's correlation for PCV with bodyweight, BCS, and haematological parameters in alpacas and llamas.

Author Contributions: Conceptualization, M.G.W.; methodology, M.G.W.; software, M.G.W.; validation, M.G.W. and S.B.W.; formal analysis, M.G.W.; investigation, M.G.W.; resources, M.G.W. and M.G.; data curation, M.G.W. and S.N.; writing—original draft preparation, M.G.W.; writing—review and editing, S.B.W., S.N., T.M.P., and M.G.; visualization, M.G.W.; supervision, M.G.; project administration, M.G.W.; funding acquisition, M.G.W. All authors have read and agreed to the published version of the manuscript.

Funding: This publication was supported by Deutsche Forschungsgemeinschaft and University of Veterinary Medicine Hannover, Foundation within the funding programme Open Access Publishing. Open Access funding enabled and organized by Projekt DEAL.

Institutional Review Board Statement: The study was conducted in accordance with the guidelines of the Declaration of Helsinki and approved by the Animal Welfare Officer of the University of Veterinary Medicine Hannover, Foundation under the registration number TVO-2018-V-66.

Data Availability Statement: The datasets supporting the conclusions of this article are included within the article and its additional file. The raw data used in this study are not publicly available since they are veterinary patient records subject to confidentiality. The raw data are located in the patient archive of the Clinic for Swine, Small Ruminants, Forensic Medicine and Ambulatory Service of the University of Veterinary Medicine Hannover Foundation and were analysed with the permission of the clinic management and the Animal Welfare Officer of the University of Veterinary Medicine Foundation.

Acknowledgments: The authors would like to thank the entire laboratory staff of the Clinic for Swine, Small Ruminants, Forensic Medicine and Ambulatory Service, University of Veterinary Medicine Hannover for their excellent work in preparing the haematological data, as well as the entire clinical staff involved in assessment of the clinical parameters and the care of the animals. The authors further gratefully acknowledge the help of Frances Sherwood-Brock, English editorial office, University of Veterinary Medicine Hannover, Foundation, Germany, for proofreading the manuscript for correct English and Martin Beyerbach of the Department of Biometry, Epidemiology and Information Processing, University of Veterinary Medicine Hannover, Foundation, Germany, for statistical advice.

Conflicts of Interest: The authors declare no conflict of interest.

References

1. Neubert, S.; von Altrock, A.; Wendt, M.; Wagener, M.G. Llama and Alpaca Management in Germany-Results of an Online Survey among Owners on Farm Structure, Health Problems and Self-Reflection. *Animals* **2021**, *11*, 102. [CrossRef]
2. Hengrave Burri, I.; Martig, J.; Sager, H.; Liesegang, A.; Meylan, M. South American Camelids in Switzerland. I. Population, Management and Health Problems. *Schweiz. Arch. Für Tierheilkd.* **2005**, *147*, 325–334. [CrossRef]
3. Davis, R.; Keeble, E.; Wright, A.; Morgan, K. South American Camelids in the United Kingdom: Population Statistics, Mortality Rates and Causes of Death. *Vet. Rec.* **1998**, *142*, 162–166. [CrossRef]
4. Van Saun, R.J. Nutritional Requirements and Assessing Nutritional Status in Camelids. *Vet. Clin. North. Am. Food Anim. Pract.* **2009**, *25*, 265–279. [CrossRef]
5. Wagener, M.G.; Grimm, L.M.; Ganter, M. Anaemia in a Llama (Lama glama): Treatment, Regeneration and Differential Diagnoses. *Vet. Rec. Case Rep.* **2018**, *6*, e000638. [CrossRef]
6. Jefferies, B. Body Condition Scoring and Its Use in Management. *Tasman. J. Agric.* **1961**, *32*, 19–21.
7. Ferguson, J.D.; Galligan, D.T.; Thomsen, N. Principal Descriptors of Body Condition Score in Holstein Cows. *J. Dairy Sci.* **1994**, *77*, 2695–2703. [CrossRef]
8. Edmonson, A.; Lean, I.; Weaver, L.; Farver, T.; Webster, G. A Body Condition Scoring Chart for Holstein Dairy Cows. *J. Dairy Sci.* **1989**, *72*, 68–78. [CrossRef]
9. Johnson, L.W. Llama Nutrition. *Vet. Clin. North Am. Food Anim. Pract.* **1994**, *10*, 187–201. [CrossRef]
10. Bath, G.F.; Malan, F.; Van Wyk, J. The "FAMACHA" Ovine Anaemia Guide to assist with the control of haemonchosis. In Proceedings of the 7th Annual Congress of the Livestock Health and Production Group of the South African Veterinary Association, Port Elizabeth, South Africa, 5–7 June 1996; p. 5.

11. Storey, B.E.; Williamson, L.H.; Howell, S.B.; Terrill, T.H.; Berghaus, R.; Vidyashankar, A.N.; Kaplan, R.M. Validation of the FAMACHA© System in South American camelids. *Vet. Parasitol.* **2017**, *243*, 85–91. [CrossRef] [PubMed]
12. Hilton, C.; Pugh, D.; Wright, J.; Waldridge, B.; Simpkins, S.; Heath, A. How To Determine and When to Use Body Weight Estimates and Condition Scores in Llamas. *Vet. Med.* **1998**, *93*, 1015–1018.
13. Bromage, G. Feeding and Nutrition. In *Llamas and Alpacas: A Guide to Management*; The Crowood Press Ltd.: Marlborough, UK, 2006; pp. 34–45.
14. Duncanson, G.R. Nutrition and Metabolic Diseases. In *Veterinary Treatment of Llamas and Alpacas*; CABI: Oxfordshire, UK, 2012; pp. 13–21.
15. Wagener, M.G.; Ganter, M. Body Condition Scoring in South American Camelids. *Der Prakt. Tierarzt* **2020**, *101*, 684–696.
16. Gauly, M. Fütterung. In *Neuweltkameliden-Haltung, Zucht, Erkrankungen*, 4th ed.; Thieme: Stuttgart, Germany, 2018; pp. 46–67.
17. Van Saun, R.J. Nutritional Diseases of Llamas and Alpacas. *Vet. Clin. North. Am. Food Anim. Pract.* **2009**, *25*, 797–810. [CrossRef]
18. Baumgartner, W.; Wittek, T.; Gauly, M. Allgemeiner Klinischer Untersuchungsgang. In *Klinische Propädeutik der Haus- und Heimtiere*, 9th ed.; Enke: Stuttgart, Germany, 2018; pp. 50–68.
19. Proost, K.; Pardon, B.; Pollaris, E.; Flahou, T.; Vlaminck, L. Dental Disease in Alpacas. Part 2: Risk Factors Associated with Diastemata, Periodontitis, Occlusal Pulp Exposure, Wear Abnor-Malities, and Malpositioned Teeth. *J. Vet. Intern. Med.* **2020**, *34*, 1039–1046. [CrossRef]
20. Frezzato, G.; Stelletta, C.; Murillo, C.E.P.; Simonato, G.; Cassini, R. Parasitological Survey To Address Major Risk Factors Threatening Alpacas in Andean Extensive Farms (Arequipa, Peru). *J. Vet. Med Sci.* **2020**, *82*, 1655–1661. [CrossRef]
21. Edwards, E.E.; Garner, B.C.; Williamson, L.H.; Storey, B.E.; Sakamoto, K. Pathology of Haemonchus contortus in New World Camelids in the Southeastern United States: A Retrospective Review. *J. Vet. Diagn. Investig.* **2016**, *28*, 105–109. [CrossRef]
22. Crosse, P.; Ayling, R.; Whitehead, C.; Szladovits, B.; English, K.; Bradley, D.; Solano-Gallego, L. First Detection of 'Candidatus Mycoplasma Haemolamae'infection in Alpacas in England. *Vet. Rec.* **2012**, *171*, 71–75. [CrossRef]
23. Kaufmann, C.; Meli, M.; Hofmann-Lehmann, R.; Riond, B.; Zanolari, P. Epidemiology of'Candidatus Mycoplasma Haemolamae'infection in South American Camelids in Central Europe. *J. Camelid Sci.* **2011**, *4*, 23–29.
24. Smith, B.B.; Pearson, E.G.; Timm, K.I. Third Compartment Ulcers in the Llama. *Vet. Clin. North. Am. Food Anim. Pract.* **1994**, *10*, 319–330. [CrossRef]
25. Foster, A.; Bidewell, C.; Barnett, J.; Sayers, R. Haematology and Biochemistry in Alpacas and Llamas. *In Practice* **2009**, *31*, 276–281. [CrossRef]
26. Andrews, A.; Cox, A. Suspected Nutritional Deficiency Causing Anaemia in Llamas (Lama glama). *Vet. Rec.* **1997**, *140*, 153–154. [CrossRef] [PubMed]
27. Morin, D.; Garry, F.; Weiser, M.; Fettman, M.; Johnson, L. Hematologic Features of Iron Deficiency Anemia in Llamas. *Vet. Pathol.* **1992**, *29*, 400–404. [CrossRef] [PubMed]
28. Luethy, D.; Stefanovski, D.; Salber, R.; Sweeney, R. Prediction of Packed Cell Volume after Whole Blood Transfusion in Small Ruminants and South American Camelids: 80 Cases (2006–2016). *J. Vet. Intern. Med.* **2017**, *31*, 1900–1904. [CrossRef] [PubMed]
29. Van Wyk, J.A.; Bath, G.F. The FAMACHA System for Managing Haemonchosis in Sheep and Goats by Clinically Identifying Individual Animals for Treatment. *Vet. Res.* **2002**, *33*, 509–529. [CrossRef]
30. Vilela, V.L.R.; Feitosa, T.F.; Linhares, E.F.; Athayde, A.C.R.; Molento, M.B.; Azevedo, S.S. FAMACHA© Method as an Auxiliary Strategy in the Control of Gastrointestinal Helminthiasis of Dairy Goats Under Semiarid Conditions of Northeastern Brazil. *Vet. Parasitol.* **2012**, *190*, 281–284. [CrossRef]
31. Gauly, M.; Schackert, M.; Erhardt, G. Use of FAMACHA© Scoring System as a Diagnostic Aid for the Registration of Distinguishing Marks in the Breeding Program for Lambs Exposed to an Experimental Haemonchus contortus Infection. *Dtsch. Tierarztl. Wochenschr.* **2004**, *111*, 430–433.
32. Koopmann, R.; Holst, C.; Epe, C. Experiences with the FAMACHA©-Eye-Colour-Chart for Identifying Sheep and Goats for Targeted Anthelmintic Treatment. *Berl. Und Müncher Tierärztliche Wochenschr.* **2006**, *119*, 436–442.
33. Papadopoulos, E.; Gallidis, E.; Ptochos, S.; Fthenakis, G. Evaluation of the FAMACHA© System for Targeted Selective Anthelmintic Treatments for Potential Use in Small Ruminants in Greece. *Small Rumin. Res.* **2013**, *110*, 124–127. [CrossRef]
34. Di Loria, A.; Veneziano, V.; Piantedosi, D.; Rinaldi, L.; Cortese, L.; Mezzino, L.; Cringoli, G.; Ciaramella, P. Evaluation of the FAMACHA System for Detecting the Severity of Anaemia in Sheep from Southern Italy. *Vet. Parasitol.* **2009**, *161*, 53–59. [CrossRef]
35. Reynecke, D.P.; Van Wyk, J.A.; Gummow, B.; Dorny, P.; Boomker, J. Validation of the FAMACHA© Eye Colour Chart Using Sensitivity/Specificity Analysis on Two South African Sheep Farms. *Vet. Parasitol.* **2011**, *177*, 203–211. [CrossRef]
36. Scheuerle, M.; Mahling, M.; Muntwyler, J.; Pfister, K. The Accuracy of the FAMACHA©-Method in Detecting Anaemia and Haemonchosis in Goat Flocks in Switzerland Under Field Conditions. *Vet. Parasitol.* **2010**, *170*, 71–77. [CrossRef]
37. Kaplan, R.; Burke, J.; Terrill, T.; Miller, J.; Getz, W.; Mobini, S.; Valencia, E.; Williams, M.; Williamson, L.; Larsen, M. Validation of the FAMACHA© Eye Color Chart for Detecting Clinical Anemia in Sheep and Goats on Farms in the Southern United States. *Vet. Parasitol.* **2004**, *123*, 105–120. [CrossRef]
38. Olah, S.; van Wyk, J.A.; Wall, R.; Morgan, E.R. FAMACHA©: A Potential Tool for Targeted Selective Treatment of Chronic Fasciolosis in Sheep. *Vet. Parasitol.* **2015**, *212*, 188–192. [CrossRef]

39. Grace, D.; Himstedt, H.; Sidibe, I.; Randolph, T.; Clausen, P.-H. Comparing FAMACHA© Eye Color Chart and Hemoglobin Color Scale Tests for Detecting Anemia and Improving Treatment of Bovine Trypanosomosis in West Africa. *Vet. Parasitol.* **2007**, *147*, 26–39. [CrossRef] [PubMed]
40. Viesselmann, L.C.; Videla, R.; Schaefer, J.; Chapman, A.; Wyrosdick, H.; Schaefer, D.M. Mycoplasma haemolamae and Intestinal Parasite Relationships With Erythrocyte Variables in Clinically Healthy Alpacas and Llamas. *J. Vet. Intern. Med.* **2019**, *33*, 2336–2342. [CrossRef] [PubMed]
41. Cocquyt, C.M.; Van Amstel, S.; Cox, S.; Rohrbach, B.; Martín-Jiménez, T. Pharmacokinetics of Moxidectin in Alpacas Following Administration of an Oral or Subcutaneous Formulation. *Res. Vet. Sci.* **2016**, *105*, 160–164. [CrossRef] [PubMed]
42. Galvan, N.; Middleton, J.R.; Nagy, D.W.; Schultz, L.G.; Schaeffer, J.W. Anthelmintic Resistance in a Herd of Alpacas (Vicugna Pacos). *Can. Vet. J.* **2012**, *53*, 1310. [PubMed]
43. Viesselmann, L.C.; Videla, R.; Flatland, B. Verification of the Heska Element Point-of-Care Blood Gas Instrument for Use with Venous Blood From Alpacas and Llamas, With Determination of Reference Intervals. *Vet. Clin. Pathol.* **2018**, *47*, 435. [CrossRef]
44. Gauly, M.; Maier, H.; Trah, M. Blood Collection, Haematological Values, Biochemical Parameters of New World Camelids. *Tierärztliche Umsch.* **1998**, *53*, 751–754.
45. Wagener, M.G.; Grossmann, T.; Stöter, M.; Ganter, M. Hematological Diagnostics in Llamas and Alpacas. *Der Prakt. Tierarzt* **2018**, *99*, 481–493.
46. Hengrave Burri, I.; Tschudi, P.; Martig, J.; Liesegang, A.; Meylan, M. South American Camelids in Switzerland. II. Reference Values for Blood Parameters. *Schweiz. Arch. Für Tierheilkd.* **2005**, *147*, 335–343. [CrossRef] [PubMed]
47. Maia, D.; Rosalinski-Moraes, F.; Van Wyk, J.A.; Weber, S.; Sotomaior, C.S. Assessment of a Hands-On Method for FAMACHA© System Training. *Vet. Parasitol.* **2014**, *200*, 165–171. [CrossRef] [PubMed]
48. Rafia, S.; Taghipour-Bazargani, T.; Khaki, Z.; Bokaie, S.; Tabrizi, S.S. Effect of Body Condition Score on Dynamics of Hemogram in Periparturient Holstein Cows. *Comp. Clin. Pathol.* **2012**, *21*, 933–943. [CrossRef]
49. Torres-Chable, O.M.; García-Herrera, R.A.; González-Garduño, R.; Ojeda-Robertos, N.F.; Peralta-Torres, J.A.; Chay-Canul, A.J. Relationships Among Body Condition Score, FAMACHA© Score and Haematological Parameters in Pelibuey Ewes. *Trop. Anim. Health Prod.* **2020**, *52*, 3403–3408. [CrossRef]
50. Munoz, A.; Riber, C.; Trigo, P.; Castejon, F. Hematology and Clinical Pathology Data in Chronically Starved Horses. *J. Equine Vet. Sci.* **2010**, *30*, 581–589. [CrossRef]
51. Weiser, M.; Fettman, M.; Van Houten, D.; Johnson, L.; Garry, F. Characterization of Erythrocytic Indices and Serum Iron Values in Healthy Llamas. *Am. J. Vet. Res.* **1992**, *53*, 1776–1779.
52. Hajduk, P. Haematological Reference Values for Alpacas. *Aust. Vet. J.* **1992**, *69*, 89–90. [CrossRef]
53. Fowler, M.; Zinkl, J. Reference Ranges for Hematologic and Serum Biochemical Values in Llamas (Lama glama). *Am. J. Vet. Res.* **1989**, *50*, 2049–2053.
54. Al-Izzi, S.; Abdouslam, O.; Al-Bassam, L.; Azwai, S. Haematological Parameters in Clinically Normal Llamas (Lama glama). *Prax. Vet.* **2004**, *52*, 225–232.
55. Husakova, T.; Pavlata, L.; Pechova, A.; Tichy, L.; Hauptmanova, K. The Influence of Sex, Age and Season on the Haematological Profile of Alpacas (Vicugna pacos) in Central Europe. *Vet. Med.* **2015**, *60*, 407–414. [CrossRef]
56. Goggs, R. Normoblasts: Not Always Normal. *Vet. Rec.* **2014**, *175*, 506–507. [CrossRef] [PubMed]
57. Wagener, M.G.; Puff, C.; Stöter, M.; Schwennen, C.; Ganter, M. Regenerative Anaemia in Alpacas (Vicugna pacos) Can Lead to a Wrong Diagnosis of Leucocytosis. *Vet. Rec. Case Rep.* **2020**, *8*, e001257. [CrossRef]
58. Newhall, D.A.; Oliver, R.; Lugthart, S. Anaemia: A Disease or Symptom. *Neth. J. Med.* **2020**, *78*, 104–110.

MDPI

Article

Assessment of Meat-Type Sheep Welfare Using Animal-Based Measures

Naceur M'Hamdi [1,*,†], Cyrine Darej [1,†], Khaoula Attia [1], Hajer Guesmi [1], Ibrahim El Akram Znaïdi [2], Rachid Bouraoui [3], Hajer M'Hamdi [4], Lamjed Marzouki [5] and Moez Ayadi [5]

1 Research Laboratory of Ecosystems & Aquatic Resources, National Agronomic Institute of Tunisia, Carthage University, 43 Avenue Charles Nicolle, Tunis 1082, Tunisia; cyrine.darej@gmail.com (C.D.); attiakhaoula@ymail.com (K.A.); guessmihajer55@gmail.com (H.G.)
2 Department of Animal Sciences, High Agronomic Institute of Chott Mariem, University of Sousse, Sousse 4000, Tunisia; akram_znaidi@hotmail.com
3 Laboratory ADIPARA, Higher School of Agriculture of Mateur, Road Tabarka-7030, Mateur, Bizerte 7030, Tunisia; bouraoui.rachid@yahoo.fr
4 Ministry of Agriculture, CRDA Ben Arous, New Medina, Ben Arous 2063, Tunisia; hajervet@gmail.com
5 Unit of Functional Physiology and Bio-Resources Valorization (BF-VBR), Higher Institute of Biotechnology of Beja, University of Jendouba, Beja 9000, Tunisia; Lamjed.marzouki@ipeis.rnu.tn (L.M.); moez_ayadi2@yahoo.fr (M.A.)
* Correspondence: naceur_mhamdi@yahoo.fr
† These authors contributed equally.

Simple Summary: There was every indication that animal welfare will continue to be a major issue affecting livestock farming in the future. The main welfare issues affecting sheep were feeding strategies, health, and diseases. The health problems of sheep are avoidable with good grazing, breeding, and stockmanship. However, sheep must be given adequate supervision to ensure that any welfare issues are quickly noticed and addressed. Assessing animal welfare can be used as management tools by farmers to identify welfare issues and recognize poor welfare.

Abstract: This study aimed to assess the welfare of Tunisian sheep in extensive sheep production systems using animal-based measures of ewe welfare. This study encompasses the first national survey of sheep welfare in which animal-based outcomes were tested. Animal-based welfare measures were derived from previous welfare protocols. Fifty-two Tunisian farms were studied and a number from 20 to 100 animals by flock were examinated. The whole flock was also observed to detect clinical diseases, lameness, and coughing. The human-animal relationship was selected as welfare indicators. It was evaluated through the avoidance distance test. The average avoidance distance was 10.47 ± 1.23 and 8.12 ± 0.97 m for a novel person and farmer, respectively. The global mean of body condition score (BCS) was 2.4 with 47% of ewes having a BCS of two, which may be associated with an increased risk of nutritional stress, disease, and low productivity. Ten farms had more than 7% of lambs with a low body condition score, which may be an indication of a welfare problem. The results obtained in the present study suggest that the used animal-based measures were the most reliable indicators that can be included in welfare protocols for extensive sheep production systems.

Keywords: animal-based measures; indicators; sheep welfare; stress

Citation: M'Hamdi, N.; Darej, C.; Attia, K.; Guesmi, H.; Znaïdi, I.E.A.; Bouraoui, R.; M'Hamdi, H.; Marzouki, L.; Ayadi, M. Assessment of Meat-Type Sheep Welfare Using Animal-Based Measures. *Animals* **2021**, *11*, 2120. https://doi.org/10.3390/ani11072120

Academic Editors: María-Luz García and María-José Argente

Received: 18 February 2021
Accepted: 9 April 2021
Published: 16 July 2021

Publisher's Note: MDPI stays neutral with regard to jurisdictional claims in published maps and institutional affiliations.

1. Introduction

Sheep farming in Tunisia occupies an important place in the economic and social levels. It is the main source of income for most of the rural population [1]. The national sheep flock size accounts for 3.7 million heads of sheep and contributes to around 42.5% of the red meat and 5% of milk production [2]. However, market demand from consumers for assurance schemes for high-quality and safe animal products is increasing [3,4]. The concept of welfare in animals has gained importance in recent years. That is due to the fact that ensuring animal welfare is not only a duty that has to be performed legally and ethically but it should also be considered as a way of direct economic contribution to the enterprise. The concept of welfare comprises physical and mental health [5]. The welfare of animals means a life away from any undesired emotions (pain, suffering, and distress). On-farm welfare assessments can be used for immediate or ongoing on-farm monitoring and to demonstrate compliance with national and international legal welfare standards and farm assurance schemes [6,7]. Parameters used for the assessment of animal welfare can refer to either the physiology, behavior, production or health of an animal [8]. Therefore, the objective of this study is to evaluate the welfare state of Tunisian extensive sheep using selected animal-based welfare measures.

2. Materials and Methods

2.1. Study Sites and Animals

This study was carried out in the North of Tunisia in the sub-humid bioclimatic area (rainfall > 550 mm). Fifty-two farms were visited during the lambing season (September–December) in 2017. They were selected through random sampling from lists of farmers data obtained from the Northwest Forestry and Pastoral Development Office of Beja, Tunisia. The farmers were contacted and asked whether they wanted to participate in the study. A total of 1040 Noire de Thibar ewes (840) and lambs (200), aged 2–6 years, from small and medium flocks of approximately 350 breeding ewes were randomly selected using systematic random sampling and examined by a one trained person at lambing, mid lactation, and weaning. Each farm was visited three times during the period of the study. The ewes were managed under extensive conditions, in a year-round outdoor system, grazing pastures and managed under commercial conditions. Natural rangelands forage (hay, barley in green, etc.) and concentrated feed were the main food resources for sheep. The ewe sample size was selected based on the following equation reported by Cochran [9] under a 95% confidence interval and precision of 10%. This number was supported by the Animal Welfare Indicators Project (AWIN) sheep protocol, which recommends a sample of 92 animals when the farm size is $\geq$2000 breeding ewes [10].

$$n = \frac{N}{1 + N(\mathrm{e})^2}$$

where n is the sample size, N is the population size, and e is the level of precision.

The flocks were observed to detect signs of clinical disease, lameness, and coughing. The human-animal relationship and fear testing were selected as welfare indicators based on behavior at pasture.

2.2. Welfare Indicator Assessments

Indicators used in this study, were inspired by AWIN [10] (Table 1). Groups of sheep, ranging in number from 20 to 100, from each selected study farm were presented by the farmer and assessed using eight indicators of welfare [11–13]. These indicators were considered to be key animal-based outcome measures to be included in the on-farm protocols [11]. The validity and feasibility of the indicators selected in this study have been previously justified and reported in other studies [13–15]. The body condition score was used as an indirect measure of good feeding. The body condition was scored on a scale from 1 to 5 [16] during lambing and at mid-lactation for ewes and weaning for lambs. A score of 1 is very thin and 5 is very fat. Good housing was evaluated through fleece cleanliness [10]. All the animals were assessed and scored on visual cleanliness of the fleece (0–3 scale). The score 0 represented a visually clean fleece, with minor fecal material or mud in the fleece. A score of 1 represented small spots of dirt under the belly, legs, and tail; a score of 2 represented a generally dirty fleece; and a score of 3 represented a very dirty fleece, stained with fecal material or mud under the belly, legs, and tail. For good health, three indicators were evaluated; lameness, respiratory disorders, and dirtiness [10]. Lameness was scored following a 3-point scale (0–2) that takes into account the smoothness of movement (score 0 for imperfect mobility, score 1 for lame, and score 2 for severely lame. The respiratory rate was determined by measuring the time (seconds) required to take ten breaths; these data were then converted to breaths per minute (bpm).The respiratory rate was scored on a scale from 1 (acute) to 3 (progressive). Dirtiness was evaluated using a 3-point scale, through a visual assessment of one side and behind the hindquarters and belly. The human-animal relationship (HAR) was defined as the degree of relatedness or distance between the animal and the human. The test was conducted by the trained person and the farmer according to the procedure reported by Waiblinger et al. [17]. The heart rate was measured electrocardiographically. In short-term measurements of 1 h on restrained standing animals two skin electrodes were adequate, placed on each side of the thorax in the plane of the heart [18]. Each indicator was assessed by observing the behavior and physical appearance of the individual sheep within the group and scored. Following each assessment, the observer recorded the number of sheep observed with each welfare indicator (count data). The total number of sheep in each assessment group was counted to determine the number of animals not affected by each welfare condition.

Table 1. Animal welfare indicators for sheep adopted from AWIN (2015).

Welfare Principles	Welfare Criteria	Welfare Indicators
Good feeding	Appropriate nutrition	Body condition score
Good housing	Comfort around resting	Fleece cleanliness
Good health	Absence of diseases, injuries, and pain	Lameness, respiratory disorders, and rear and dirtiness
Appropriate behavior	Expression of social behavior	Familiar human approach

2.3. Statistical Analysis

Data analysis was performed using the SAS statistical package [19]. A descriptive exploratory analysis was carried out to summarize the main characteristics of the assessments performed. The overall proportion (%) of affected sheep recorded was calculated by dividing the total number of affected animals by the total number of sheep in the sample group or the flock. The significant difference between proportions was calculated by chi-square. Results were expressed as means ± SD. Results with an associated probability less than or equal to 0.05 were considered significant.

3. Results

3.1. Body Condition Score

Assessing body condition was an important animal-based measurement. Descriptive statistics for BCS were presented in Figure 1. More than 75% of the BCS of ewes recorded in the study ranged from 2.5 to 3.5 at lambing. Ewes were considered lean if the score was less than 2 and fat if the score was equal to 5. In our study, we noticed that 17% of lambs and 24% of ewes had a body condition score of less than 2. Therefore, the highest number of animals (57 and 64%) had a BCS of 3. However, only 2.5% of ewes were considered fat.

3.2. Cleanliness

For good housing, a flock can be considered clean, since 60% of ewes had a cleanliness score of 1 and only 14% were considered dirty (Figure 2). There was no significant difference between lambs and ewes ($p > 0.05$), this can be explained by the fact that during our visits to farms, lambs were not separated from their mothers.

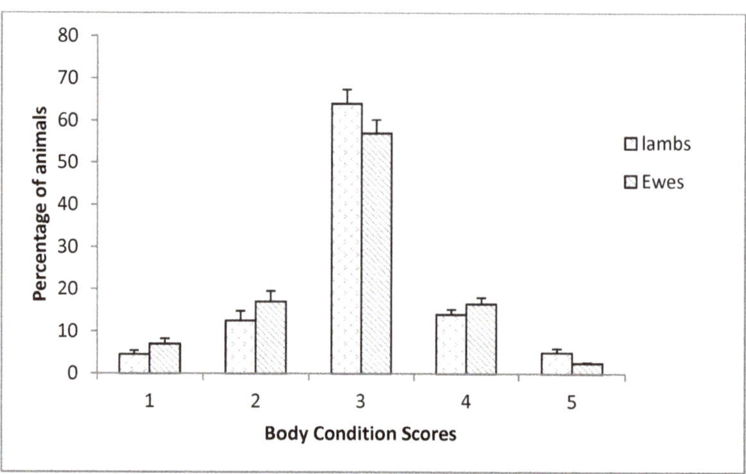

Figure 1. Distribution of the average scores of body condition for the studied sheeps at lambing, mid-lactation, and weaning.

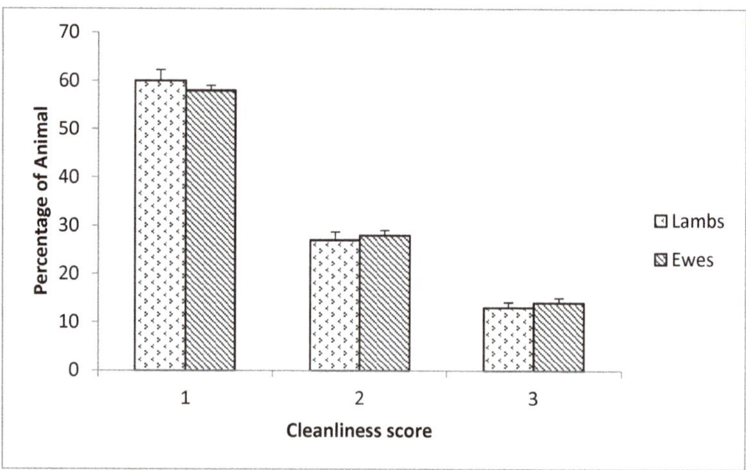

Figure 2. Distrubition of the average of scores of cleanliness for the studied sheeps at lambing, mid-lactation, and weaning.

3.3. Health

Good health was an important component of animal welfare and it can be defined as the absence of injuries, disease, and pain [20]. For health, four indicators were assessed: Respiratory rate, lesions and dirtiness, and finally lameness (Table 2). We reported in our study an average percentage of moderate lameness ewes with values of 10.5% and 4.65% with severe lameness. Lameness was considered the most common sign of limb injury, which compromises the animals' welfare. In our study, the significant percentage of sheep with no lameness (84.87%) may be indicative of good overall welfare within the flock. For the respiratory rate, 89.75% of animals had no respiratory problem and only 8% had a moderate problem.

Table 2. Mean ± standard deviation (SD) of the selected health indicators measured in 52 farms for ewes at lambing and mid-lactation.

Variables	Percentage of Ewes with		
	No Problem	Moderate	Severe
Respiratory rate	89.75 ± 7.24	8.1 ± 0.21	2.15 ± 0.14
Lesions	86.95 ± 6.13	9.7 ± 0.33	3.35 ± 0.23
Dirtiness	17.2 ± 3.2	57.3 ± 5.17	12.75 ± 1.33
Lameness	84.87 ± 5.27	10.48 ± 1.12	4.65 ± 0.57

3.4. *Animal-Human Interactions*

The quality of the human-animal relationship can be one of the most important factors in determining the welfare of an animal. The nature and frequency of this relationship can vary markedly in different sheep farming systems. The mean withdrawal distance was all greater ($p < 0.01$) for a novel person (10.47 m) than the farmer (8.12 m) (Table 3). Moreover, sheep withdrew from the advancing person at distances exceeding 20 m in the approach. For heart rates, the average value was 128.4 bpm for a novel person for a test during the 10 s. HAR was significantly ($p < 0.05$) higher compared to the HAR reported for the farmer (97.8 bpm).

Table 3. Mean ± SD of avoidance distance and heart rate measured in 52 farms for ewes at lambing and mid-lactation.

Approach Test	Withdrawal Distance (m)			Heart Rate (bpm)		
	Min	Mean	Max	Min	Mean	Max
Novel person	7.5	10.47 ± 1.23	13.78	63.5	128.4 ± 1.42	208.7
Farmer	5.24	8.12 ± 0.97	10.17	58.3	97.8 ± 6.45	197.6

4. Discussion

4.1. *Body Condition*

This study constitutes the first evaluation of the welfare of sheep conducted in Tunisia. The body condition score assesses the amount of fat and muscle overlying the spine. Overall, the body condition was considered good. Previous studies have also shown similar results [21]. In other studies [22,23], the general average score for the body conditions of Norduz ewes was 2.9. The descriptive analysis of BCS reveals that the median of BCS was 3.0 and only 5.5% of animals were scored with a BCS lower than 2 and 15% greater than 4. These results agreed with those of Keinprecht et al. [24]. Most of the ewes in this study were within the recommended body condition. However, thin ewes were observed within flocks, suggesting that some farmer's nutritional management was not and or were not identifying/treating individual thin ewes. A low BCS, at mid-lactation indicated prior long term poor welfare. In addition, low values of BCS occured when energy expenditure surpasses the intake and body fat was mobilized to meet the animal's needs, whereas high values indicated over-feeding. Indeed, ensuring that all sheep in a flock meet their nutritional requirements was not easily achievable in extensive systems.

4.2. *Cleanliness*

Fleece cleanliness had previously been proposed as an important welfare measure for sheep, as it can provide information about the quality of the environment. In our study, animals were reared in simple shelters and a large backyard. A higher score for fleece cleanliness was reported in this study (Figure 2). Previous studies [15,25] reported similar results and explained the higher score to the good housing conditions. Several authors [8,15,26] suggested that looking at the degree of dirtiness/cleanliness of a sheep in a flock can give a good insight into the housing conditions. However, in an Italian study [14], authors assessed animals and found significant differences between lambs and ewes.

4.3. Health

It was well acknowledged that health and disease were important aspects of welfare. The indicators selected in this study were respiratory disorder, lameness, and lesions. Lameness is one of the major welfare concerns of sheep. Lame ewes were found across all farms, while moderate and severe lameness were notified to farmers. In our study, the significant percentage of sheep with no lameness (84.87%) may be indicative of good overall welfare within the flock. However, Winter and Arsenos [27] reported a high prevalence (up to 75% of sheep). For respiratory disorder, we noticed that 8% of ewes have moderate disorders and 2.5% with a severe disorder which is considered higher than other findings [12]. For the respiratory rate, our results agreed with those of Lees et al. [28]. For lesions, we reported 86.95% with no problem and 57% with moderate lesions. We can conclude that the flocks required more care and attention to avoid health problems, lesions, and injuries [12,25,29].

4.4. Animal-Human Interactions

The assessment of an animal's reaction to humans was a good indicator of the human-animal relationship. In this study, the presence of the farmer had a strong calming effect (Table 3). Our results were highly consistent with previous results that affirmed that the presence of a familiar person can calm the animal [30]. Furthermore, the presence of a familiar person reduced stress and fear of humans in sheep [31,32]. However, for a novel person, sheep showed a higher avoidance distance [26]. HAR was a good indocator of the stockpersons' attitudes towards farm animals. Moreover, the low average approach distance of our study was explained by the fact that sheeps recognize individual humans and are more likely to approach those who treat them well than those who act in an aggressive way. Then, the presence of a familiar human may calm the animals in potentially aversive situations. HAR was a major determinant of sheep welfare and particularly pertinent to extensive systems with limited interactions with people. The heart rates of the sheep in this study were similar to the previous values [5,33]. In other studies, higher heart rates in Scottish Blackface lambs were found [8]. The attitude a stockperson holds about animals will strongly influence their behavior towards animals [34]. Conversely, the regular experience of positive human-animal interactions can decrease the animals' general level of stress [35] and enhance the reproductive performance and the presence of a familiar person can calm the animal in potentially aversive situations [30]. The reaction to an approaching human may be best suited for use when assessing extensively managed animals as it most closely resembles the situations the animals experience regularly.

5. Conclusions

This was a first and preliminary study in Tunisia. We can conclude that on-farm welfare assessments can be used for immediate or ongoing on-farm monitoring by farmers. Animal-based measures often reflect the outcome of resource inputs and management practice. The use of behavioral principles should improve the efficiency of livestock handling and reduce stress on animals. The present study reveals the extent to which this species is capable of habituating to common human-related stimuli. The respiratory rate was considered a serious welfare consequence for lambs. The human-animal relationship (HAR) is a major determinant of sheep welfare since it is an important source of fear in farmed sheep.

Author Contributions: Conceptualization, N.M., C.D. and K.A.; Formal analysis, N.M., K.A. and M.A.; Investigation, R.B.; Methodology, N.M.; Resources, H.M.; Software, N.M., C.D., K.A., H.G. and R.B.; Supervi-sion, L.M. and M.A.; Validation, N.M., K.A., H.G. and L.M.; Visualization, I.E.A.Z.; Writing—original draft, N.M. and C.D.; Writing—review & editing, I.E.A.Z. All authors have read and agreed to the published version of the manuscript.

Funding: This research received no external funding.

Institutional Review Board Statement: All procedures related to animal care, handling, and sampling were conducted under the approval of the Official Animal Care and Use Committee of the Higher School of Agriculture of Mateur (protocol no. 05/15) before the initiation of research and followed the Tunisian guidelines.

Data Availability Statement: Data will be published under demand.

Acknowledgments: The authors acknowledge the Pasture and Livestock Office (OEP) of Tunisia administrative and technical support and farmers for the animals used.

Conflicts of Interest: The authors declare no conflict of interest.

References

1. Ibidhi, R.; Frija, A.; Jaouad, M.; Ben Salem, H. Typology analysis of sheep production, feeding systems and farmers strategies for livestock watering in Tunisia. *Small Rumin. Res.* **2018**, *160*, 44–53. [CrossRef]
2. OEP. Available online: http://www.oep.nat.tn/index.php/en/donnees-sectorielles/41-productions (accessed on 17 February 2020).
3. Blokhuis, H.J.; Jones, R.B.; Geers, R.; Miele, M.; Veissier, I. Measuring and monitoring animal welfare: Transparency in the foodproduct quality chain. *Anim. Welf.* **2003**, *12*, 445–455.
4. Main, D.C.J.; Kent, J.P.; Wemelsfelder, F.; Ofner, E.; Tuyttens, F.A.M. Applications for methods of on-farm welfare assessment. *Anim. Welf.* **2003**, *12*, 523–528.
5. Cook, C. Basal and stress response cortisol levels and stress avoidance learning in sheep (Ovis ovis). *N. Z. Veter. J.* **1996**, *44*, 162–163. [CrossRef]
6. Destrez, A.; Deiss, V.; Leterrier, C.; Boivin, X.; Boissy, A. Long-term exposure to unpredictable and uncontrollable aversive events alters fearfulness in sheep. *Animal* **2013**, *7*, 476–484. [CrossRef] [PubMed]
7. Dwyer, C.; Lawrence, A. Introduction to animal welfare and the sheep. In *The Welfare of Sheep*; Dwyer, C., Ed.; Springer: Berlin/Heidelberg, Germany, 2008; pp. 1–40.
8. Goddard, P. Welfare assessment in sheep. *InPractice* **2011**, *33*, 508–516. [CrossRef]
9. Cochran, W.G. *Sampling Techniques*; John Wiley & Sons, Inc.: New York, NY, USA, 1977.
10. AWIN. AWIN Welfare Assessment Protocol for Sheep. Available online: http://www.animal-welfare-indicators.net/site/flash/pdf/AWINProtocolSheep.pdf (accessed on 17 February 2020).
11. Phythian, C.J.; Mullan, S.; Butterworth, A.; Lambton, S.; Ilić, J.; Burazerović, J.; Burazerović, E.; Leach, K.A. A pilot survey of farm animal welfare in Serbia, a country preparing for EU accession. *Vet. Med. Sci.* **2017**, *3*, 208–226. [CrossRef]
12. Phythian, C.J.; Toft, N.; Cripps, P.J.; Michalopoulou, E.; Winter, A.C.; Jones, P.H. Inter-observer agreement, diagnostic sensitivity and specificity of animal-based indicators of young lamb welfare. *Animal* **2013**, *7*, 1182–1190. [CrossRef] [PubMed]
13. Phythian, C.J.; Toft, N.; Cripps, P.J.; Michalopoulou, E.; Winter, A.C.; Jones, P.H. Reliability of indicators of sheep welfare as-sessed by a group observation method. *Vet. J.* **2012**, *193*, 257–263. [CrossRef] [PubMed]
14. Napolitano, F.; De Rosa, G.; Ferrante, V.; Grasso, F.; Braghieri, A. Monitoring the welfare of sheep in organic and con-ventional farms using an ANI 35L derived method. *Small Rumin. Res.* **2009**, *1*, 49–57. [CrossRef]
15. Caroprese, M.; Casamassima, D.; Rassu, S.P.G.; Napolitano, F.; Sevi, A. Monitoring the on-farm welfare of sheep and goats. *Ital. J. Anim. Sci.* **2009**, *8*, 343–354. [CrossRef]
16. Calavas, D.; Sulpice, P.; Lepetitcolin, E.; Bugnard, F. Assessing the accuracy of BCS in ewes under field conditions. *Vet. Res.* **1998**, *29*, 129–138. [PubMed]
17. Waiblinger, S.; Menke, C.; Fölsch, D. Influences on the avoidance and approach behaviour of dairy cows towards humans on 35 farms. *Appl. Anim. Behav. Sci.* **2003**, *84*, 23–39. [CrossRef]
18. Webster, A.J.F. Continuous measurement of heart rate as an indicator of the energy expenditure of sheep. *Br. J. Nutr.* **1967**, *21*, 769–785. [CrossRef] [PubMed]
19. *Statistical Analysis System*; Release 9.4; SAS Institute Inc.: Cary, NC, USA, 2012.
20. Keeling, L.J. Defining a framework for developing assessment systems. In *An Overview of the Development of the Welfare Quality Project Assessment Systems. Welfare Quality Report n: 12*; Keeling, L.J., Ed.; Cardiff University: Cardiff, UK, 2009; pp. 1–7.
21. Verbeek, E.; Waas, J.R.; Oliver, M.H.; McLeay, L.M.; Ferguson, D.M.; Matthews, L.R. Motivation to obtain a food reward of pregnant ewes in negative energy balance: Behavioral, metabolic and endocrine considerations. *Horm. Behav.* **2012**, *62*, 162–172. [CrossRef] [PubMed]
22. Karaku, F.; Atmaca, M. The effect of ewe body condition at lambing on the growth of lambs and colostral specific gravity. *Arch. Anim. Breed.* **2016**, *59*, 107–112. [CrossRef]
23. Banchero, G.; Quintans, G. Lamb vigour of merino ewes in high and low body condition with or without a lupin supple-ments during the last two weeks of pregnancy. In *Reproduction in Domestic Ruminants V, Proceedings of The Seventh International Symposium on Reproduction in Domestics Ruminants, Wellington, New Zealand, 7 Auguest 2006*; Nottingham University Press: Nottingham, UK, 2007; p. 495.

24. Keinprecht, H.; Pichler, M.; Pothmann, H.; Huber, J.; Iwersen, M.; Drillich, M. Short term repeatability of body fat thickness measurement and body condition scoring in sheep as assessed by a relatively small number of assessors. *Small Rumin. Res.* **2016**, *139*, 30–38. [CrossRef]
25. Stubsjoen, S.M.; Hektoen, L.; Valle, P.S. Assessment of sheep welfare using on-farm registrations and performance data. *Anim. Welf.* **2011**, *20*, 239–251.
26. Sevi, A. Animal-based measures for welfare assessment. *Ital. J. Anim. Sci.* **2009**, *8*, 904–911. [CrossRef]
27. Winter, A.C.; Arsenos, G. Diagnosis of white line lesions in sheep. *InPractice* **2009**, *31*, 17–21. [CrossRef]
28. Lees, A.M.; Sullivan, M.L.; Olm, J.C.W.; Cawdell-Smith, A.J.; Gaughan, J.B. A panting score index for sheep. *Int. J. Biometeorol.* **2019**, *63*, 973–978. [CrossRef] [PubMed]
29. Lovatt, F. Clinical examination of sheep. *Small Rumin. Res.* **2010**, *92*, 72–77. [CrossRef]
30. Waiblinger, S.; Boivin, X.; Pedersen, V.; Tosi, M.V.; Janczak, A.M.; Visser, E.K.; Jones, R.B. Assessing the human-animal rela-tionship in farmed species: A critical review. *Appl. Anim. Behav. Sci.* **2006**, *101*, 185–242. [CrossRef]
31. Boivin, X.; Nowak, R.; Desprès, G.; Tournadre, H.; Le Neindre, P. Discrimination between shepherds by lambs reared under artificial conditions. *J. Anim. Sci.* **1997**, *75*, 2892–2898. [CrossRef] [PubMed]
32. Hemsworth, P.H.; Rice, M.; Karlen, M.G.; Calleja, L.; Barnett, J.L.; Nash, J.; Coleman, G.J. Human–animal interactions at abattoirs: Relationships between handling and animal stress in sheep and cattle. *Appl. Anim. Behav. Sci.* **2011**, *135*, 24–33. [CrossRef]
33. Komesaroff, P.A.; Esler, M.; Clarke, I.J.; Fullerton, M.J.; Funder, J.W. Effects of estrogen and estrous cycle on glucocorticoid and catecholamine responses to stress in sheep. *Am. J. Physiol. Content* **1998**, *275*, E671–E678. [CrossRef]
34. Hemsworth, P.H.; Coleman, G.J. *Human-Livestock Interactions: The Stockperson and the Productivity and Welfare of Intensively Farmed Animals*; CAB International: Wallingford, UK, 1998.
35. Seabrook, M.F.; Bartle, N.C. Environmental factors influencing the production and welfare of farm animals-human factors. In *Farm Animals and the Environment*; Phillips, C.J.C., Piggins, D., Eds.; CAB International: Wallingford, UK, 1992; pp. 111–130.

MDPI

Article

Pre- and Post-Slaughter Methodologies to Estimate Body Fat Reserves in Lactating Saanen Goats

Leonardo Sidney Knupp [1], Mondina Francesca Lunesu [2], Roberto Germano Costa [1,*], Mauro Ledda [3], Sheila Nogueira Ribeiro Knupp [4], Marco Acciaro [5], Mauro Decandia [5], Giovanni Molle [5], Ana Helena Dias Francesconi [2] and Antonello Cannas [2]

1 Departamento de Zootecnia, Universidade Federal da Paraíba, Rodovia BR 079 km 12, Areia 58397-000, Brazil; LeonardoKnupp@hotmail.com

2 Dipartimento di Agraria, Università degli Studi di Sassari, Viale Italia 39, 07100 Sassari, Italy; mflunesu@uniss.it (M.F.L.); france@uniss.it (A.H.D.F.); cannas@uniss.it (A.C.)

3 Dipartimento di Medicina Veterinaria, Università degli Studi di Sassari, Via Viena 2, 07100 Sassari, Italy; vetleddamauro@gmail.com

4 Centro de Saúde e Tecnologia Rural, Universidade Federal de Campina Grande, Av. Universitária s/n, Patos 58708-110, Brazil; sheilanribeiro@hotmail.com

5 Agricultural Research Agency of Sardinia—AGRIS Sardegna, Loc. Bonassai, 07100 Sassari, Italy; macciaro@agrisricerca.it (M.A.); mdecandia@agrisricerca.it (M.D.); gmolle@agrisricerca.it (G.M.)

* Correspondence: betogermano@hotmail.com

Simple Summary: In this study, we present the results of a trial on which we compared pre- and post-slaughter methodologies to estimate body fat reserves in dairy goats. Our results evidenced that fat thickness measured with ultrasound in the perirenal region was the best pre-slaughter measurement for estimating fat reserves in lactating Saanen goats, whereas empty body weight and hot carcass weight were the best post-slaughter predictors for estimating fat reserves. Body condition score could be a useful tool, but it seems that it needs to be re-evaluated to predict adequately fat depots in lactating Saanen goats.

Abstract: This work aimed to compare pre- and post-slaughter methodologies to estimate body fat reserves in dairy goats. Twenty-six lactating Saanen goats ranging from 43.6 to 69.4 kg of body weight (BW) and from 1.84 to 2.96 of body condition score (BCS; 0–5 range) were used. Fifteen pre-slaughter and four post-slaughter measurement values were used to estimate the weight of fat in the omental (OM), mesenteric (MES), perirenal (PR), organ (ORG), carcass (CARC), and non-carcass components (NC) and total (TOT, calculated as the sum of CARC and NC) depots in goats. The pre-slaughter measurements were withers height; rump height; rump length; pelvis width; chest depth; shoulder width; heart girth; body length; sternum height; BW; BCS assessed in the lumbar (BCSl) and sternal (BCSs) regions; and fat thickness measured by ultrasound in the lumbar (FTUSl), sternal (FTUSs), and perirenal (FTUSpr) regions. The post-slaughter measurements were hot carcass weight (HCW), empty body weight (EBW), and fat thickness measured by digital caliper in the lumbar (FTDCl) and sternal (FTDCs) regions. Linear and multiple regressions were fit to data collected. BW, BCS (from lumbar and sternal regions), all somatic measurements, and fat thickness measured by ultrasound in the lumbar and sternal regions were not adequate to estimate the weight of total fat in lactating Saanen goats ($R^2 \leq 0.55$). The best pre-slaughter and post-slaughter estimators of OM, MES, PR, ORG, NC, and TOT fat were FTUSpr and EBW, respectively. Among pre- and post-slaughter measurements, BCSl (R^2 = 0.63) and HCW (R^2 = 0.82) provided the most accurate predictions of CARC fat, respectively. Multiple regression using the pre-slaughter variables FTUSpr, BW, and BCSl yielded estimates of TOT fat with an R^2 = 0.92 (RSD = 1.14 kg). On the other hand, TOT fat predicted using the post-slaughter variables HCW and FTDCs had an R^2 = 0.83 (RSD = 1.41 kg). These results confirm that fat reserves can be predicted in lactating Saanen goats with high precision using multiple regression equations combining in vivo measurements.

Keywords: body condition score; body measurements; fat depots; goats; prediction equation; ultrasound

Citation: Knupp, L.S.; Lunesu, M.F.; Costa, R.G.; Ledda, M.; Knupp, S.N.R.; Acciaro, M.; Decandia, M.; Molle, G.; Francesconi, A.H.D.; Cannas, A. Pre- and Post-Slaughter Methodologies to Estimate Body Fat Reserves in Lactating Saanen Goats. *Animals* **2021**, *11*, 1440. https://doi.org/10.3390/ani11051440

Academic Editors: María-Luz García and María-José Argente

Received: 23 March 2021
Accepted: 15 May 2021
Published: 18 May 2021

1. Introduction

In most goat production systems under harsh conditions, the ability of the animal to retain and mobilize body reserves is of considerable importance in determining goat productivity and survival. Such relevance is due to the fact that the nutritional status of goats fluctuates throughout the year [1] because of changes in the amount and quality of nutrients in the diet [2] and physiological state of the animal [3]. Accurate and precise determination of nutritional status in lactating goats is important to avoid depletion of energy stored and to minimize tissue protein mobilization, thus increasing milk production [2].

The nutritional status of animals can be estimated by direct and indirect methods. The comparative slaughter is the most accurate direct method, but it is expensive, because at least half of the carcass is lost [4], it is destructive and laborious, and it does not allow for the use of the same animal more than once [5]. Therefore, indirect methods are preferable because most of them are not complex and can be applied to live animals [6].

Throughout the years, researchers have developed many indirect methods to estimate nutritional status, such as body weight (BW) and body measurements [7,8], body condition score (BCS, [9,10]), urea space [11], adipocyte diameter [12], real-time ultrasonography [13], computed tomography [1], dual-energy X-ray absorptiometry, and magnetic resonance imaging [6]. Some of these methods are very expensive and difficult to use in many farm animals. Others, such as BCS and body measurements, have basically no cost and can be performed in experimental and field conditions.

The BCS method was developed by Russel et al. [10] for meat lambs, which accumulate fat in the subcutaneous region, whereas it might not be appropriate for dairy goats, which deposit body fat mostly as visceral fat [1,14]. In lactating Alpine does, Ngwa et al. [2] noted that the amount of fat in non-carcass components (visceral and renal fat) is almost twice that in carcass and a considerable amount of internal fat is mobilized in early lactation. Härter et al. [15] developed equations to predict abdominal fat depots in pregnant non-lactating Saanen goats using ultrasound measurements of the *Longissimus* muscle area (LMA) and kidney fat thickness (KFT). The authors reported high coefficient of determination for non-carcass fat and total body fat (R^2 = 0.77 and 0.80, respectively) when using LMA and KFT associated with BW. However, to our knowledge, there are no studies comparing different pre- and post-slaughter methodologies to estimate fat reserves in lactating Saanen goats. Thus, this work aimed to (i) study the relationship between BCS and body measurements with BW and body fat, (ii) compare pre- and post-slaughter methodologies as predictors to estimate body fat depots, and (iii) develop equations that could be used as an indicator of nutritional status in lactating Saanen goats.

2. Materials and Methods

2.1. Animals

The study was carried out using 26 mature lactating Saanen goats randomly selected from the experimental flock of Agris Research farm of Bonassai in Olmedo (Northwestern Sardinia, Italy, 40°40′16.215″ N, 8°22′0.392″ E, 32 m a.s.l.). Animals were chosen from a larger group fed a high-starch diet, homogeneous for lambing date, age (6–7 years), and milk yield. Goats were clinically healthy and had mean BW of 56.4 ± 6.8 kg. Animals were milked twice a day and had access to feed and water until slaughter. Their care and use followed the Italian national law and ethic regulations (DL. no. 116, 27/01/1992). The animal protocol described below was performed in compliance with the EU and Italian regulations on animal welfare, and all measurements were taken by personnel previously trained and authorized by the institutional authorities managing ethical issues at the University of Sassari. Experimental procedures with animals (goats) were approved by the Animal Care and Use Committee of the University of Sassari and Agris, Italy (CIBASA 10.12.2014).

2.2. Pre-Slaughter Measurements

2.2.1. Somatic Measurements and Body Condition Score

The following somatic measurements, based on Cam et al. [8], were made on all goats 16 h before slaughter: withers height (WH), the distance between the top of the withers to the ground; rump length (RL), distance from hip to pin; rump height (RH) vertical distance from top of pelvic girdle and the ground; pelvis width (PW), distance between trochanters; chest depth (CD), the distance between the withers and the sternum; shoulder width (SW), the horizontal distance between the processes on the left and on the right shoulder blade; heart girth (HG), the smallest circumference around the animal just behind the foreleg; body length (BL), the distance between the withers and the cross; and sternum height (SH), the distance between the sternum and the ground. The measurements of WH, RH, CD, SW, and SH were taken with a Lydtin metric stick (metal tube of 80 to 230 cm length). Pelvis width was measured with a thickness compass, and RL, HG, and BL with a linear meter.

For practical reasons, i.e., for the lack of an appropriate precision scale suitable for live animals in the site of slaughtering, body weight was measured with an electronic scale immediately after slaughter (blood was collected and weighed). Two experienced workers evaluated the BCS in the lumbar and sternal region by using the Hervieu et al. [9] reference scale (0 to 5 score). In both cases, the BCS intervals were of 0.25 units. The BCS was assessed at the moment of the selection of the animals and at the end of the trial, just before slaughtering.

2.2.2. Measurement of Fat Thickness Using Ultrasound

Fat thickness was measured, simultaneously with the previous measures, using a real time MyLab One ultrasound system (Esaote S.p.A., Genova, Italy). Trichotomy was performed in the area to be measured and gel was used as a coupling agent to improve the quality of the images. Ultrasound pictures were taken twice on three different anatomical sites: (1) lumbar fat thickness (FTUSl), measured in the area of the longissimus muscle around the 13th thoracic vertebrae (last rib), by using an ultrasound probe SL3323 VET (array of 13-6 Mhz and 40-mm length; Esaote S.p.A., Genova, Italy); (2) perirenal fat thickness (FTUSpr), measured behind the 13th rib on the right side of the body using an ultrasound probe SV3513 VET (array of 10-5 Mhz and 50-mm length; Esaote S.p.A., Genova, Italy), according to Härter et al. [15]; and (3) sternal fat thickness (FTUSs), measured using an ultrasound probe SL3323 VET (array of 13-6 Mhz and 40-mm length; Esaote S.p.A., Genova, Italy) positioned perpendicularly to the third sternebra on the sternum. Images were obtained with a linear probe (transducer) of 6 Mhz and silicone acoustic attachment (standoff) for FTUSl and FTUSs measurements and an 8 Mhz convex transducer for FTUSpr measurements. The pictures were stored on a computer and, subsequently, analyzed with the software MyLab Desk™/Desk (Esaote S.p.A., Genova, Italy) to obtain the fat thickness measurements.

2.3. Post-Slaughter Measurements

2.3.1. Slaughter Procedures, Hot Carcass, and Empty Body Weight

The animals were slaughtered under general anesthesia and exsanguinated from the jugular vein in the facilities of the Hospital of the Veterinary Department of the University of Sassari (Sassari, Sardinia, Italy). The weights of blood, head, skin, feet, tail, empty viscera (rumen–reticulum, omasum, abomasum, small intestine, and large intestine), mesentery, internal fat, liver, heart, kidneys, spleen, lungs, tongue, esophagus, trachea, and reproductive system, and hot carcass weight (HCW) were recorded. The digestive tract compartments were isolated, weighed, emptied, and weighed again. The empty body weight (EBW) was calculated by difference of live weight and the content of the gastrointestinal tract, bladder, and gallbladder empty. The fat tissue surrounding the digestive tract, omental (OM) fats, and mesenteric (MES) fats was removed, along with any associated connective tissue and weighed. Perirenal fat (PR) was removed from the kidneys and weighed. Organ fat from heart, liver, and lungs was removed from each organ and weighed together (ORG).

2.3.2. Carcass Measurements

Carcasses were stored at 4 °C for 24 h in a cooler. Then, carcasses were split down the backbone with a band saw into two halves (right and left). The right half of each carcass was ribbed at the 12th and 13th thoracic vertebrae at the same anatomical points where measurements had been taken on the live animal using ultrasound. Lumbar fat thickness was measured by using a digital caliper (FTDCl). Similarly, a transversal cut was performed at the third sternebra on the sternum vertebra, and sternal fat was then measured using a digital caliper (FTDCs).

2.3.3. Fat Content on Carcass and Non-Carcass Components

The left side of each carcass was frozen until subsequent determination of chemical composition, whereas all non-carcass components (digestive tract, pluck, reproductive tract, and mammary gland), including head and skin, were stored in separate polyethylene bags at −20 °C until preparation for analysis. All frozen components (carcass and non-carcass) were cut into pieces of 5–6 cm^3 while still frozen, and then minced and ground by using a mill grinder (TC 42 Golia HP 10 HS, La Felsinea S.R., Padova, Italy). After the ground material was mixed thoroughly with a mechanical mixer (ME 30, La Felsinea S.R., Padova, Italy), samples were taken in three replicates. The samples were weighed, frozen at −80 °C, and subsequently analyzed for dry matter by liophilization (Lyolab 3000, Jouan Nordic, Allerød, Denmark). Then, samples were reground in a blender (Knifetec Mill 1095, Foss, Höganäs, Sweden) and analyzed for fat. Carcass (CARC) and non-carcass (NC) fat was determined by continuously extracting the samples with petroleum ether for 6 h by using the AOAC method 920.39 (AOAC International, 2005).

2.4. *Statistical Analysis*

The statistical analyses were performed using linear single variable with the GLM procedure of SAS software (version 9.2, SAS System Inc., Cary, NC, USA) for the weights of fat in the different depots as dependent variables (y), and BW, BCSl, BCSs, somatic measurements, FTUSl, FTUSs, FTUSpr, HCW, EBW, FTDCl, and FTDCs as independent variables (x). The variables included in the multiple regressions were selected using the REG procedure with the STEPWISE method of SAS. Since ultrasound is not so cheap and requires more time than BCS, BW, and somatic measures to be used under field conditions, additional simplified equations were developed without the use of ultrasound also using the REG procedure with the STEPWISE method of SAS.

3. Results

3.1. *Pre-Slaughter Measurements*

3.1.1. Somatic Measurements

Heart girth ranged from 86 to 104 cm, with a mean of 94 cm (Table 1). Among all somatic measurements, only HG had regression coefficients significantly different from zero ($p < 0.05$) in all fat depots analyzed. The relationship between HG and BW showed a mean HG change of 1.2 cm per unit (kg) of BW (BW = 1.2 HG − 57.7; R^2 = 0.75; Figure 1).

Table 1. Somatic measurements; body weight (BW); lumbar and sternal body condition scores (BCSl and BSs, respectively); hot carcass weight (HCW); empty body weight (EBW); lumbar, sternal, and perirenal fat thickness depth measured by ultrasound (FTUSl, FTUSs, and FTUSpr, respectively); lumbar and sternal fat thickness depth measured by digital caliper (FTDCl and FTDCs, respectively); and fat depot weights.

Item	Mean	Minimum	Maximum	Standard Deviation
Somatic measurements (cm)				
Wither height	70.9	65.0	77.0	3.4
Rump height	73.6	69.0	82.0	3.3
Rump length	21.6	17.5	28.0	3.0
Pelvis width	18.7	16.0	25.0	1.9
Chest depth	35.1	29.0	39.0	2.1
Shoulder width	18.1	13.5	24.0	2.5

Table 1. *Cont.*

Item	Mean	Minimum	Maximum	Standard Deviation
Heart girth	94.2	86.0	104.0	4.8
Body length	74.7	68.0	90.0	4.7
Sternum height	35.3	28.0	44.0	3.9
BW (kg)	56.4	43.6	69.4	6.8
BCSl (scale 0–5)	2.64	1.84	2.96	0.3
BCSs (scale 0–5)	2.64	1.75	3.00	0.3
HCW (kg)	24.3	18.1	30.1	3.4
EBW (kg)	47.6	36.1	59.9	6.8
FTUSl (mm)	2.27	1.08	3.50	0.6
FTDCl (mm)	2.13	1.47	3.60	0.5
FTUSs (mm)	22.68	9.70	27.90	4.5
FTDCs (mm)	22.10	9.79	29.82	4.7
FTUSpr (cm)	1.21	0.44	2.34	0.1
Fat depot weight (kg)				
Omental	1.83	0.30	4.38	1.3
Mesenteric	0.94	0.46	1.54	0.3
Perirenal	0.72	0.07	2.02	0.6
Organ fat (heart, liver, and lungs)	0.22	0.08	1.06	0.2
Total fat (kg)				
Carcass	4.61	1.33	6.58	1.3
Non-carcass	5.56	2.16	9.38	2.1
Carcass and non-carcass	10.17	3.49	15.93	3.2

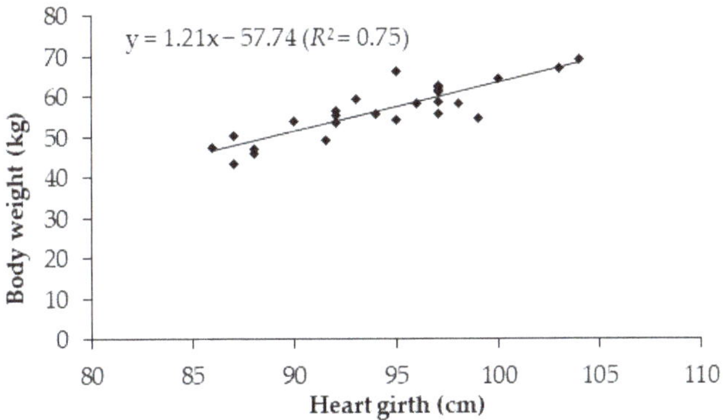

Figure 1. Relationship between heart girth and body weight in lactating Saanen goats.

3.1.2. Body Weight

Body weight after slaughter (summed with blood from exsanguinations) ranged from 44 to 69 kg, with a mean of 56 kg (Table 1). The regressions between the weight of fat in each of the different fat depots and BW (Table 2) had determination coefficients (R^2) that varied between 0.21 for the organ (ORG) depot (RSD = 0.17 kg) and 0.58 for carcass depot (RSD = 0.86 kg). The determination coefficient for the relationship between the total weight of fat (TOT, sum of fat on carcass and non-carcass components) and BW (Table 2) was 0.55 (RSD = 2.25 kg).

3.1.3. Body Condition Score

Body condition score assessed at lumbar or sternal region averaged 2.6, but sternal BCS detected a lower fatness level compared to BCSl (1.75 versus 1.84, respectively) (Table 1). The Pearson correlation between lumbar and sternal BCS was 0.852, with $p < 0.001$. The regression of sternal BCS on lumbar BCS had a non-significant intercept, with BCS sternal = 0.999 BCS lumbar. The regression equations for prediction of the weights of fat depots using both BCS, from lumbar and sternal region, had low accuracy. The R^2 values for BCSl varied between 0.10 for organs and 0.63 for carcass fat, and those for BCSs varied between 0.07 for organs and 0.54 for carcass fat (Table 2).

Table 2. Regression equations (y = a + bx), coefficient of determination (R^2), and residual standard deviation (RSD) values for estimate of the different fat depot weights (kg) and the total weight of all fat combined (y variables) based on the pre-slaughter measurement values of heart girth, body weight (BW), and lumbar and sternal body condition scores (x variables).

Item [1]	Intercept ± Standard Error	b ± Standard Error	R^2	RSD	*p*-Value
Heart girth					
OM	−9.26 ± 4.58	0.12 ± 0.05	0.20	1.17	0.023
MES	−2.96 ± 1.06	0.04 ± 0.01	0.36	0.27	0.001
PR	−3.83 ± 2.17	0.05 ± 0.02	0.16	0.56	0.046
ORG	−1.47 ± 0.67	0.02 ± 0.01	0.21	0.17	0.019
CARC	−12.49 ± 3.82	0.18 ± 0.04	0.46	0.98	0.001
NC	−19.27 ± 6.58	0.26 ± 0.07	0.37	1.68	0.001
TOT	−31.76 ± 9.95	0.44 ± 0.10	0.43	2.55	0.001
BW at slaughter					
OM	−3.82 ± 1.87	0.10 ± 0.03	0.28	1.11	0.005
MES	−0.86 ± 0.43	0.03 ± 0.01	0.43	0.26	0.001
PR	−1.89 ± 0.86	0.05 ± 0.01	0.28	0.51	0.005
ORG	−0.49 ± 0.29	0.01 ± 0.005	0.21	0.17	0.020
CARC	−3.65 ± 1.45	0.15 ± 0.02	0.58	0.86	<0.001
NC	−6.55 ± 2.57	0.21 ± 0.04	0.49	1.52	<0.001
TOT	−10.20 ± 3.80	0.36 ± 0.07	0.55	2.25	<0.001
BCS lumbar (scale 0–5)					
OM	−3.95 ± 2.38	2.21 ± 0.89	0.20	1.17	0.022
MES	−0.42 ± 0.63	0.52 ± 0.24	0.17	0.31	0.038
PR	−1.73 ± 1.12	0.93 ± 0.42	0.17	0.55	0.037
ORG	−0.38 ± 0.37	0.23 ± 0.14	0.10	0.18	0.116
CARC	−5.79 ± 1.64	3.94 ± 0.62	0.63	0.81	<0.001
NC	−6.30 ± 3.57	4.50 ± 1.34	0.32	1.76	0.003
TOT	−12.08 ± 5.07	8.44 ± 1.91	0.45	2.50	0.001
BCS sternal (scale 0–5)					
OM	−3.08 ± 2.37	1.87 ± 0.89	0.16	1.20	0.047
MES	−0.49 ± 0.60	0.55 ± 0.22	0.20	0.30	0.023
PR	−1.45 ± 1.10	0.83 ± 0.42	0.14	0.56	0.057
ORG	−0.28 ± 0.37	0.19 ± 0.14	0.07	0.19	0.183
CARC	−4.76 ± 1.76	3.55 ± 0.66	0.54	0.90	<0.001
NC	−4.95 ± 3.59	3.99 ± 1.35	0.27	1.82	0.007
TOT	−9.71 ± 5.21	7.54 ± 1.96	0.38	2.64	0.001

[1] OM = omental fat; MES = mesenteric fat; PR = perirenal fat; ORG = organ fat (heart, liver, and lungs); CARC = carcass fat; NC = non-carcass fat; TOT = total fat depot (TOT = CARC + NC).

The relationship between lumbar and sternal BCS and BW in lactating Saanen goats provided low R^2 (Figure 2). For BCSl, the equation was BW (kg) = 12.29 BCSl + 24.20 (R^2 = 0.22) and for BCSs, the equation was BW (kg) = 14.12 BCSs + 19.36 (R^2 = 0.32).

3.1.4. Ultrasound Measurements

Thickness of fat in the lumbar region measured using ultrasound ranged from 1.1 to 3.5 mm, with a mean of 2.3 mm (SD = 0.6). Fat in the sternal region was much thicker, ranging from 9.7 to 27.9 mm, with a mean of 22.7 mm (SD = 4.5). Perirenal fat was also high, ranging from 44 to 234 mm, with a mean of 121 mm (SD = 53) (Table 1). The determination coefficients of the equations were slightly lower using FTUSl, ranging between 0.10 for organ fat (RSD = 0.18 kg) and 0.33 for non-carcass fat (RSD = 1.76 kg), than using FTUSs thickness, ranging from 0.05 for organ fat (RSD = 0.19 kg) and 0.55 for carcass fat (RSD = 0.89 kg) (Table 3). Nevertheless, higher R^2 were found using FTUSpr, with values ranging between 0.05 for organ fat (RSD = 0.19 kg) and 0.86 for PR fat (RSD = 0.22 kg) (Table 3).

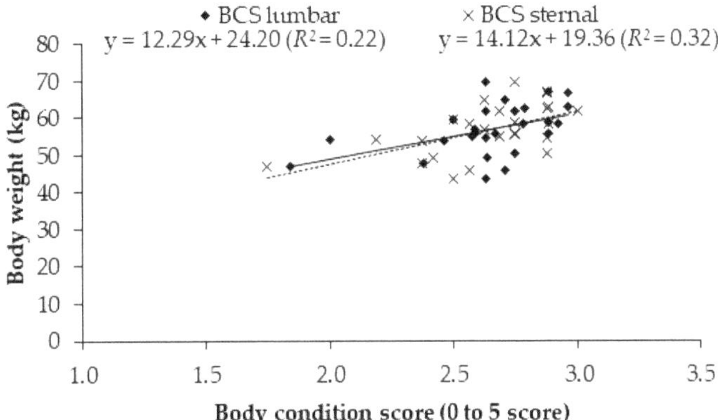

Figure 2. Relationship between lumbar or sternal body condition score (BCS, scale 0–5) and body weight in lactating Saanen goats.

Table 3. Regression equations (y = a + bx), coefficient of determination (R^2), and residual standard deviation (RSD) values for estimate of the different fat depot weights (kg) and the total weight of all fat combined (y variables) based on the pre-slaughter measurement values lumbar, sternal, and perirenal fat thickness depths measured by ultrasound (FTUSl, FTUSs, and FTUSpr, respectively) (x variables).

Item [1]	Intercept ± Standard Error	b ± Standard Error	R^2	RSD	*p*-Value
Lumbar fat thickness (FTUSl)					
OM	−0.46 ± 0.92	1.04 ± 0.39	0.23	1.16	0.012
MES	0.37 ± 0.24	0.27 ± 0.10	0.23	0.30	0.012
PR	−0.38 ± 0.42	0.50 ± 0.18	0.25	0.53	0.008
ORG	−0.003 ± 0.14	0.10 ± 0.06	0.10	0.18	0.120
CARC	2.07 ± 0.92	1.13 ± 0.39	0.27	1.16	0.007
NC	1.10 ± 1.40	2.01 ± 0.59	0.33	1.76	0.002
TOT	3.16 ± 2.23	3.14 ± 0.95	0.32	2.81	0.002
Sternal fat thickness (FTUSs)					
OM	−1.81 ± 1.05	0.16 ± 0.05	0.35	1.06	0.001
MES	0.17 ± 0.29	0.03 ± 0.01	0.24	0.29	0.011
PR	−1.01 ± 0.48	0.08 ± 0.02	0.37	0.48	0.001
ORG	0.02 ± 0.19	0.01 ± 0.01	0.05	0.19	0.278
CARC	−0.04 ± 0.88	0.21 ± 0.04	0.55	0.89	<0.001
NC	−1.11 ± 1.57	0.30 ± 0.07	0.44	1.59	0.001
TOT	−1.15 ± 2.33	0.51 ± 0.10	0.51	2.36	<0.001
Perirenal fat thickness (FTUSpr)					
OM	−0.53 ± 0.26	2.23 ± 0.22	0.81	0.57	<0.001
MES	0.48 ± 0.11	0.45 ± 0.09	0.49	0.24	<0.001
PR	−0.40 ± 0.10	1.06 ± 0.09	0.86	0.22	<0.001
ORG	0.14 ± 0.09	0.08 ± 0.07	0.05	0.19	0.283
CARC	2.88 ± 0.46	1.64 ± 0.49	0.42	1.01	0.001
NC	1.94 ± 0.51	3.44 ± 0.43	0.73	1.11	<0.001
TOT	4.82 ± 0.93	5.07 ± 0.78	0.63	2.03	<0.001

[1] OM = omental fat; MES = mesenteric fat; PR = perirenal fat; ORG = organ fat (heart, liver, and lungs); CARC = carcass fat; NC = non-carcass fat; TOT = total fat depot (TOT = CARC + NC)

3.1.5. Multiple Regressions

To increase the accuracy of the equation that predicted fat depot using only one independent variable (Tables 2–4), we calculated multiple regressions (Table 5). The inclusion of heart girth in the regression using FTUSpr to predict omental fat improved the R^2 from 0.79 to 0.85 (RSD = 0.65 and 0.57 kg, respectively), and, when predicting mesenteric fat, it improved the R^2 from 0.46 to 0.62 (RSD = 0.26 and 0.23 kg, respectively). The weight of perirenal fat was best predicted by an equation that included both FTUSpr and BW, increasing the R^2 value from 0.84 to 0.88 (RSD = 0.26 and 0.23 kg, respectively) compared to FTUSpr alone. The carcass fat weight was best predicted by an equation with three variables, FTUSpr, BW, and BCSl (R^2 = 0.92, RSD = 0.46 kg). Similarly, non-carcass fat weight was best predicted by an equation with three variables (FTUSpr, HG, and BCSl; R^2 = 0.91, RSD = 0.71 kg). The best equation to predict total fat weight included FTUSpr, BW, and BCSl (R^2 = 0.92, RSD = 1.14 kg).

3.2. Post-Slaughter Measurements

3.2.1. Hot Carcass Weight

Hot carcass weight varied between 18.1 and 30.1 kg with a mean of 24.3 kg (SD = 3.4 kg) (Table 1), corresponding to a mean killing out percentage (100 × HCW/BW) of 42.7 ± 3.0 (data not shown). The values of R^2 for equations using HCW varied from 0.17 for organ fat (RSD = 0.18 kg) to 0.82 for carcass fat (RSD = 0.57 kg), with a value of 0.74 for total fat (RSD = 1.73 kg) (Table 4).

Table 4. Regression equations (y = a + bx), coefficient of determination (R^2), and residual standard deviation (RSD) values for estimate of the different fat depot weights (kg) and the total weight of all fat combined (y variables) based on the post-slaughter measurement values hot carcass weight (HCW), empty body weight (EBW), and lumbar and sternal fat thickness depths measured by a digital caliper (FTDCl and FTDCs, respectively) (x variables).

Item [1]	Intercept ± Standard Error	b ± Standard Error	R^2	RSD	p-Value
Hot carcass weight (HCW)					
OM	−4.00 ± 1.39	0.24 ± 0.06	0.43	0.99	0.001
MES	−0.81 ± 0.31	0.07 ± 0.01	0.58	0.22	<0.001
PR	−1.95 ± 0.65	0.11 ± 0.03	0.42	0.46	0.001
ORG	−0.32 ± 0.25	0.02 ± 0.01	0.17	0.18	0.037
CARC	−3.55 ± 0.80	0.34 ± 0.03	0.82	0.57	<0.001
NC	−5.89 ± 1.83	0.47 ± 0.07	0.63	1.30	<0.001
TOT	−9.44 ± 2.43	0.81 ± 0.10	0.74	1.73	<0.001
Empty body weight (EBW)					
OM	−4.30 ± 1.34	0.13 ± 0.03	0.47	0.95	0.001
MES	−0.86 ± 0.30	0.04 ± 0.01	0.61	0.21	<0.001
PR	−2.08 ± 0.63	0.06 ± 0.01	0.46	0.44	0.001
ORG	−0.42 ± 0.24	0.01 ± 0.01	0.23	0.17	0.012
CARC	−3.20 ± 0.95	0.16 ± 0.02	0.74	0.67	<0.001
NC	−6.44 ± 1.69	0.25 ± 0.03	0.68	1.20	<0.001
TOT	−9.64 ± 2.40	0.42 ± 0.05	0.74	1.70	<0.001
Lumbar fat thickness (FTDCl)					
OM	1.52 ± 1.12	0.16 ± 0.51	0.01	1.31	0.760
MES	0.75 ± 0.29	0.10 ± 0.13	0.02	0.33	0.473
PR	0.72 ± 0.52	0.01 ± 0.24	0.01	0.61	0.976
ORG	−0.10 ± 0.15	0.15 ± 0.07	0.17	0.18	0.036
CARC	3.42 ± 1.11	0.56 ± 0.51	0.05	1.29	0.279
NC	4.65 ± 1.82	0.45 ± 0.83	0.01	2.12	0.591
TOT	8.07 ± 2.86	1.02 ± 1.31	0.02	3.32	0.444
Sternal fat thickness (FTDCs)					
OM	−1.97 ± 1.03	0.17 ± 0.04	0.37	1.03	0.001
MES	0.09 ± 0.29	0.04 ± 0.01	0.28	0.29	0.005
PR	−1.09 ± 0.47	0.08 ± 0.02	0.40	0.47	0.001
ORG	−0.08 ± 0.18	0.01 ± 0.01	0.11	0.18	0.099
CARC	−0.34 ± 0.82	0.22 ± 0.04	0.61	0.82	<0.001
NC	−1.55 ± 1.50	0.32 ± 0.07	0.50	1.51	<0.001
TOT	−1.89 ± 2.20	0.54 ± 0.10	0.57	2.21	<0.001

[1] OM = omental fat; MES = mesenteric fat; PR = perirenal fat; ORG = organ fat (heart, liver, and lungs); CARC = carcass fat; NC = non-carcass fat; TOT, total fat depot (TOT = CARC + NC).

Table 5. Multiple regression equations and coefficient of determination (R^2) and residual standard deviation (RSD) values for estimates of the different fat depot weights (kg) and the total weight of all the fat depots combined (y variables) based on the pre-slaughter measurement values body weight (BW); lumbar and sternal body condition scores (BCSl and BCSs, respectively); heart girth (HG); and lumbar, sternal, and perirenal fat thickness depths measured by ultrasound (FTUSl, FTUSs, and FTUSpr, respectively) (× variables).

Step	Dependent Variable (y) [1]	Independent Variables (x)	Intercept ± Standard Error [2]	b ± Standard Error	R^2	RSD
1	OM	FTUSpr	−7.77 ± 2.85	2.10 ± 0.27	0.79	0.65
2		HG		0.08 ± 0.03	0.85	0.57
1	MES	FTUSpr	−2.43 ± 1.13	0.35 ± 0.11	0.46	0.26
2		HG		0.03 ± 0.01	0.62	0.23
1	PR	FTUSpr	−1.66 ± 0.50	0.97 ± 0.11	0.84	0.26
2		BW		0.02 ± 0.01	0.88	0.23
1	ORG	FTUSl	−0.21 ± 0.15	0.07 ± 0.03	0.09	0.19
2		BW		0.004 ± 0.003	0.40	0.07
1	CARC	FTUSpr	−7.63 ± 1.14	0.80 ± 0.23	0.47	1.09
2		BW		0.09 ± 0.02	0.79	0.70
3		BCSl		2.25 ± 0.47	0.92	0.46
1	NC	FTUSpr	−18.21 ± 3.57	2.59 ± 0.36	0.70	1.25
2		HG		0.19 ± 0.04	0.89	0.76
3		BCSl		1.32 ± 0.75	0.91	0.71
1	TOT	FTUSpr	−15.80 ± 2.85	3.22 ± 0.58	0.62	2.28
2		BW		0.22 ± 0.05	0.86	1.44
3		BCSl		3.79 ± 1.18	0.92	1.14

All regressions are significant at $p < 0.05$. [1] OM = omental fat; MES = mesenteric fat; PR = perirenal fat; ORG = organ fat (heart, liver, and lungs); CARC = carcass fat; NC = non-carcass fat; TOT = total fat depot (TOT = CARC + NC). [2] The intercept is the same within each group of equations predicting the same dependent variable.

3.2.2. Empty Body Weight

Empty body weight mean was 47.6 kg, varying between 36.1 and 59.9 kg (Table 1). The values of R^2 for equations using EBW varied from 0.23 for organ fat (RSD = 0.17 kg) to 0.74 for carcass fat (RSD = 0.67 kg) and total fat (RSD = 1.70 kg) (Table 4).

3.2.3. Digital Caliper Measurements

The mean depths of the fat measured by digital caliper in the lumbar and sternal regions were 2.1 mm (range 1.5–3.6 mm) and 22.1 mm (range 9.8–29.8 mm), respectively (Table 1). The determination coefficients of the equations using FTDCl as a predictor were all extremely low (varying between 0.01 and 0.17) and were not significant ($p > 0.05$, except ORG). Predictions using FTDCs had R^2 values between 0.11 for organ fat (RSD = 0.18 kg) and 0.61 for carcass fat (RSD = 0.82 kg) (Table 4).

3.2.4. Multiple Regressions

The inclusion of FTDCs in in the equation using EBW to predict the weight of the OM fat resulted in an improvement in the accuracy (R^2 value increased from 0.47 to 0.52; Table 6). In contrast, the prediction of the MES fat weight, where EBW was the best single predictor, was not improved by the addition of any other post-slaughter variables. The addition of FTDCs, in combination with EBW, increased the R^2 value from 0.46 to 0.53 (RSD = 0.44 and 0.43 kg, respectively) in the equation to predict PR fat weight and from 0.68 to 0.74 (RSD = 1.20 and 1.12 kg, respectively) in the equation to predict non-carcass fat. For the prediction of organ fat, the equation obtained had a coefficient of determination very low (R^2 = 0.32) with the use of the EBW and FTDCl.

Table 6. Multiple regression equations and coefficient of determination (R^2) and residual standard deviation (RSD) values for estimates of the different fat depot weights (kg) and the total weight of all the fat depots combined (y variables) based on the post-slaughter measurement values hot carcass weight (HCW), empty body weight (EBW), and lumbar and sternal fat thickness depths measured by a digital caliper (FTDCl and FTDCs, respectively) (x variables).

Step	Dependent Variable (y) [1]	Independent Variable(s) (x)	Intercept ± Standard Error [2]	b ± Standard Error	R^2	RSD
1	OM	EBW	−4.46 ± 1.31	0.09 ± 0.03	0.47	0.95
2		FTDCs		0.08 ± 0.05	0.52	0.92
1	MES	EBW	−0.86 ± 0.30	0.04 ± 0.01	0.61	0.21
1	PR	EBW	−2.16 ± 0.60	0.04 ± 0.02	0.46	0.44
2		FTDCs		0.04 ± 0.02	0.53	0.43
1	ORG	EBW	−0.54 ± 0.24	0.01 ± 0.005	0.23	0.17
2		FTDCl		0.11 ± 0.07	0.32	0.16
1	CARC	HCW	−3.83 ± 0.55	0.36 ± 0.07	0.82	0.57
2		FTDCs		0.12 ± 0.02	0.91	0.40
3		EBW		-0.06 ± 0.04	0.92	0.38
1	NC	EBW	−6.70 ± 1.58	0.19 ± 0.04	0.68	1.20
2		FTDCs		0.14 ± 0.06	0.74	1.12
1	TOT	HCW	−10.44 ± 2.00	0.60 ± 0.10	0.74	1.73
2		FTDCs		0.27 ± 0.08	0.83	1.41

All regressions are significant at $p < 0.05$. [1] OM = omental fat; MES = mesenteric fat; PR = perirenal fat; ORG = organ fat (heart, liver, and lungs); CARC = carcass fat; NC = non-carcass fat; TOT = total fat depot (TOT = CARC + NC). [2] The intercept is the same within each group of equations predicting the same dependent variable.

The predictions of carcass and total fat were markedly more precise than those of the internal organs. The addition of the FTDCs with HCW increased the precision from 0.82 to 0.91 (RSD = 0.57 and 0.40 kg, respectively) to predict carcass fat weight (Table 6), while the addition of the FTDCs with HCW increased the precision from 0.74 to 0.83 (RSD = 1.73 and 1.41 kg, respectively) to predict total fat weight (Table 6).

4. Discussion

As mentioned in the method, for practical reasons, i.e., for the lack of an appropriate precision scale suitable for live animals in the site of slaughtering, body weight was measured with an electronic scale immediately after slaughter (blood was collected and weighed). Since this technique avoided the errors associated with animal movement during weighing, it is likely that BW measurement immediately post-mortem was at least as accurate and precise as when carried out on live animals.

Among all body dimension characters evaluated, HG was the most related trait to BW (BW = 1.2 HG − 57.7; R^2 = 0.75; Figure 1). In a recent work, McGregor [16] observed a moderate correlation (R^2 = 0.60) in Angora goats, with a 1 kg increase in live weight for each 1 cm increase in heart girth, which was very similar to the present work. Slippers et al. [17] reported that body weight was highly correlated with heart girth in Nguni goats (R^2 > 0.88). In contrast to what observed for HG, BCS was not a good predictor to estimate live weight (Figure 2), probably because of the moderate correlation between BW and lumbar BCS (r = 0.50) and sternum BCS (r = 0.56). McGregor [16] reported a correlation of 66% between BW and lumbar BCS, corroborating that it is difficult to estimate the BW using BCS in goats. Although the level of precision obtained when using BW to predict weights of fat depots such as organ fat and omental fat was not high (R^2 = 0.21 and 0.28, respectively), a higher precision was achieved when predicting carcass and total fat content (R^2 = 0.58 and 0.55, respectively).

When using the lumbar BCS method, Russel et al. [10] in Scottish Blackface ewes and Teixeira et al. [18] in Rasa Aragonesa ewes obtained R^2 values close to 0.90 for BCS as a predictor of the amount of body fat. However, the distribution of body fat in goats differs appreciably from that in ewes [19]. The data of the present study in Saanen goats confirmed that subcutaneous fat deposits are not highly noticeable in the dorsal region of this species. In fact, according to Hervieu et al. [9], large amounts of accumulated fat are deposited in the sternal region in goats. Although Mendizabal et al. [12] reported that the precision to estimate total fat in Spanish Blanca Celtibérica goats using sternal BCS was much better (R^2 = 0.90) than those achieved using lumbar BCS (R^2 = 0.59), in the present study, sternal BCS did not estimate fat reserves satisfactorily (R^2 < 0.55). Furthermore, in the same region, both ultrasound (R^2 = 0.51) and digital caliper (R^2 = 0.57) had low precision in the estimation of fat reserves. These differences can be attributed, at least in part, to the much greater ranges of BW and BCS (33.0 to 80.5 kg and 0.75 to 4.25, respectively) evaluated by Mendizabal et al. [12] compared to those obtained in the present work (43.6 to 69.4 kg and 1.75 to 3.00, respectively). This is plausible considering the mathematical and statistical approaches used because the regression fit of the model is dependent on the range of the dataset. The utilization of high ranges of BCS is scientifically correct but tends to overestimate the ability of the method to predict the actual body reserves and visceral fat of the animals, since it includes a range of BCS and body reserves values rarely seen in commercial goat flocks (e.g., Eknaes et al. [1] estimated a total body fat and protein content in goats in different stages of lactation raised intensively and extensively lower than that in our experiment, reported in Table 1), while a method to estimate body reserves should work within the values commonly observed in commercial flocks. Another possible explanation is that Saanen goats do not deposit fat in the lumbar or sternal region proportionally to the visceral fat depots.

Among all somatic measurements taken, only heart girth presented a significant correlation with all fat depots. However, the determination coefficients of the equations using heart girth were consistently low, with values ranging between 0.16 for organ fat (RSD = 0.56 kg) and 0.46 for carcass fat (RSD = 0.98 kg). Differently, in Pelibuey ewes, Bautista-Díaz et al. [7] observed that abdominal circumference was the best somatic measurement taken to estimate the weights of carcass fat (R^2 = 0.73), visceral fat (R^2 = 0.64), and total body fat (R^2 = 0.71). In fact, these results confirm that sheep, especially meat breeds, have a higher deposition of fat in the subcutaneous region, whereas dairy goat breeds deposit a major part of fat in the visceral internal cavity [14].

When estimating fat depots using ultrasound, we attained higher levels of precision when measuring the perirenal fat thickness (R^2 values between 0.05 and 0.86) compared to the lumbar region (R^2 values from 0.10 to 0.33) or the sternal region (R^2 values from 0.05 to 0.55). These findings confirmed that the perirenal fat thickness measured with ultrasound can adequately estimate fat reserves in lactating Saanen goats (except ORG fat). In fact, in a previous study carried out on Saanen goats, Härter et al. [15] found that abdominal fat was the main energy reserve and that perirenal fat thickness measured by ultrasound was significantly correlated with BW and renal, omental, and non-carcass fat.

Considering the post-slaughter measurements evaluated in this study, we found that hot carcass weight and empty body weight were the best predictors of the amount of total fat stored by the goats (R^2 = 0.74 and RSD = 1.7 kg, for both). The use of HCW or EBW removes the large effect that differences in gastrointestinal contents, which varied from 5.6 to 12.7 kg, have on BW. Similarly, Mendizabal et al. [2] found that EBW and, especially, HCW were the best post-slaughter predictors of the weights of fat depots in Spanish Blanca Celtibérica goats.

Lumbar fat thickness measured by a digital caliper was the worst predictor of the weights of individual fat depots, with R^2 values lower than 0.2, likely because the very thin layer of fat that lost its firmness after cutting the muscle and, therefore, made measurements difficult. Sternal fat thickness measured by a digital caliper was not a good predictor of fat depots either, although R^2 values were higher (0.11–0.61 range) compared to the lumbar region. In Spanish Blanca Celtibérica goats, Delfa et al. [20] dissected the lumbar and sternal region and found that the fat percentage of the lumbar square joint was only 15% compared to 41% of fat in the sternal triangle joint. Therefore, it is evident that the BCS scales proposed by Hervieu et al. [9] for Alpine and Saanen goats need to be re-evaluated. A BCS method based on body palpations is difficult to adopt in goats because of a lack of subcutaneous adipose tissue in this species. Firstly, it would be necessary to evaluate if there is a correlation between the fat located in the lumbar or sternal region and the total fat of the animals. If findings show a high correlation, this could mean that the BCS methods can be used to predict the body fat of dairy goats, although some adjustments might still be necessary. However, if studies show that the fat thickness located in the lumbar and sternal region is not highly correlated with the total fat, mainly located in the visceral region, then new methods should be developed. Hervieu et al. [9] confirmed that there is a significant correlation between the fat scores given by BCS and their respective fat fractions (in the lumbar and sternal regions). However, the authors did not evaluate whether this correlation also regarded the body composition as a whole.

On the basis of the different regressions of the pre-slaughter measurements tested for each fat depot, we found that FTUSpr yielded the most precise estimates of body fat in lactating Saanen goats, with the exception of organ fat depot estimation, and the addition of BW and BCSl substantially improved the precision of the estimates of total body fat (R^2 increasing from 0.62 to 0.92). We hypothesized that there is a wide variability of body fat on equal BCS (low precision of estimation). However, the estimation of carcass and total fat could be improved if BW and BCSl were added to FTUSpr, as shown in Table 5. These results suggest that goats were of different sizes (large and small) and possibly in some cases with similar BCS. In addition, BW was a discrete indicator of carcass and total fat and was moderately accurate indicator for MES fat; BCSl was a particularly good predictor of carcass fat and FTUSpr predicted with high accuracy omental, perirenal, and non-carcass fat. Therefore, the addition of these three variables (BW, BCSl, and FTUSpr) seems to be complementary in predicting total fat.

On the other hand, on the basis of the multiple regression analysis using post-slaughter measurements, we found that EBW was the first variable and gave the best predictions of OM, MES, PR, ORG, and NC fat depots, whereas HCW was the first variable in CARC and TOT fat. The addition of FTDCs as a second variable was helpful when estimating the fat reserves in OM, PR, NC, and TOT fat (R^2 increasing from 0.74 to 0.83). These results agree with those obtained by Mendizabal et al. [12], who found that HCW and EBW were the most used post-slaughter variables to predict fat depots in Spanish Blanca Celtibérica goats, confirming the importance of these measurements to predict fat depots in goats.

When the main results obtained with the multiple regression analysis are considered, the recommended equations to be used at field level, when ultrasound is not available, might be summarized as

(1) OM: $4.79 + 0.13 \times \text{BW} + 1.57 \times \text{BCSl} - 0.16 \times \text{WH} - 0.15 \times \text{RL}$ ($R^2 = 0.55$);
(2) MES: $1.57 + 0.05 \times \text{BW} - 0.04 \times \text{WH}$ ($R^2 = 0.56$);
(3) PR: $2.79 + 0.08 \times \text{BW} - 0.07 \times \text{WH} - 0.09 \times \text{PW}$ ($R^2 = 0.48$);
(4) ORG: $1.21 + 0.02 \times \text{BW} - 0.06 \times \text{CD}$ ($R^2 = 0.60$);
(5) CARC: $-3.31 + 0.12 \times \text{BW} + 3.19 \times \text{BCSl} - 0.07 \times \text{RH} - 0.12 \times \text{SW}$ ($R^2 = 0.87$);
(6) NC: $1.65 + 0.25 \times \text{BW} + 2.61 \times \text{BCSl} - 0.16 \times \text{WH} - 0.31 \times \text{PW}$ ($R^2 = 0.69$);
(7) TOT: $-4.21 + 0.36 \times \text{BW} + 6.12 \times \text{BCSl} - 0.21 \times \text{WH} - 0.38 \times \text{SW}$ ($R^2 = 0.79$).

5. Conclusions

Fat thickness measured with ultrasound in the perirenal region was the best pre-slaughter measurement for estimating fat reserves in lactating Saanen goats, whereas empty body weight and hot carcass weight were the best post-slaughter predictors for estimating fat reserves. Body condition score could be a useful tool, but it seems that it needs to be re-evaluated to predict adequately fat depots in lactating Saanen goats. The best variable to predict carcass and total fat content was hot carcass weight, but methodologies able to predict weights of fat reserves in live animals are preferable for practical and economic reasons.

Author Contributions: Conceptualization, L.S.K., M.D., G.M., M.F.L. and A.C.; methodology, L.S.K., R.G.C., M.D. and A.C.; software, L.S.K.; validation, L.S.K., R.G.C. and A.C.; formal analysis, L.S.K.; investigation, L.S.K., S.N.R.K., M.A., M.F.L., M.D., A.H.D.F., G.M. and A.C.; resources, A.C.; data curation, L.S.K.; writing—original draft preparation, L.S.K.; writing—review and editing, L.S.K., M.F.L., R.G.C., M.L., S.N.R.K., M.A., M.D., G.M., A.H.D.F. and A.C.; visualization, M.L., A.H.D.F. and R.G.C.; supervision, A.C.; project administration, A.C. and M.D.; funding acquisition, A.C. All authors have read and agreed to the published version of the manuscript.

Funding: This research was funded by the REGIONE AUTONOMA DELLA SARDEGNA, Legge Regionale 7 agosto 2007, n. 7, annualità 2012 for the project "Modulazione della partizione dell'energia alimentare fra latte e riserve corporee in ovini e caprini".

Institutional Review Board Statement: The study was conducted according to the European Union and Italian regulations on animal welfare, and all measurements were taken by personnel previously trained and authorized by the institutional authorities managing ethical issues at the University of Sassari and AGRIS. Experimental procedures with animals (goats) were approved by the Animal Care and Use Committee of the University of Sassari and Agris, Italy (CIBASA 10.12.2014).

Data Availability Statement: The data presented in this study are available on request from the corresponding author.

Acknowledgments: The authors thank the technical assistance of Gesumino Spanu, Antonio Fenu, Antonio Mazza, Roberto Rubattu, and Alessandra Marzano from the University of Sassari (Sassari, Italy) and the technical support of Salvatore Puggioni, Savatore Contini, Stefano Picconi, and Roberto Mura from the Argis Sardegna (Sassari, Italy). The authors are grateful to CNPq (Conselho Nacional de Desenvolvimento Científico e Tecnológico) for giving a doctoral scholarship to the first author and to CAPES (Coordenação de Aperfeiçoamento de Pessoal de Nível Superior) for giving an internship ("Sandwich Program") to the first author in Italy (BEX: 1865/14-5).

Conflicts of Interest: The authors declare no conflict of interest.

References

1. Eknæs, M.; Kolstad, K.; Volden, H.; Hove, K. Changes in body reserves and milk quality throughout lactation in dairy goats. *Small Rumin. Res.* **2006**, *63*, 1–11. [CrossRef]
2. Ngwa, A.T.; Dawson, L.J.; Puchala, R.; Detweiler, G.; Merkel, R.C.; Wang, Z.; Tesfai, K.; Sahlu, T.; Ferrell, C.L.; Goetsch, A.L. Effects of stage of lactation and dietary forage level on body composition of Alpine dairy goats. *J. Dairy Sci.* **2009**, *92*, 3374–3385. [CrossRef] [PubMed]
3. Cannas, A.; Tedeschi, L.O.; Fox, D.G.; Pell, A.N.; Van Soest, P.J. A mechanistic model for predicting the nutrient requirements and feed biological values for sheep. *J. Anim. Sci.* **2004**, *82*, 149–169. [CrossRef] [PubMed]
4. Silva, T.S.; Chizzotti, M.L.; Busato, K.C.; Rodrigues, R.T.S.; da Silva, I.F.; Queiroz, M.A.Á. Indirect methods for predicting body composition of Boer crossbreds and indigenous goats from the Brazilian semiarid. *Trop. Anim. Health Prod.* **2015**, *47*, 1217–1220. [CrossRef] [PubMed]
5. Akdag, F.; Teke, B.; Meral, Y.; Arslan, S.; Ugurlu, M. Prediction of carcass composition by ultrasonic measurement and the effect of region and age on ultrasonic measurements. *Small Rumin. Res.* **2015**, *133*, 82–87. [CrossRef]
6. Scholz, A.M.; Bünger, L.; Kongsro, J.; Baulain, U.; Mitchell, A.D. Non-invasive methods for the determination of body and carcass composition in livestock: Dual-energy X-ray absorptiometry, computed tomography, magnetic resonance imaging and ultrasound: Invited review. *Animal* **2015**, *9*, 1250–1264. [CrossRef] [PubMed]
7. Bautista-Díaz, E.; Salazar-Cuytun, R.; Chay-Canul, A.J.; Garcia Herrera, R.A.; Piñeiro-Vázquez, Á.T.; Magaña Monforte, J.G.; Tedeschi, L.O.; Cruz-Hernández, A.; Gómez-Vázquez, A. Determination of carcass traits in Pelibuey ewes using biometric measurements. *Small Rumin. Res.* **2017**, *147*, 115–119. [CrossRef]
8. Cam, M.A.; Olfaz, M.; Soydan, E. Possibilities of using morphometrics characteristics as a tool for body weight prediction in Turkish hair goats (Kilkeci). *Asian J. Anim. Vet. Adv.* **2010**, *5*, 52–59. [CrossRef]
9. Hervieu, J.; Morand-Fehr, P.; Schmidely, P.; Fedele, V.; Delfa, R. Mesures anatomiques permettant d'expliquer les variations des notes sternales, lombaires et caudales utilisées pour estimer l'état corporel des chèvres laitières. In *Etat Corporel des Brebis et Chèvres, Proceedings of CIHEAM, Zaragoza*; Purroy, A., Ed.; 1991; pp. 43–56. Available online: https://om.ciheam.org/article.php?IDPDF=92605094 (accessed on 18 May 2021).
10. Russel, A.J.F.; Doney, J.M.; Gunn, R.G. Subjective assessment of body fat in live sheep. *J. Agric. Sci. (Camb.)* **1969**, *72*, 451–454. [CrossRef]
11. Ngwa, A.T.; Dawson, L.J.; Puchala, R.; Detweiler, G.; Merkel, R.C.; Tovar-Luna, I.; Sahlu, T.; Ferrell, C.L.; Goetsch, A.L. Urea space and body condition score to predict body composition of meat goats. *Small Rumin. Res.* **2007**, *73*, 27–36. [CrossRef]
12. Mendizabal, J.A.; Delfa, R.; Arana, A.; Purroy, A. A comparison of different pre- and post-slaughter measurements for estimating fat reserves in Spanish Blanca Celtibérica goats. *Can. J. Anim. Sci.* **2010**, *90*, 437–444. [CrossRef]
13. Teixeira, A.; Joy, M.; Delfa, R. In vivo estimation of goat carcass composition and body fat partition by real-time ultrasonography. *J. Anim. Sci.* **2008**, *86*, 2369–2376. [CrossRef] [PubMed]

14. Colomer-Rocher, F.; Kirton, A.H.; Mercer, G.J.K.; Duganzich, D.M. Carcass composition of New Zealand Saanen goats slaughtered at different weights. *Small Rumin. Res.* **1992**, *7*, 161–173. [CrossRef]
15. Härter, C.J.; Silva, H.G.O.; Lima, L.D.; Castagnino, D.S.; Rivera, A.R.; Neto, O.B.; Gomes, R.A.; Canola, J.C.; Resende, K.T.; Teixeira, I.A.M.A. Ultrasonographic measurements of kidney fat thickness and Longissimus muscle area in predicting body composition of pregnant goats. *Anim. Prod. Sci.* **2014**, *54*, 1481–1485. [CrossRef]
16. McGregor, B.A. Relationships between live weight, body condition, dimensional and ultrasound scanning measurements and carcass attributes in adult Angora goats. *Small Rumin. Res.* **2017**, *147*, 8–17. [CrossRef]
17. Slippers, S.C.; Letty, B.A.; de Villiers, J.F. Prediction of the body weight of Nguni goats. *S. Afr. J. Anim. Sci.* **2000**, *30* (Suppl. 1), 127–128. [CrossRef]
18. Teixeira, A.; Delfa, R.; Colomer-Rocher, F. Relationships between fat depots and body condition score or tail fatness in the Rasa Aragonesa breed. *Anim. Sci.* **1989**, *49*, 275–280. [CrossRef]
19. Colomer-Rocher, F.; Morand-Fehr, P.; Kirton, A.H. Standard methods and procedures for goat carcass evaluation, jointing and tissue separation. *Livest. Prod. Sci.* **1987**, *17*, 149–159. [CrossRef]
20. Delfa, R.; González, C.; Teixeira, A.; Gosalvez, L.F.; Tor, M. Relationships between body fat depots, carcass composition, live weight and body condition scores in Blanca Celtibérica goats. In *Body Condition of Sheep and Goats: Methodological Aspects and Applications, Proceeding of CIHEAM, Zaragoza*; Purroy, A., Ed.; 1995; pp. 109–119. Available online: https://om.ciheam.org/article.php?IDPDF=96605600 (accessed on 18 May 2021).

 animals

MDPI

Article

Health Status and Stress in Different Categories of Racing Pigeons

Marjan Kastelic [1], Igor Pšeničnik [2], Gordana Gregurić Gračner [3], Nina Čebulj Kadunc [4], Renata Lindtner Knific [2], Brigita Slavec [2], Uroš Krapež [2], Aleksandra Vergles Rataj [5], Olga Zorman Rojs [2], Barbara Pulko [2], Maša Rajšp [2], Nina Mlakar Hrženjak [2] and Alenka Dovč [2,*]

1 BUBA d.o.o.-Veterinary Clinic and Pet Supply Store, Rožna dolina 5, 1290 Grosuplje, Slovenia; vet.kastelic@siol.net

2 Institute for Poultry, Birds, Small Mammals and Reptiles, Veterinary Faculty, University of Ljubljana, Gerbičeva 60, 1000 Ljubljana, Slovenia; igorpsenicnik88@gmail.com (I.P.); renata.lindtnerknific@vf.uni-lj.si (R.L.K.); brigita.slavec@vf.uni-lj.si (B.S.); uros.krapez@vf.uni-lj.si (U.K.); olga.zorman-rojs@vf.uni-lj.si (O.Z.R.); barbara.pulko@gmail.com (B.P.); masa.rajsp95@gmail.com (M.R.); nina.mlakar@vf.uni-lj.si (N.M.H.)

3 Department of Animal Hygiene, Behavior and Animal Welfare, Faculty of Veterinary Medicine, University of Zagreb, Heinzelova 55, 10000 Zagreb, Croatia; ggracner@gmail.com

4 Institute of Preclinical Sciences, Veterinary Faculty, University of Ljubljana, Gerbičeva 60, 1000 Ljubljana, Slovenia; nina.cebulj.kadunc@vf.uni-lj.si

5 Institute of Microbiology and Parasitology, Veterinary Faculty, University of Ljubljana, Gerbičeva 60, 1000 Ljubljana, Slovenia; aleksandra.verglesrataj@vf.uni-lj.si

* Correspondence: alenka.dovc@vf.uni-lj.si; Tel.: +386-1-477-9250

Simple Summary: Corticosterone is the most important "stress" hormone in birds. Stress response is influenced by different factors, such as phylogeny, feed supply, age, body condition, health status, climate, predators. Pigeon races over long distances, 500 km or more, can lead to the "exploitation" of animals if not strictly regulated and observed, jeopardizing their welfare status. Animals should be in good health and body condition, and health monitoring must be implemented. In stressful situations such as races, the possibility of infection increases. Clinically asymptomatic infections can flare up later in the breeding season and can cause high offspring mortality. For example, infections with circoviruses are particularly important because of their ability to weaken the immune system. The purpose of this work is to identify the critical stress points during the active training season of racing pigeons for the improvement of their condition and the preservation of their welfare during races. The aim of our study was to determine the serum corticosterone levels in different categories of racing pigeons exposed to severe stress factors. At the time of racing, some parameters of stress, including environmental factors, or the presence of infectious diseases or parasites, were recorded. It was found that participation in the race significantly increased serum corticosterone levels and remained high even one month after the race. Therefore, training and races should be properly managed and planned.

Abstract: The influence of different stress parameters in racing pigeon flocks, such as the presence of diseases and environmental conditions at the time of the races, were described. A total of 96 racing pigeons from 4 pigeon flocks were examined, and health monitoring was carried out. No helminth eggs and coccidia were found. *Trichomonas* sp. was confirmed in subclinical form. Paramyxoviruses and avian influenza viruses were not confirmed, but circovirus infections were confirmed in all flocks. *Chlamydia psittaci* was confirmed in one flock. Blood samples were collected, and HI antibody titers against paramyxoviruses before and 25 days after vaccination were determined. To improve the conditions during racing and the welfare of the pigeons, critical points were studied with regard to stress factors during the active training season. Serum corticosterone levels were measured in the blood serum of four different categories of pigeons from each flock. Corticosterone levels were almost twice as high in pigeons from the category that were active throughout the racing season, including medium- and long-distance racing, compared to the other three categories that were not racing actively. Within five hours of the finish of a race, the average serum corticosterone level was 59.4 nmol/L in the most physically active category. The average serum corticosterone level in this category remained at 37.5 nmol/L one month after the last race.

Citation: Kastelic, M.; Pšeničnik, I.; Gregurić Gračner, G.; Čebulj Kadunc, N.; Lindtner Knific, R.; Slavec, B.; Krapež, U.; Vergles Rataj, A.; Zorman Rojs, O.; Pulko, B.; et al. Health Status and Stress in Different Categories of Racing Pigeons. *Animals* **2021**, *11*, 2686. https://doi.org/10.3390/ani11092686

Academic Editors: María-Luz García and María-José Argente

Received: 11 August 2021
Accepted: 8 September 2021
Published: 13 September 2021

Keywords: *Columba livia domestica*; infectious diseases; serum corticosterone; welfare

1. Introduction

Pigeons are naturally gifted with the ability to find "home" from distant places, relying on abilities beyond memory. Humans discovered this capability by accident and then began to breed them selectively [1]. The breeding of domestic pigeons is one of the most rapidly developing areas of the animal world. The great races attract breeders from dozens of countries every year [1,2].

Shows and races with pigeons often lead to exploitation, injury, and death if not strictly regulated and observed [3]. Good body condition and stress resistance usually depend on feed supply, but health status throughout the year is also of great importance. Transport to the race and environmental factors during the race can be very stressful for pigeons, and sometimes a high number of pigeons perish. Warzecha [4] indicated that these problems can affect many animals, and government veterinarians should be actively involved in these activities.

In many species, including birds, rodents, reptiles, and amphibians, corticosterone (CORT) is the main glucocorticoid involved in the regulation of energy, immune, and stress responses. Responses to chronic stressor exposure and chronically elevated glucocorticoids include reduced growth, immunocompetence, reproduction, and survival. The effects of elevated glucocorticoids have an influence on survival, physiological, behavioral, reproductive, and intergenerational responses in wild vertebrates [5,6].

CORT is the major "stress" hormone in birds, with short-term changes mediating adaptive behavioral and metabolic responses to adverse environmental events (increased effort, transport, predators, fasting, and climatic conditions) as well as health status [7–9].

Because of diurnal rhythms of plasma corticosterone levels, the time of sampling is important [10,11]. However, in birds, the maximum level of CORT naturally occurs at daybreak [10]. Lumeij et al. [12] determined that baseline serum corticosterone concentrations in racing pigeons varied from less than 0.2 to 1.24 μg/dL (5.77–35.77 nmol/L) after 24 h of rest.

Various stressors can also occur during transport to and during the race, especially if it is not strictly regulated. Inappropriate environmental factors during the race, such as air temperature, relative humidity, air velocity, and magnetic radiation, can lead to additional stress [13,14].

Stressful situations during races also increase the possibility of infections. Clinically asymptomatic infections can flare up later in the breeding season and cause great losses of offspring [15]. The major bacterial pathogens in a racing flock are *Salmonella typhimurium* var. Copenhagen, *Escherichia coli*, and a group of bacteria that cause chronic respiratory disease and lead to poor performance, mostly caused by *Chlamydia psittaci* (CP), *Pasteurella*, and *Mycoplasma* species. In addition, some fungi and yeasts (e.g., *Aspergillus*, *Candida*, *Cryptococcus*), endoparasites (e.g., *Eimeria*, *Haemoproteus*, *Trichomonas*), and ectoparasites (e.g., Mallophaga, Hippoboscid pigeon flies) molested birds, stressing them and causing various diseases [16,17]. Zigo [2] found an increased incidence of endoparasite infestation and respiratory syndrome at the time of racing.

Diseases caused by viruses, whether clinical or subclinical, play an important role in the occurrence of stress. The most often detected groups of viruses include paramyxoviruses (avian paramyxovirus 1) (APMV-1), circoviruses (pigeon circoviruses) (PiCV), adenoviruses, herpesviruses, and poxviruses [18]. Avian influenza viruses (AIV) are less often detected in pigeons. They play a minor role in the epidemiology of H5 influenza. In pigeons, influenza A virus of subtype H7 can cause conjunctivitis, tremor, paresis of wings and legs, and wet droppings. Nevertheless, free-flying domestic pigeons can act as mechanical vectors and vehicles for long-distance transmission of any influenza A virus, if plumage or feet are contaminated [19,20].

Among viruses, PiCV is the most frequently detected among pigeons [21], which is important due to its ability to weaken the immune system. The main consequences of PiCV infection are atrophy and other pathophysiological changes to organs of the immune system (e.g., bursa of Fabricius, thymus, spleen gut-associated lymphoid tissue, bronchus-associated lymphoid tissue, bone marrow, liver, kidney, larynx, trachea, lung, small and large intestine, pancreas). It has been established that infection with pigeon circovirus leads to apoptosis of lymphocytes. For the reasons mentioned above, PiCV is considered as a potential immunosuppressive agent [22]. PiCV infections are capable of predisposing birds to concomitant infections with other pathogens [23].

The aim of this study was to assess CORT levels in different categories of racing pigeons exposed to severe stress factors in order to determine critical stress points during the active season, improve conditions during racing flights and improve pigeon welfare by tracking certain stress parameters, such as the presence of infectious diseases or parasites, and determining environmental factors during the racing season.

2. Materials and Methods

Four flocks of racing pigeons from different breeders were included in the study. Each pigeon flock consisted of 100 to 150 parent racing pigeons. All flocks had similar husbandry conditions and were fed with diets produced by the same manufacturer. Active pigeons, which participated in training and races, always flight together. Transport to the trainings and games was also similar.

2.1. Pigeons; Sampling

A total of 96 racing pigeons from 4 breeders (24 per breeder) were examined for various stress parameters during the racing season. Four different categories (6 pigeons from each group) were compared in each pigeon flock. The first category/group (G1) consisted of sexually mature breeding pigeons not included in training or races. The second group (G2) contained young pigeons (less than 1-year-old) that did not participate in training. The third group (G3) contained pigeons that participated in training but not in medium- or long-distance flights. In the fourth group (G4), racing pigeons used in training and on the medium- and long-distance flights were included. These birds were active throughout the racing season. The pigeons from G3 and G4 were 2 to 7 years old. All the pigeons had identification rings.

A total of 24 pigeons from G4 group (6 from each flock) were tested for CORT levels twice: within 3 to 5 h after returning from the last race (G4a) and 30 days after the last race in the season (G4b). Pigeons from all categories were also tested for CORT concentration in both samplings.

To cheque the health status before the last race, the pigeons were clinically examined, and various samples were taken for further laboratory analyses. Common fecal samples were collected for each group separately (N = 16, 4 from each flock, 1 for each group) for intestinal parasites testing. Samples were transported to the laboratory in transport bags at 4 °C. Additionally, 6 oropharyngeal samples were collected for *Trichomonas* sp. testing per group. Transport of samples to the laboratory was carried out in transport bags at room temperature, and samples were examined within 6 h.

Cloacal and oropharyngeal samples (Copan swabs, Brescia, Italy) and blood samples (blood collection tubes, BD Microtainer®, SST™, Monroe, LA, USA) were collected from each bird individually before (G1, G2, G3) the last race and immediately after the race was over (G4). In all categories/groups (G1, G2, G3, and G4), 6 cloacal and 6 oropharyngeal samples were collected for PiCV, APMV, AIV, and CP determination. Additionally, 6 oropharyngeal samples (for each group) for *Trichomonas* sp. and 6 pools of feces (for each group) for endoparasites were collected. Blood samples in a volume of 0.5 mL were obtained by venipuncture of the ulnar cutaneous vein and collected in microtainer tubes using a serum separator (Becton Dickinson, Heidelberg, Germany). Swabs, feces, and blood samples were transported to the laboratory at 4 °C. Swabs were stored at −20 °C.

A hemagglutination inhibition assay (IHA) was performed within 48 h after collection. Serum was used to determine HI antibody titers against paramyxoviruses and the immunity status of the presumably vaccinated pigeons. The first sampling results did not show satisfactory protection. Thus, all 4 flocks were revaccinated 5 days after the last race was finished. The Chevivac-P200 vaccine (Chevita GmbH, Pfaffenhofen, Germany) was used for vaccination. The vaccine was administered strictly subcutaneously dorsally in the neck toward the tail but not immediately behind the head, according to the manufacturer's instructions. The effect and responsiveness of vaccination were assessed after 25 days (i.e., 30 days after the last race) and is reported in the text as group G4b.

2.2. Laboratory Tests

Cloacal and oropharyngeal swabs were used for molecular detection of pathogenic viruses and bacteria. Swabs were vortexed individually in 2 mL of PBS for 2 min, and 100 μL aliquots of each swab were pooled to produce 300 μL samples for DNA and RNA extraction. Pools were prepared in sterile PBS from 3 samples of cloacal or oropharyngeal swabs collected from each pigeon group (G1, G2, G3, and G4; 2 pools for each group) and for each flock separately (N = 64: 32 cloacal and 32 oropharyngeal pools). Total DNA and RNA were extracted from 140 μL of the pooled samples using the QIAamp Viral RNA Mini Kit (Qiagen, Hilden, Germany) according to the manufacturer's instructions. Previously published molecular methods were used to detect pathogens in the samples collected in the study: PiCV [24], APMV-1 [25], AIV [26] and CP [27]. A species-specific real-time PCR assay was used for further determination of CP [28].

Pooled fecal samples (for each group) were examined for the presence of endoparasites using the flotation and sedimentation method [29].

Trichomonas sp. was detected microscopically in freshly prepared wet mounts. If no trichomonas was present in the observed sample, a drop of iodine solution was added, and the sample was re-examined.

IHA for HI antibody titers against APMV-1 was performed as described in a previous study [30].

CORT was measured in serum using a commercial enzyme immunoassay Corticosterone ELISA (Demeditec Diagnostics GmbH, Kiel, Germany) according to the manufacturer's instructions.

2.3. Last Race

The last race was from Timisoara, Romania, to Ljubljana and its surroundings, Slovenia, 155 pigeons participated. The average flight distance for pigeons from G4 was 504.91 km. The pigeons were released at 12:20 on 9 July 2017 (18°45′43″ E, 45°49′03″ N), and the race was completed at 6:41 on 10 July 2017. Speed results were calculated based on the speed of the first 5 pigeons. According to the data provided by breeders through internet applications [31] the weather was clear, and the air temperature was very high (34 °C at 2 m above the ground).

During the last race, an animal welfare expert followed the procedures during transport, and he was present with the driver.

2.4. Data Analysis

After data collection, mean values of serum corticosterone levels in different categories of racing pigeons were calculated. Next, non-parametric tests were applied to find possible statistically significant differences between individual groups (Mann–Whitney test, $p < 0.05$), and in the case of G4a and G4b groups, the differences immediately and 30 days after the race for the same group of pigeons (Wilcoxon test, $p < 0.05$). For those analyses, the Statistical Package for the Social Sciences, version 25 was utilized.

3. Results

3.1. Health Status of the Pigeon Flocks

All the flocks were examined before the last race started. No helminth eggs were found in the fecal samples. Examination of the beak cavity confirmed the presence of *Trichomonas* sp. in all four flocks. However, none of them was found to have diarrhea or debris in the beak cavity. In all categories, the pigeons were clinically healthy and showed no presence of ectoparasites.

The presence of viruses (PiCV, APMV-1, AIV) and *Chlamydia psittaci* was checked by molecular methods in cloacal and oropharyngeal samples.

PiCV was detected in cloacal and oropharyngeal samples in all breeders but not in all categories. Among the adult breeding group (G1) and racing pigeons included in training and in races (G4), the virus was detected in two flocks. Among young pigeons (from G2 and G3), the virus was detected in three flocks. The results for PiCV are shown in Table 1.

Table 1. The presence of PiCV in individual flocks and categories.

Flock	Category Inside the Flock [1]	Circovirus Cloacal Sample	Circovirus Oropharyngeal Sample
Flock 1	G1	negative	negative
	G2	negative	negative
	G3	positive	positive
	G4	negative	negative
Flock 2	G1	positive	negative
	G2	positive	positive
	G3	negative	positive
	G4	positive	positive
Flock 3	G1	positive	positive
	G2	positive	positive
	G3	positive	positive
	G4	negative	negative
Flock 4	G1	negative	negative
	G2	positive	positive
	G3	negative	negative
	G4	positive	negative

[1] G1: sexually mature breeding pigeons; G2: young pigeons, less than one year old, not participating in training; G3 pigeons participating in training but not in medium- or long-distance flights; G4: racing pigeons participating in training on medium- and long-distance flights.

APMV-1 and AIV were not detected. CP was confirmed in one cloacal sample in flock 2.

To determine paramyxovirus immune status in the flocks, sera samples were examined by the IHA for HI-antibody titers against APMV-1 (Table 2).

3.2. Serum Corticosterone Levels in Different Categories of Racing Pigeons

Within five hours after the race was finished, the average CORT level in G4a was 59.4 nmol/L, almost two-fold higher compared to those in the other three categories (G1, G2, and G3). In the same category, one month after the last race (G4b), CORT levels remained higher in two flocks compared to levels before the final race (Tables 3 and 4).

Pairwise comparisons showed statistically significant differences for the following groups: G1 vs. G3, G2 vs. G3, G1 vs. G4a, G2 vs. G4a, and G3 vs. G4a (Table 5). Although there was no statistically significant difference found between G4a vs. G4b in the total sample, some differences emerged after we made comparisons for the individual flock separately (Table 5). Namely, significant differences in the measurements emerged for flocks 2 and 4 (in both cases $Z = -2.201$, $p = 0.028$).

Table 2. HI-antibody titers (log[2]) against APMV-1.

Flock	Category Inside the Flock [1]	Average of Antibody Titer (Range)
Flock 1	G1	1.7 (0–4)
	G2	0.0 (0)
	G3	0.0 (0)
	G4a	2.3 (0–4)
	G4b	5.3 (4–7)
Flock 2	G1	4.0 (1–7)
	G2	0.0 (0)
	G3	2.3 (0–7)
	G4a	0.7 (0–4)
	G4b	1.2 (0–4)
Flock 3	G1	1.8 (0–6)
	G2	0.0 (0)
	G3	0.7 (0–4)
	G4a	0.0 (0)
	G4b	0.0 (0)
Flock 4	G1	2.7 (0–5)
	G2	0.0 (0)
	G3	1.3 (0–2)
	G4a	0.8 (0–4)
	G4b	1.2 (0–3)
All flocks together	G1	3.5 (0–7)
	G2	0.0 (0)
	G3	1.7 (0–7)
	G4a	1.1 (0–4)
	G4b	2.2 (0–7)

[1] G1: sexually mature breeding pigeons; G2: young pigeons, less than one-year-old, not participating in training; G3 pigeons participating in training but not in medium- or long-distance flights; G4a: racing pigeons participating in training on medium- and long-distance flights; G4b: racing pigeons participating in training on medium- and long-distance flights 25 days after immunization (i.e., 30 days after the last race). Pigeons from G4a and G4b were the same pigeons.

Table 3. Serum CORT levels in different groups of pigeons in all flocks together.

Category/Group [1]	Corticosterone Level nmol/L (SE) [2]
G1	33.8 (6.04)
G2	28.3 (3.92)
G3	16.9 (3.78)
G4a	59.4 (10.62)
G4b	37.5 (7.89)

[1] G1: sexually mature breeding pigeons; G2: young pigeons, less than one-year-old, they do not participate in training; G3 pigeons participated in training but not in medium- or long-distance flights; G4a: racing pigeons participated in training on the medium- and long-distance flights—tested within three to five hours after returning from the last race; G4b: racing pigeons participated in training on the medium- and long-distance flights tested 25 days after immunization (i.e., 30 days after the last race). Pigeons from G4a and G4b were the same pigeons.
[2] SE: Standard error of the mean.

3.3. *Results Achieved in the Last Race*

Transport from Ljubljana to Timisoara was controlled, and welfare was provided by using specially adapted trailers for pigeons. Climatic conditions were appropriate, and pigeons were observed every six hours, and the water supply was checked.

The average speed of the first five pigeons was 1037.94 m/min in flock 1, 1042.43 m/min in flock 2, 855.13 m/min in flock 3, and 1158.74 m/min in flock 4.

Table 4. Serum CORT levels in different groups of pigeons are presented separately by individual flocks.

Flock (CORT Average nmol/L)	Category Inside the Flock [1]	Corticosterone Level nmol/L (SE) [3]
Flock 1 (40.50/6.72) [2]	G1	36.5 (11.31)
	G2	34.3 (4.86)
	G3	9.1 (1.50)
	G4a	65.5 (18.34)
	G4b	57.1 (20.73)
Flock 2 (37.92/5.19) [2]	G1	42.3 (7.67)
	G2	31.8 (8.10)
	G3	25.4 (14.04)
	G4a	69.0 (10.77)
	G4b	21.1 (7.15)
Flock 3 (44.02/8.85) [2]	G1	43.1 (18.78)
	G2	32.6 (10.90)
	G3	13.0 (3.99)
	G4a	66.0 (34.53)
	G4b	65.4 (13.53)
Flock 4 (18.31/3.99) [2]	G1	13.2 (3.79)
	G2	14.5 (4.84)
	G3	20.1 (3.94)
	G4a	37.3 (17.23)
	G4b	6.5 (1.52)

[1] G1: sexually mature breeding pigeons; G2: young pigeons, less than one-year-old, they do not participate in training; G3 pigeons participated in training but not in medium- or long-distance flights; G4a: racing pigeons participated in training on the medium- and long-distance flights—tested within three to five hours after returning from the last race; G4b: racing pigeons participated in training on the medium- and long-distance flights tested 25 days after immunization (i.e., 30 days after the last race was finished). Pigeons from G4a and G4b were the same pigeons; [2] In the average estimation of the CORT levels immediately after the last race, only G4b was not included. [3] SE: Standard error of the mean.

Table 5. Pairwise comparisons of categories (groups).

	Mann-Whitney Test									Wilcoxon Test
Sig.	G1 vs. G2	G1 vs. G3	G2 vs. G3	G1 vs. G4a	G2 vs. G4a	G3 vs. G4a	G1 vs. G4b	G2 vs. G4b	G3 vs. G4b	G4a vs. G4b
Z	−0.196	−2.753	−2.784	−2.052	−2.567	−4.228	−0.454	−0.227	−1.588	−1.867
p	0.845	0.006	0.005	0.040	0.010	<0.001	0.650	0.821	0.112	0.062

4. Discussion

4.1. Health Status of Pigeon Flocks

Pigeons are exposed to various stressors during the racing season. Increased stress during transport to the race and the race itself is an important factor that significantly affects their health status [2]. In stressful situations, the possibility of infection increases [13]. Clinically asymptomatic infections may flare up later in the breeding season and lead to major losses in the flock. Therefore, regular monitoring of flock health status is very important and should be conducted before and after the active season. In our study, antibodies against APMV-1 were assessed, and the presence of certain pathogens (PiCV, APMV-1, AIV, CP) that could cause disease or act as a stress factor in its subclinical form was investigated. It is also important to know the epidemiological situation of these pathogens in rural areas and surrounding countries. Pathogens, such as PiCV [32], APMV-1 [33,34], adenoviruses [34,35], AIV [34], CP [31,34], and *Trichomonas* sp. [36], were confirmed in different categories of pigeons in Slovenia. Transmission of the above-mentioned pathogens from feral pigeons to racing pigeons has been frequently noted in clinical practice.

The specificity of pigeon training and racing significantly impedes the principles of biosecurity [37]. Breeders in Slovenia train and race together, and, therefore, the flocks have closer contact, which could lead to transmission of viral pathogens (e.g., PiCV, APMV-1), bacterial pathogens (e.g., *Chlamydia psittaci*), or parasites (e.g., *Trichomonas* sp.). Another critical biosecurity issue is contact between racing pigeons and feral pigeons. Contact commonly occurs during long race flights, when pigeons have to rest, drink, and feed; these situations could result in the transmission of infections from feral pigeons.

In a previous study, 74.3% of cloacal and 54.1% of oropharyngeal samples collected from feral pigeons in Slovenia were positive for PiCV [32]. In another study of Slovenian racing pigeons, results showed that 93.3% of cloacal and 96.7% of oropharyngeal samples were positive for PiCV [34] and are comparable with results obtained in other countries [21]. The findings showed a high prevalence of infections with PiCV, a pathogen that could cause immunosuppression, which may favor secondary infections [38]. In the present study, PiCV was detected in all flocks but not in all categories in each flock. In the adult breeding group (G1) and pigeons included in training and racing (G4), the virus was detected in two flocks and in young pigeons (G2 and young pigeons from G3) in three flocks. Based on our limited results, we can only speculate that there is no direct correlation between PiCV infection and CORT levels.

The results for AIV were negative in our previous study and this study, which coincides with the results of other authors [19,39,40].

Chlamydial infections in feral pigeons in Europe and the focus on public health implications are described by Magnino et al. [41]. They found 19.4% to 95.6% seropositive pigeons and 3.4–50% PCR positive pigeons, indicating high importance of transmission to racing pigeons and also to humans. In our previous study of chlamydial infections in racing pigeons, CP was confirmed in cloacal swabs in 16.7% of samples. The determination of CP also indicates a high risk of infection to humans [39]. In this study, CP was confirmed in only one cloacal sample (2.1%). All oropharyngeal samples were negative.

Endoparasites could cause subclinical or even clinical disease. To prevent serious disease and consequently stress, pigeons must be treated regularly [9,16]. Zigo [2] found an increased incidence of coccidiosis (40.4%), trichomoniasis (17.3%), and other endoparasitoses (11.5%) at the time of racing. The results of parasitological examinations, performed just before the last race, showed that all samples were negative for helminths and coccidia, though *Trichomonas* sp. was confirmed in all flocks.

Due to the fact that APMV-1 is a serious viral pathogen that is endemic and is common in feral pigeons [17,37], breeders should regularly vaccinate pigeons to prevent infection and transmission. Vaccination is usually performed before and after the active flying season. Inactivated vaccines are used for the prophylaxis of APMV-1 in pigeons. Vaccines are based on different strains of paramyxoviruses, with the LaSota strain being one of the most commonly used [37]. Inactivated vaccines are administered by subcutaneous injection in various prevention programs. After the vaccination of 4- to 6-week-old seronegative pigeons with the inactivated LaSota vaccine in aqueous suspension, mean antibody titers were between 3 (log_2) and 5 (log_2) after 1 to 2 months. Titers between 2 (log_2) and 3 (log_2) were detected for 6 months, and thereafter, a decrease in titers was observed [42].

In our study, APMV-1 was not detected, but we did find that pigeons in all flocks, especially young pigeons less than one-year-old (G2), were rather poorly protected against this virus based on antibody titers detected by the IHA test (Table 1). The first IHA test found a titer of 3.5 (log_2) in one flock, below 2.0 in two flocks, and as low as 0.0 in another flock. The flocks were vaccinated again 5 days after the last race, but only the active fliers (G4) were checked 25 days after immunization (i.e., 30 days after the last race). Titers were slightly elevated in flocks 1, 2, and 4 but still remained below 2.0 in flock 2 and 4. In flock 3, the titer remained 0.0. In this flock, the CORT level remained high after 30 days (65.4 nmol/L). A high CORT level (57.1 nmol/L) after 30 days was also found in flock 1, but the titer, in this case, increased from 2.3 to 5.3, and the minimum titer was 4. We can assume that there was no direct relationship between the vaccination response and stress. The reason for the low titer increase was probably due to the short testing interval after revaccination, which should be 1 to 2 months [42], or due to some other unknown stressors.

4.2. Serum Corticosterone Levels in Different Categories of Racing Pigeons

In all bird species studied, CORT is considered the most important glucocorticoid [12]. CORT measurements have been proven to be useful in measuring the welfare status of racing pigeons [43]. Romero and Wingfield [44] determined 2–9 ng/mL (5.77–25.96 nmol/L) for baseline serum CORT levels and 14–15 ng/mL (40.38–43.26 nmol/L) for stress-induced levels in free-living pigeons. Lumeij et al. [12] measured baseline concentrations of CORT in the serum of 30 racing pigeons after they had been kept quiet for 24 h. The average level was 0.34 µg/dL (9.81 nmol/L), and maximum level was 1.24 µg/dL (35.77 nmol/L).

Our results of CORT in serum were within the range of results obtained by other authors [12,44]. In three out of four flocks, the lowest level of CORT was found in the serum of pigeons that were only trained (G3). The average CORT level was 16.9 nmol/L (range from 9.1 to 25.4 nmol/L). In the fourth flock, the lowest level of CORT was found in the adult breeding group (G1) (13.2 nmol/L) and slightly higher in pigeons that were only trained (G3) (14.5 nmol/L). Active racing pigeons (G4) were the most stressed group. The average measured CORT level was 59.4 nmol/L (range from 37.3 to 69.0 nmol/L), and it seems that exhaustion during the race significantly increased the CORT level in serum.

Comparing active racing pigeons (G4) one month after the end of the last race, we found that in two flocks (flock 2 and flock 4), the levels of CORT dropped to levels lower than the levels of the whole flock during the active season, and in two flocks (flock 1 and flock 3), the levels were still higher than the levels before the last race. The levels remained high at 57.1 nmol/L in flock 1 and 56.4 nmol/L in flock 2. The reason could be further stress in the flock during this period or the values remaining high all the time. However, measurements need to be recorded more often to obtain the right answer.

The average values obtained from CORT in serum were the highest in flock 3 (44.02 nmol/L) and the lowest in flock 4 (18.31 nmol/L). When comparing the speed results of the last race, the pigeons from flock 3 had the worst race results, and the best results were obtained in flock 4. The average speed of the first five pigeons from flock 4 was 1037.94 m/min and 855.13 m/min from flock 3. We know that flock 4 (G3 and G4) had the most intensive training during the whole active season, which reflected in slightly higher CORT values compared to those in G1 and G2. The CORT level in the flock dropped below the pre-race level one month after the race. The breeder maintained a high level of health care after the race and used pills, electrolytes, and herbal teas at the time of the active season. This was also described in the literature [1] as a method to maximize the performance of pigeons and protect breeders' investment.

4.3. Stress Factors during Transport to the Race and during the Race

Fast and appropriate transport to the race is very important. The number of pigeons in a box, supply of food and water during transport and at the launching place, as well as the supply of fresh air, should be strictly controlled and implemented [14]. The transport our pigeons received reached a high level of maximum support in terms of pigeon welfare. The only negative potential is that the common transport of several different flocks could lead to a possible transmission of diseases.

In addition, the duration and direction of the race should be included in the planning of each race, and environmental factors should be followed to ensure that the race occurs under suitable conditions that do not impact the welfare of pigeons. For these purposes, our breeders used the internet application called Ventusky by Mojzik and Prantl [24]. The chosen location for the last race was Timisoara, Romania, with an average flight distance of 504.91 km. During the race, the weather was clear, but the air temperature was very high, 34 °C at 2 m above the ground. Most, but not all of the pigeons that participated in the race returned to the pigeon houses within six days. Consequently, pigeon breeders prepared an internal protocol that allows racing only at temperatures up to 30 °C. The protocol was also adopted at the conclusion of a joint meeting of the Slovenian Pigeon Federation. The intention of this decisions is to improve the welfare of racing pigeons.

5. Conclusions

Race flights commonly result in elevated stress as measured by CORT levels and influence the welfare of pigeons that participate in or train for such events. Therefore, races and training should be properly managed and planned with pigeons thoroughly prepared for such challenges.

Only pigeons in good condition and those that are clinically healthy should be allowed to participate in the race. The presence of infectious diseases or parasites should be assessed before the start of each racing season. At the end of the racing season, pigeons may have elevated levels of stress hormones, such as CORT.

The distance, duration, and direction of the race should be planned according to environmental factors in order to reduce stress. It is necessary to avoid temperatures above 30 °C and to predict adverse weather conditions (storms, strong winds), pigeon exposure to predators, and unfavorable magnetic waves.

As training and racing can be very stressful for animals, strictly regulating the factors and circumstances that could jeopardize of racing animal welfare of racing pigeons should be a priority for those involved in the above-mentioned activities.

Author Contributions: Authors M.K., I.P., G.G.G., R.L.K., B.P., O.Z.R., and A.D. collaboratively discussed and prepared the concept, methodology, investigation, resources, curation of data, written review and editing. Preparation of the original written draft was performed by R.L.K., M.R., N.M.H., and A.D. Laboratory tests were performed by N.Č.K. (corticosterone detection), U.K. and B.S. (serology and molecular diagnostics), and A.V.R. (parasitology). The figures were prepared by I.P. and B.P. Please see the CRediT taxonomy for the explanation of terms. All authors have read and agreed to the published version of the manuscript.

Funding: This work was financially supported by the Slovenian Ministry of Agriculture, Forestry and Food and the Slovenian Research Agency (Project number V-4 2024 and Program Group number P4–0092).

Institutional Review Board Statement: The national Animal Protection Act (Official Gazette of RS 38/2013; 21/18; 92/20) defines in 21 a. article that non-experimental clinical veterinary practices are not considered as a procedure on animals for scientific purposes.

Informed Consent Statement: Not applicable.

Data Availability Statement: Not applicable.

Acknowledgments: The authors would like to thank the pigeon breeders from the Breeder Association of Sport Carrier Pigeon "SEL Ljubljana" and the Breeder Association of Sport Carrier Pigeon "LET Tržič" for their cooperation in this study.

Conflicts of Interest: The authors declare no conflict of interest. The funders had no influence on the design of the study; the collection, analysis, or interpretation of the data; the writing of the manuscript; or the decision to publish the results. None of the authors of this paper has a financial or personal relationship with any other person or organization that could inappropriately influence or bias the content of the paper.

References

1. Jerolmack, C. Animal archeology: Domestic pigeons and the nature-culture dialectic. *Qual. Soc. Rev.* **2007**, *3*, 74–95.
2. Zigo, F. Changes in Bacterial Microflora in Young Carrier Pigeons during the Race Season. *Int. J. Avian Wildl. Biol.* **2017**, *2*, 41–44. [CrossRef]
3. Wells, D.L.; Hepper, P.G. Pet ownership and adults' views on the use of animals. *Soc. Anim.* **1997**, *5*, 45–63.
4. Warzecha, M. Pigeon sport and animal rights. *Dtsch. Tierarztl. Wochenschr.* **2007**, *114*, 108–113.
5. Sopinka, N.M.; Patterson, L.D.; Redfern, J.C.; Pleizier, N.K.; Belanger, C.B.; Midwood, J.D.; Crossin, G.T.; Cooke, S.J. Manipulating glucocorticoids in wild animals: Basic and applied perspectives. *Conserv. Physiol.* **2015**, *3*, cov031. [CrossRef]
6. Romero, L.M.; Dickens, M.J.; Cyr, N.E. The Reactive Scope Model—A new model integrating homeostasis, allostasis, and stress. *Horm. Behav.* **2009**, *55*, 375–389. [CrossRef] [PubMed]
7. Angelier, F.; Parenteau, C.; Trouvé, C.; Angelier, N. Does the stress response predict the ability of wild birds to adjust to short-term captivity? A study of the rock pigeon (*Columbia livia*). *R. Soc. Open Sci.* **2016**, *3*, 160840. [CrossRef] [PubMed]

8. Harvey, S.; Merry, B.J.; Phillips, J.G. Influence of stress on the secretion of corticosterone in the duck (*Anas platyrhynchos*). *J. Endocrinol.* **1980**, *87*, 161–171. [CrossRef]
9. Humphreys, E.M.; Carrington-Cotton, A.; Wernham, C.V. *A Review of an Exploratory Tial on Sparrowhawk Attack Rates at Pigeon Racing Lofts in Relation to Some Management Practices*; BTO Scotland, School of Biological and Environmental Sciences: Stirling, Scotland, UK, 2010.
10. Beuving, G.; Vonder, G.M. Daily rhythm of corticosterone in laying hens and the influence of egg laying. *J Reprod. Fertil.* **1977**, *51*, 169–173. [CrossRef]
11. Westerhof, I.; Mol, J.A.; Van den Brom, W.E.; Lumeij, J.T.; Rijnberk, A. Diurnal rhythms of plasma corticosterone concentrations in racing pigeons (*Columba livia domestica*) exposed to different light regimens, and the influence of frequent blood sampling. *Avian Dis.* **1994**, *38*, 428–434. [CrossRef]
12. Lumeij, J.T.; Boschma, Y.; Mol, J.; De Kloet, E.R.; Van den Brom, W.E. Action of ACTH1-24 upon plasma corticosterone concentrations in racing pigeons (*Columba livia domestica*). *Avian Pathol.* **1987**, *16*, 199–204. [CrossRef] [PubMed]
13. Scope, A.; Filip, T.; Gabler, C.; Resch, F. The influence of stress from transport and handling on hematologic and clinical chemistry blood parameters of racing pigeons (*Columba livia domestica*). *Avian Dis.* **2002**, *46*, 224–229. [CrossRef]
14. Geraldy, K. Transport of racing pigeons and animal welfare. *Dtsch. Tierarztl. Wochenschr.* **2007**, *114*, 114–115. [PubMed]
15. Dovč, A.; Krapež, U.; Slavec, B.; Pintarič, Š.; Jereb, G.; Tomašič, A.; Dobeic, M. Diagnostics of Some Viral and Bacterial Diseases in Pigeon Flocks Using Air Sampler. In Proceedings of the XXII Counseling Disinfection, Disinsection and Deratisation with International Participation—One World One Health, Kosmaj, Serbia, 23–26 May 2013; Radenković-Damnjanović, B., Ed.; Faculty of Veterinary medicine Beograd: Beograd, Serbia, 2013; pp. 221–229.
16. Rupiper, D.J. Diseases that affect race performance of homing pigeons. Part II: Bacterial, fungal and parasitic diseases. *J. Avian Med. Surg.* **1998**, *12*, 138–148.
17. Harlin, R.; Wade, L. Bacterial and parasitic diseases of Columbiformes. *Vet. Clin. N. Am. Exot. Anim. Pract.* **2009**, *12*, 453–473. [CrossRef] [PubMed]
18. Rupiper, D.J. Diseases that affect race performance of homing pigeons. Part I: Husbandry, diagnostic strategies, and viral diseases. *J. Avian Med. Surg.* **1998**, *12*, 70–77.
19. Kaleta, E.F.; Hönicke, A. Review of the literature on avian influenza A viruses in pigeons and experimental studies on the susceptibility of domestic pigeons to influenza A viruses of the haemagglutinin subtype H7. *Dtsch. Tierarztl. Wochenschr.* **2004**, *111*, 467–472.
20. Stenzel, T.A.; Pestka, D.; Tykalowski, B.; Smialek, M.; Koncicki, A. Epidemiological investigation of selected pigeon viral infections in Poland. *Vet. Rec.* **2012**, *171*, 562. [CrossRef] [PubMed]
21. Stenzel, T.; Koncicki, A. The epidemiology, molecular characterization and clinical pathology of circovirus infections in pigeons—Current knowledge. *Vet. Q.* **2017**, *37*, 166–174. [CrossRef]
22. Abadie, J.; Nguyen, F.; Groizeleau, C.; Amenna, N.; Fernandez, B.; Guereaud, C.; Guigand, L.; Robart, P.; Lefebvre, B.; Wyers, M. Pigeon circovirus infection: Pathological observations and suggested pathogenesis. *Avian Pathol.* **2001**, *30*, 149–158. [CrossRef]
23. Raue, R.; Schmidt, V.; Freick, M.; Reinhardt, B.; Johne, R.; Kamphausen, L.; Kaleta, E.F.; Muller, H.; Krautwald-Junghanns, M.E. A disease complex associated with pigeon circovirus infection, young pigeon disease syndrome. *Avian Pathol.* **2005**, *34*, 418–425. [CrossRef] [PubMed]
24. Halami, M.Y.; Nieper, H.; Müller, H.; Johne, R. Detection of a novel circovirus in mute swans (*Cygnus olor*) by using nested broad-spectrum PCR. *Virus Res.* **2008**, *132*, 208–212. [CrossRef]
25. Wise, M.G.; Suarez, D.L.; Seal, B.S.; Pedersen, J.C.; Senne, D.A.; King, D.J.; Kapczynski, D.R.; Spackman, E. Development of a real-time reverse-transcription PCR for detection of Newcastle disease virus RNA in clinical samples. *J. Clin. Microbiol.* **2004**, *42*, 329–338. [CrossRef]
26. Spackman, E.; Senne, D.A.; Myers, T.J.; Bulaga, L.L.; Garber, L.P.; Perdue, M.L.; Lohman, K.; Daum, L.T.; Suarez, D.L. Development of a real-time reverse transcriptase PCR assay for type A influenza virus and the avian H5 and H7 hemagglutinin subtypes. *J. Clin. Microbiol.* **2002**, *40*, 3256–3260. [CrossRef] [PubMed]
27. Ehricht, R.; Slickers, P.; Goellner, S.; Hotzel, H.; Sachse, K. Optimized DNA microarray assay allows detection and genotyping of single PCR-amplifiable target copies. *Mol. Cell. Probes* **2006**, *20*, 60–63. [CrossRef]
28. Pantchev, A.; Sting, R.; Bauerfeind, R.; Tyczka, J.; Sachse, K. New real-time PCR tests for species-specific detection of *Chlamydophila psittaci* and *Chlamydophila abortus* from tissue samples. *Vet. J.* **2009**, *181*, 145–150. [CrossRef] [PubMed]
29. Thienpont, D.; Rochette, F.; Vanparijs, O.F.J. *Diagnosing Helminthiasis through Coprological Examination*; Janssen Research Foundation: Beerse, Belgium, 1979; pp. 31–34.
30. Dovč, A.; Zorman-Rojs, O.; Vergles Rataj, A.; Bole-Hribovšek, V.; Krapež, U.; Dobeic, M. Health status of free-living pigeons (*Columba livia domestica*) in the city of Ljubljana. *Acta Vet. Hung.* **2004**, *52*, 219–226. [CrossRef] [PubMed]
31. Mojzik, M.; Prantl, M. Ventusky. InMeteo: Meteorologické Informace. Plzen. 2013. Available online: https://www.ventusky.com (accessed on 26 June 2021).
32. Krapež, U.; Slavec, B.; Steyer, A.F.; Pintarič, S.; Dobeic, M.; Rojs, O.Z.; Dovč, A. Prevalence of pigeon circovirus infections in feral pigeons in Ljubljana, Slovenia. *Avian Dis.* **2012**, *56*, 432–435. [CrossRef]

33. Krapež, U.; Steyer, A.F.; Slavec, B.; Barlič-Maganja, D.; Dovč, A.; Račnik, J.; Rojs, O.Z. Molecular characterization of avian paramyxovirus type 1 (Newcastle disease) viruses isolated from pigeons between 2000 and 2008 in Slovenia. *Avian Dis.* **2010**, *54*, 1075–1080. [CrossRef]
34. Dovč, A.; Jereb, G.; Krapež, U.; Gregurić-Gračner, G.; Pintarič, S.; Slavec, B.; Knific, R.L.; Kastelic, M.; Kvapil, P.; Mićunović, J.; et al. Occurrence of bacterial and viral pathogens in common and noninvasive diagnostic sampling from parrots and racing pigeons in Slovenia. *Avian Dis.* **2016**, *60*, 487–492. [CrossRef]
35. Zadravec, M. Adenoviruses in Birds in Slovenia. Ph.D. Thesis, University of Ljubljana, Veterinary Faculty, Ljubljana, Slovenia, 2012.
36. Zadravec, M.; Zorman Rojs, O.; Račnik, J.; Vergles Rataj, A.; Vlahović, K.; Dovč, A. The prevalence of trichomoniasis in ornamental and free-living pigeons (*Columbia livia*) in Slovenia. *Vet. Nov.* **2006**, *32*, 5–10.
37. Pestka, D.; Stenzel, T.; Koncicki, A. Occurrence, characteristics and control of pigeon paramyxovirus type 1 in pigeons. *Pol. J. Vet. Sci.* **2014**, *17*, 379–384. [CrossRef] [PubMed]
38. Harzer, M.; Heenemann, K.; Sieg, M.; Vahlenkamp, T.; Freick, M.; Ruckner, A. Prevalence of pigeon rotavirus infections: Animal exhibitions as a risk factor for pigeon flocks. *Arch. Virol.* **2021**, *166*, 65–72. [CrossRef] [PubMed]
39. Panigrahy, B.; Senne, D.A.; Pedersen, J.C.; Shafer, A.L.; Pearson, J.E. Susceptibility of pigeons to avian influenza. *Avian Dis.* **1996**, *40*, 600–604. [CrossRef] [PubMed]
40. Yamamoto, Y.; Nakamura, K.; Yamada, M.; Mase, M. Limited susceptibility of pigeons experimentally inoculated with H5N1 highly pathogenic avian influenza viruses. *J. Vet. Med. Sci.* **2012**, *74*, 205–208. [CrossRef]
41. Magnino, S.; Haag-Wackernagel, D.; Geigenfeind, I.; Helmecke, S.; Dovc, A.; Prukner-Radovčić, E.; Residbegović, E.; Ilieski, V.; Laroucau, K.; Donati, M.; et al. Chlamydial infections in feral pigeons in Europe: Review of data and focus on public health implications. *Vet. Microbiol.* **2009**, *135*, 54–67. [CrossRef]
42. Duchatel, J.P.; Flore, P.H.; Hermann, W.; Vindevogel, H. Efficacy of an inactivated aqueous-suspension Newcastle disease virus vaccine against paramyxovirus type 1 infection in young pigeons with varying amounts of maternal antibody. *Avian Pathol.* **1992**, *21*, 321–325. [CrossRef]
43. Dovč, A.; Pulko, B.; Pšeničnik, I.; Kalin, A.; Raduha, D.; Čebulj-Kadunc, N.; Slavec, B.; Krapež, U.; Lindtner Knific, R.; Kastelic, M.; et al. Exposure to stress in different categories of racing pigeons at the time of racing. In Proceedings of the First Annual Meeting of the European Congress of Veterinary Behavioural Medicine and Animal Welfare (ECVBMAW) in Association with 24th Annual Meeting of the European Society of Veterinary Clinical Ethology (ESVCE), Berlin, Germany, 27–29 September 2018; Mastorakis, N., Ed.; DVG Service: Geisen, Germany, 2018; pp. 188–190.
44. Romero, L.M.; Wingfield, J.C. Regulation of the hypothalamic-pituitary-adrenal axis in free-living pigeons. *J. Comp. Physiol. B* **2001**, *171*, 231–235. [CrossRef]

Article

Genetic Analysis of the Fatty Acid Profile in Gilthead Seabream (*Sparus aurata* L.)

Antonio Vallecillos [1], María Marín [1], Martina Bortoletti [2], Javier López [1], Juan M. Afonso [3], Guillermo Ramis [4], Marta Arizcun [5], Emilio María-Dolores [1] and Eva Armero [1,*]

1 Department of Agronomic Engineering, Technical University of Cartagena, Paseo Alfonso XIII 48, 30202 Cartagena, Spain; antonio.vallecillos@edu.upct.es (A.V.); mirenamm@gmail.com (M.M.); jlaela2@hotmail.com (J.L.); emilio.mariadolores@carm.es (E.M.-D.)
2 Department of Comparative Biomedicine and Food Science (BCA), University of Padova, Viale dell'Universitá 16, 35020 Legnaro, Italy; martina.bortoletti@studenti.unipd.it
3 Institute of Sustainable Aquaculture and Marine Ecosystems (GIA-ECOAQUA), Carretera de Taliarte s/n, 35214 Telde, Spain; juanmanuel.afonso@ulpgc.es
4 Department of Animal Production, University of Murcia, Avenida Teniente Flomesta 5, 30860 Murcia, Spain; guiramis@um.es
5 Spanish Institute of Oceanography, Oceanographic Center of Murcia, Carretera de la Azohía s/n, 30860 Puerto de Mazarrón, Spain; marta.arizcun@ieo.es
* Correspondence: eva.armero@upct.es; Tel.: +34-968-325-538; Fax: +34-968-325-433

Simple Summary: Humans require essential fatty acids in their diet and marine fish are a source of them, especially omega3 fatty acids that present high benefits on diverse vascular diseases and the immune system. Breeding programs in gilthead seabream usually include growth as the first criterion in the selection process of the fish. However, that could lead to fish with a higher fillet fat content and a fatty acid profile with a lower polyunsaturated fatty acids percentage. Fillet fat content and its fatty acids profile have been revealed as heritable traits. Therefore, further studies to go deeper in the selection process are advisable.

Abstract: The gilthead seabream is one of the most valuable species in the Mediterranean basin both for fisheries and aquaculture. Marine fish, such as gilthead seabream, are a source of n3 polyunsaturated fatty acids, highly appreciated for human food owing to their benefits on the cardiovascular and immune systems. The aim of the present study was to estimate heritability for fatty acid (FA) profile in fillet gilthead seabream to be considered as a strategy of a selective breeding program. Total of 399 fish, from a broodstock Mediterranean Sea, were analysed for growth, flesh composition and FA profile. Heritabilities for growth traits, and flesh composition (fat, protein, and moisture content) were medium. Heritability was moderate for 14:0, 16:0 and 18:1n9 and for sum of monounsaturated FA and n6/n3 ratio, and it was low for 20:1n11 and 22:6n3 and the ratio unsaturated/saturated FA. Breeding programs in gilthead seabream usually include growth as the first criterion in the selection process of the fish. However, other quality traits, such as fillet fat content and its fatty acids profile should be considered, since they are very important traits for the consumer, from a nutritional point of view and the benefits for the health.

Keywords: fatty acid profile; heritability; gilthead seabream; body weight; moisture; fat; collagen; protein

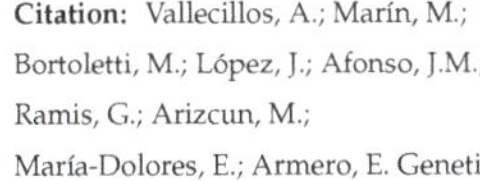

Citation: Vallecillos, A.; Marín, M.; Bortoletti, M.; López, J.; Afonso, J.M.; Ramis, G.; Arizcun, M.; María-Dolores, E.; Armero, E. Genetic Analysis of the Fatty Acid Profile in Gilthead Seabream (*Sparus aurata* L.). *Animals* **2021**, *11*, 2889. https://doi.org/10.3390/ani11102889

Academic Editors: María-Luz García and María-José Argente

Received: 30 July 2021
Accepted: 29 September 2021
Published: 3 October 2021

1. Introduction

The gilthead seabream is one of the most valuable species in the Mediterranean basin both for fisheries and aquaculture. Total production in Mediterranean countries reached 253,000 metric tons in 2019. The three most important countries producing gilthead sea bream were Greece, Turkey and Spain, in that order [1]. In the Mediterranean area, notable success has been achieved in the production of diverse species, such as sea bream and bass.

Now that production technologies have been established, interest has been redirected to increasing the quality of the product offered [2].

Fatty acids (FA), especially n3 polyunsaturated fatty acids (n3 PUFA), eicosapentaenoic (EPA, 20:5n3) and docosahexaenoic acids (DHA, 22:6n3), are recognised as being beneficial for human health, controlling a wide range of human pathologies including cardiovascular, diabetes, rheumatoid arthritis, osteoporosis, asthma, cognitive decline, neurological dysfunction and possible cancers, whilst also playing an important role in neural development and in the immune and inflammatory processes [3–6]. Arachidonic acid (20:4n6, ARA), EPA and DHA are considered as the precursors for the synthesis of eicosanoids, regulators of cell signalling and gene expression and the most powerful modulators of cell membrane fluidity [7,8]. Alteration in the PUFA content of the immune cells was demonstrated to be associated with the alteration of non-specific immunity (e.g., phagocytosis, respiratory burst and serum lysozyme), specific immunity (e.g., antibody production and resistance to pathogens), eicosanoid production and immune-related gene expression [9].

Humans, and probably all vertebrates, require essential FA in their diet that cannot be biosynthesised or interconverted, such as 18:2n6 (linoleic acid, LA) and α18:3n3 (αlinolenic acid, ALA) PUFA [10–12]. These essential fatty acids are primarily derived from plants that should be included in the vertebrates' diet. LA and ALA have vital functions in themselves, and in turn act as precursors for the long-chain PUFA (LCPUFA) ARA, EPA and DHA. Their biosynthesis from LA and ALA can be carried out by mammals, although the process of EPA and particularly DHA biosynthesis from ALA is very low in humans and in marine fish [6,13]. The biosynthetic pathway involves consecutive desaturation and elongation reactions that convert LA to ARA and ALA to EPA and DHA. The two main enzyme families involved in these conversions are the elongases of very long fatty acids (Elovl) and the fatty acyl desaturases (Fad) [6]. The EPA pathway in teleost fish is often incomplete, primarily due to a lack of Δ5 desaturase, and so synthesis of EPA from ALA is not possible in many, particularly marine, carnivorous species [14]. However, the DHA pathway from EPA is probably functional in most teleost fish, including marine species, at least in some tissues.

Modern Western diets have an excess of n6 PUFA, primarily LA, and because n6 FA and n3 FA cannot be interconverted in vertebrates, this has led to an increase in the tissue ratio of n6 to n3 LCPUFA, linked to cardiovascular, inflammatory, and neurological problems [7]. One way of addressing this n6/n3 imbalance is to increase the levels of n3 PUFA and especially n3 LCPUFA in the diet of humans. Marine fish, such as the gilthead seabream (*Sparus aurata*), are a source of LCPUFA [15]. However, vegetable feed in the fish diet has increased in recent years. Fillets of gilthead sea bream fed diets rich in plant oils show increased levels of LA and ALA, with a concurrent decrease in EPA and DHA [16,17].

Dietary and fillet FA composition are closely associated. However, Ballester-Lozano et al. [18] observed that the FA composition (%) depends not only on the diet but also on the fillet lipid content (FLC). In general, monounsaturated FA (MUFA) increased with the increase of FLC, whereas the trend for saturated FA (SFA) and PUFA was the opposite. In the case of ARA and DHA, they decreased when the FLC increased. Thus, FLC could partly explain saturated, monounsaturated and some polyunsaturated (ARA and DHA) FA but not LA, ALA, EPA and docosapentaenoic acid (22:5n3, DPA). This is likely due to marine fish showing a limited ability to convert C18 FA into LCPUFA of n6 and n3 series [19,20]. In addition, while triacylglycerols (TAG) are in fat deposits, and usually contain MUFA (C16-C18); polar lipids, in membrane cells are composed mainly of phospholipids that accumulate LCPUFA [21]. An FLC increase is related to a higher amount of fat deposits and, concurrently, a higher proportion of TAG and MUFA and a lower proportion of polar lipids and LCPUFA.

Other factors besides diet (e.g., genotype, gender, age, and production system) have a significant influence on the fillet lipid level and thus on the FA composition of most animals [22].

Hence, advances in breeding programs are essential to contribute to the profits and competitiveness of the companies, as well as to improve the quality of product, including its fatty acid profile.

Selective breeding programs have been initiated in gilthead seabream to improve growth performance and morphology traits; other objectives (feed efficiency, product quality and disease resistance) have been considered later [18,23–25]. Heritabilities, genetic and phenotypic correlations between different traits are key indicators in the success of such breeding programs.

To date, there have been few heritability studies in terms of the fillet fat content in gilthead seabream, in which medium heritability in fish and medium correlation with respect to weight were observed [26,27]. In addition, to the best of our knowledge, there has been no work about FA genetic variation in sea bream, but it has been studied in Atlantic salmon [28] and in Nile Tilapia [29]. Flesh n3 LCPUFA composition was a highly heritable trait in Atlantic salmon [28] however individual FA heritability varied from zero to medium in Nile tilapia [29]. In Atlantic salmon, families with a high percentage of n3 LCPUFA in flesh showed higher expression of lipid transport genes, cell cycle, and growth-related genes and increased activity of a transcription factor, hepatic nuclear factor 4α (HNF4 α). Dong et al. [30,31] demonstrated that HNF4α is a transcription factor of the vertebrate *Fad* gene involved in the transcription regulation of LCPUFA biosynthesis. Therefore, it is nonetheless a sensible strategy to select for this trait to improve it and optimise the efficiency of n3 LCPUFA metabolism and flesh levels, irrespective of likely dietary levels.

The aim of the present study was to estimate genetic parameters for the FA profile in fillet sea bream, for the first time in gilthead seabream, to be considered as a strategy in a selective breeding program.

2. Materials and Methods

To ensure that animal welfare standards are maintained, anaesthetic was used within the sampling procedure. All animal experiments described in this manuscript fully comply with the recommendations in the Guide for Care and Use of Laboratory Animals of the European Union Council (2010/63/EU) and, whenever necessary, fish were anesthetized.

2.1. Fish and Rearing Conditions

For growth, flesh composition (fillet fat, moisture, protein, and collagen percentages) and FA profile were analysed for 399 gilthead seabream fish. The fish came from a broodstock (n = 133; 57 males and 76 females) that had been captured in the Mediterranean Sea and maintained in Instituto Español de Oceanografía, Mazarrón, Murcia (IEO). The broodstock had never been subjected to genetic selection.

In the broodstock, the female/male ratio was approximately 2:1 in the tanks, they were under a controlled photoperiod (8L:16D) to synchronize maturation; and egg release was initiated at the beginning of February 2016. During that period, the animals were fed on Vitalis Cal (Skretting, Stavanger, Norway), and egg production was monitored daily. When the total egg production stabilized, one egg batch was established at the end of April 2016. Therefore, eggs from the broodstock were collected and pooled for four consecutive days (4 DL model) to maximize family representation. Incubation was carried out in cylindrical conical tanks (1000 L) at a density of 500–1000 larvae/L. Water conditions were as follows: Temperature 19.0 °C, salinity 34‰, and dissolved oxygen was 6.4 mg/L.

At 251 days post-hatching (dph), the fish were individually tagged in the abdominal cavity for individual identification with passive integrated transporter (PIT; Trovan Daimler-Benz, United Kingdom), following the tagging protocol described by Navarro et al. [32]; initial body weight (BW_{251dph}) and total length (TL_{251dph}) were measured; and a sample of caudal fin was collected and preserved in absolute ethanol at room temperature for future DNA extraction. Ten days later, the fish were moved to the facilities of the company Servicios Atuneros del Mediterraneo S.L. (San Pedro del Pinatar, Murcia, Spain), where they were reared in a cage in the Mediterranean Sea under intensive conditions: a cage 11 m in diameter which is anchored at a depth of 38 m in the Mediterranean Sea (average water temperature = 18.2 ± 0.9 °C, dissolved oxygen: 7.4 mg/L, 100% oxygen-saturation, salinity: 37.9‰; data estimated from open sea conditions). Fish were fed over the course of the study with extruded pellets (Dibaq S.A, Fuentepelayo-Segovia, Spain), with two different commercial diets. The first 15 months diet D4 was used (46.5% protein, 19% fat, 7% ashes, 2.75% cellulose, 17.9 MJ/kg digestible energy), and subsequently when the fish were around 220 g in weight, they were fed diet D6 (44% protein, 20% fat, 7.17% ashes, 3.07% cellulose, 17.6 MJ/kg digestible energy) until slaughter time. The FA composition of each diet was analysed in duplicate; the mean is shown in Table 1.

Table 1. Fatty acid (FA) composition of the diets (% total FAME), mean ± standard error.

FA	D4	D6
14:0	2.91 ± 0.09	1.89 ± 0.04
16:0	15.5 ± 0.11	11.4 ± 0.01
18:0	5.16 ± 0.34	4.31 ± 0.02
SFA	24.3 ± 0.09	17.5 ±0.08
16:1	4.05 ± 0.08	2.66 ± 0.00
18:1 *	29.9 ± 0.18	42.9 ± 0.30
C 20:1	3.03 ± 0.08	2.38 ± 0.08
C 22:1	2.05 ± 0.32	1.17 ± 0.15
MUFA	39.2 ± 0.13	49.7 ± 0.54
18:2n6	18.9 ± 0.21	20.1 ± 0.31
18:3n3	4.61 ± 0.26	6.05 ± 0.03
20:4n6	1.18 ± 0.59	0.00 ± 0.00
20:5n3	4.58 ± 0.03	2.81 ± 1.40
22:6n3	6.19 ± 0.52	4.01 ± 0.06
PUFA	36.4 ± 0.04	32.6 ± 0.45

SFA: saturated fatty acids, MUFA: monounsaturated fatty acids, PUFA: polyunsaturated fatty acids. 18:1 * refers only to n9c, because n9t was almost null.

At harvest size (980 dph), fish were slaughtered by immersion in ice cold water (hypothermia); final body weight (BW_{980dph}) and total length (TL_{980dph}) were measured. Fish were manually skinned and filleted without including the nape and the belly flap. Two pieces of fillet were vacuum packaged and frozen at −80 °C for further analysis.

2.2. Flesh Composition Quality

One piece of fillet from each fish was homogenized and analysed by indirect method of near-infrared spectroscopy (near infrared spectroscopy, NIR), using FOODSCAN LAB equipment (FOSS IBERIA, Barcelona, España), to obtain total collagen of the muscle, and chemical components of the muscle (fat, moisture, and protein), as a percentage of flesh.

2.3. Chromatographic Analysis of Fatty Acid Methyl Esters

Fatty acid methyl esters (FAMEs) were prepared using a solution of KOH in methanol [33], 17:0 acid was used as internal standard FA, then separated and analysed by gas chromatography. Analyses were performed on a 6890-gas chromatograph (Agilent Technologies, Palo Alto, CA, USA) equipped with a GERSTEL MultiPurpose Sampler (MPS2) and a mass spectrometer 5975 with a hyperbolic quadrupole (Agilent Technologies, Palo Alto, CA, USA). Extract (0.8 µL) of FAME was injected and separated on a DB-23 capillary column (Agilent Technologies, Palo Alto, CA, USA) of 60 m (length) × 0.25 mm (internal diameter) × 0.25 µm (film) in constant pressure mode. Chromatographic-grade helium was used as the carrier gas. The temperature of the injector was 240 °C. The inlet operated in split mode with a split ratio of 1:20. The initial oven temperature was 50 °C which was held for 1 min, then increased to 175 °C at 25 °C per min and thereafter increased to 235 °C at 4 °C per min, with a holding time of 10 min. Mass spectra were collected in the scan range m/z 40–400. The measurements were performed using an electron bombardment ion source with electron energy of 70 eV. The transfer line, source, and quadrupole temperatures were set at 280, 230, and 150 °C, respectively. The chromatograms and mass spectra were evaluated using the ChemStation software (G1791CA, Version D.03.00, Agilent Technol.). Peaks were identified by comparison of retention times with FAME standards (Supelco 37 Component FAME mix, Sigma Aldrich, St. Louis, Missouri, USA) and their mass spectra. The individual FAs are expressed as a percentage of the total FA detected.

2.4. Microsatellite Genotyping and Parental Assignment

The broodstock and offspring were genetically characterised. To this end, DNA was extracted from the caudal fin using the DNeasy kit (QIAGEN®, Hilden, Germany), and then kept at 4 °C. Next, DNA quantity and quality were determined with a NanoDrop™ 2000 spectrophotometer v.3.7 (Thermo Fisher Scientific, Wilmington, NC, USA). The multiplex SMsa1 (Super Multiplex *Sparus aurata*) was used as described in [34] for genotyping the broodstock and offspring. The electropherogram was analysed using Microsatellite analysis cloud (Thermo Fisher Scientific, Waltham, MA, USA). Direct count of heterozygosity in the offspring was calculated with the Excel package called Gene Alex [35]. For parental assignment, the exclusion method as implemented in VITASSING (v.8_2.1) software [36] was used. The number of fish assigned to a single couple was 399 and they were used to estimate the genetic parameters.

2.5. Statistical Analysis

All data were tested for normality and homogeneity of variances using SPSS (v.25.0) [37]. For growth trait (BW and TL) arithmetic means and standard errors were calculated.

Flesh composition (fillet fat, moisture, protein, and collagen percentages) and FA profile were analysed with the following general linear model (GLM):

$$Y_{ij} = \mu + b^{*}covariate_j + e_{ij} \quad (1)$$

in which,

Y_{ij} is an observation of an individual j from the origin i,

μ is the overall mean,

b is the regression coefficient between the analysed variable and the covariate *BW* for flesh composition or fillet fat percentage for FA profile,

e_{ij} is a random residual error.

The level of significant difference was set at $p < 0.05$.

Genetic parameters were estimated under a Bayesian approach using a bivariate mixed model. The model was,

$$Y = X\beta + Zu + e \quad (2)$$

where Y is the recorded data on the studied traits, β includes covariate body weight (not included for BW and TL traits), u the random animal effect and e the error. This was performed using gibbs3f90 program for all traits, as developed by Misztal et al. [38]. The analysis was carried out between two traits each time. The following multivariate normal distributions were assumed a priori for random effects:

$$\begin{gathered} p(\beta) \sim k; \\ p(u \mid G) \sim (0; G \otimes A); \\ p(e \mid R) \sim (0; R \otimes A); \end{gathered} \quad (3)$$

where A is the relationship matrix and k is a constant,

$$G = \begin{bmatrix} \sigma_{u1} & \sigma_{u1,u2} \\ \sigma_{u2,u1} & \sigma_{u2} \end{bmatrix}, R = \begin{bmatrix} \sigma_{e1} & \sigma_{e1,e2} \\ \sigma_{e2,e1} & \sigma_{e2} \end{bmatrix}. \quad (4)$$

Bounded uniform priors were assumed for the systematic effects and the (co)variance components (G, A). A single chain of 200,000 iterations was run. The first 50,000 iterations of each chain were discarded, and samples of the parameters of interest were saved every five iterations. Density plots to represent posterior marginal distribution of heritabilities, posterior means (PM) and the 95% interval of the highest posterior density (HPD 95%) were obtained through R Development Core Team [39].

3. Results and Discussion

3.1. Phenotyping

The phenotypic results for growth at 251 dph and 980 dph (BW and TL), flesh composition (fat, collagen, moisture and protein percentages) and FA profile at 980 dph in gilthead seabream are shown in Tables 2 and 3.

Table 2. Phenotypic results (least square means ± standard error) for body weight, fork length, and flesh composition for gilthead seabream.

	Offspring			Covariate BW_{980dph}	
Traits	*n*	Mean	S.E.	b	S.E.
BW_{251dph} (g)	392	43.7	0.95	NI	
TL_{251dph} (cm)	392	13.8	0.09	NI	
BW_{980dph} (g)	392	450.2	4.14	NI	
TL_{980dph} (cm)	392	28.7	0.11	NI	
Fat (%)	389	10.1	0.12	0.006 *	0.002
Collagen (%)	388	1.74	0.02	0.001 *	0.000
Moisture (%)	389	64.6	0.11	−0.007 *	0.001
Protein (%)	389	19.8	0.06	<0.000	0.001

BW_{251dph} and BW_{980dph} = body weight at 251 dph and 980 dph respectively; TL_{251dph} and TL_{980dph} = total length at 251 dph and 980 dph respectively; * = covariate was significant ($p < 0.05$). NI = not included.

Table 3. Main fatty acids (FA, expressed as %) of gilthead seabream flesh adjusted to 10.1 fat percentage.

FA	Offspring			Covariate Fat Percentage	
	n	LSM	S.E.	b	S.E.
14:0	394	3.29	0.03	0.011	0.015
16:0	371	17.9	0.23	0.018	0.090
18:0	392	4.86	0.11	−0.019	0.045
SFA	397	28.3	0.30	−0.002	0.130
16:1	397	0.32	0.04	−0.018	0.015
18:1n9c	392	33.6	0.70	0.232	0.277
18:1n9t	333	3.55	0.11	−0.005	0.047
20:1	392	2.38	0.03	−0.004	0.014
22:1	397	0.80	0.02	−0.014	0.008
MUFA	397	44.38	0.57	0.291	0.209
18:2n6	334	15.6	0.21	−0.054	0.084
18:3n3	389	3.77	0.05	0.055 *	0.022
20:4n6	399	0.63	0.02	−0.017	0.006
20:5n3	381	2.88	0.04	0.008	0.016
22:6n3	384	6.84	0.12	−0.097 *	0.048
PUFA	397	29.84	045	−0.400	0.165
$\sum$ n6/n3	394	1.05	0.35	0.249	0.140
UFA/SFA	392	2.97	0.08	−0.029	0.034

LSM: least square mean, SE: standard error; SFA: saturated fatty acids, MUFA: monounsaturated fatty acids, PUFA: polyunsaturated fatty acids, UFA: unsaturated fatty acids; * = covariate was significant ($p < 0.05$).

Regarding flesh composition, the fat percentage was high in comparison with that found by other authors [26,27] who observed less fillet fat percentage when BW was lower (In García-Celdrán et al. [26] 4.64% for BW_{690dph} = 271 g, in Elalfy et al. [27] 6.55% for BW_{700dph} = 313 g). However, when fish were raised in an estuary [27] and reached higher BW_{700dph} (440 g), the fillet fat percentage increased (8.71%). In addition, a pronounced seasonality has been observed on fillet fat that reached a maximum with the replenishment of body fat stores in early autumn [18] when our fish were slaughtered. In addition, BW had a positive significant effect on fat percentage, and this effect was less pronounced for collagen percentage (when fish weight increased 100 g the fat percentage increased 0.6% and collagen percentage decreased 0.1%). Contrary to the fat, in our study the moisture percentage was low in comparison with that in García-Celdrán et al. [26] (73%) and Elalfy et al. [27] (73.1% in the cage and 68.8% in the estuary) and BW had a negative significant effect on moisture (when fish weight increased 100 g the moisture percentage decreased 0.7%). This result was logical due to the high negative correlation between fat and moisture [25,26].

Gilthead seabream fillet showed the highest percentage of MUFA, followed by PUFA and SFA with similar percentages (Table 3). Fillet FA composition was closely related to the diet composition but not totally. In fact, in comparison with the diet, fillet showed higher SFA especially for 16:0, and lower MUFA and PUFA percentages, mainly due to the lower percentages of 18:1n9, 18:2n6 and 18:3, although for 22:6n3 the percentage increased notably in comparison to the diet. This difference in FA composition is largely explained by variations in the level of fattening, especially intramuscular fat, the percentage of PUFA, one of them DHA, decreased when FLC increased [40]. In our study, although the fish showed high FLC, the BW was much lighter than in Ballester-Lozano et al. [18] and it is likely that they had not finished their development and fat deposition, and concurrently the DHA was high.

The main fillet FA were 18:1n9, 16:0, LA, and DHA, in accordance with that described by other authors [18,41,42].

3.2. *Microsatellite Genotyping and Parental Assignment*

The use of multiplex SMSa1 PCR using the exclusion method, with a maximum of two tolerated errors, provided successful parental assignment for 91.4% of the offspring. After the assignment, six out of 76 females contributed with 52.1% of the offspring and 29 females did not produce any offspring, whilst six out of 57 males contributed with 60.9% of the offspring and 19 males did not contribute. Pedigree construction using selected highly informative microsatellite markers yielded 66 full-sib families with a mean of 3.86 sibs (range 2–28 sibs).

Regarding the study of genetic variation considering the microsatellites genotypes, high heterozygosity was observed, reaching 0.75. This value is consistent with the fact that the population came from a broodstock that had never been subjected to selection, and reveals that, at that moment, there was no danger of inbreeding.

3.3. *Genetic Parameters*

3.3.1. Heritability for Growth Traits

Heritability for BW_{251dph} and BW_{980dph} was moderate (PM = 0.22 and HPD = [0.06–0.40]; PM = 0.24 and HPD = [0.06–0.48] respectively). For TL, heritability was moderate (0.19 [0.04–0.36] at 251 dph and high (0.48 [0.18–0.80]) at 980 dph (Figure 1), in accordance with other authors [24,26,42].

In our study, TL at advanced age was presented as more heritable than BW; however, other studies [24,26,42] observed similar heritability for both traits and high genetic correlation between them, as also happened in our study (0.94 ± 0.06 genetic correlation BW-TL_{980dph}). In addition, García-Celdrán et al. [43] pointed out that heritability estimates for growth traits increased with age when they compared juveniles with commercial size fish. In our study, genetic correlation (rg) for BW or TL at different age were practically null but a safe interpretation of the rg is made difficult by the large standard errors (rg $BW_{251dph} - BW_{980dph} = 0.13 \pm 0.38$, rg $TL_{251dph} - TL_{980dph} = 0.04 \pm 0.39$).

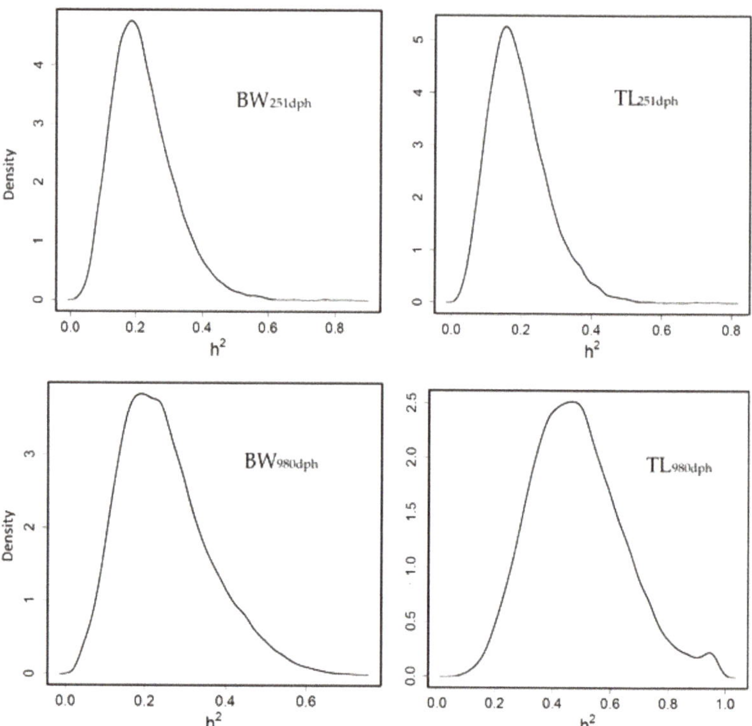

Figure 1. Posterior marginal distribution of heritabilities of body weight (BW) and total length (TL) gilthead seabream. h2 = heritability (*n* = 399).

3.3.2. Heritability of Flesh Composition

Moderate heritability was obtained for fillet fat percentage (0.24 [0.06–0.44] PM and HPD in brackets, Figure 2) which agrees with Elalfy et al. [27] and García-celdrán et al. [26], who showed 0.27 and 0.31, respectively. In our study, protein percentage heritability was moderate (0.21 [0.03–0.41]), however García-celdrán et al. [26] and Elalfy et al. [27] reported low protein heritability (0.03 and 0.08, respectively). Regarding the collagen percentage, the heritability was low (0.06 [0.002–0.19]) in this study, similar to that in García-celdrán et al. [26] (0.03) and Navarro et al. [43] (0.02). The moisture percentage showed a medium genetic heritability (0.15 [0.015–0.32]) in the present investigation, however it has been reported with considerable variation between other studies ranging from medium to low heritabilities, such as Garcia-celdrán et al. [26] and Elalfy et al. [27] (0.24 and 0.29, respectively) and Navarro et al. [43] (0.09).

It is interesting to know the genetic correlation between BW and fat percentage, since most breeding programs select fish to improve their growth. In our study, the genetic correlation between both traits was not estimated with precision because of the limited data available. When this correlation was estimated [26,27], a positive medium-high genetic correlation was observed, indicating that when fish are selected by growth, their fillet fat percentage increased indirectly.

3.3.3. Heritability of Fatty Acid Profile

Heritability was moderate for 14:0 (0.24 [0.04–0.48]), 16:0 (0.15 [0.01–0.33]) and 18:1n9c (0.20 [0.005–0.43]); it was low for 20:1 (0.12 [0.01–0.26]), and 22:6n3 (0.11 [0.017–0.25]); and practically null for 18:0 (0.02 [0.00–0.07]), 18:1n9t (0.03 [0.00–0.10]), 18:2n6t (0.09 [0.00–0.21]),

18:3n3 (0.05 [0.00–0.15]) and 20:5n3 (0.05 [0.00–0.15]) (Figure 3) and 20:4n6 (0.03 [0.00–0.10]), 16:1 (0.04 [0.00–0.12]) and 22:1 (0.06 [0.00–0.16]) although the density plots for these last three FA are not represented.

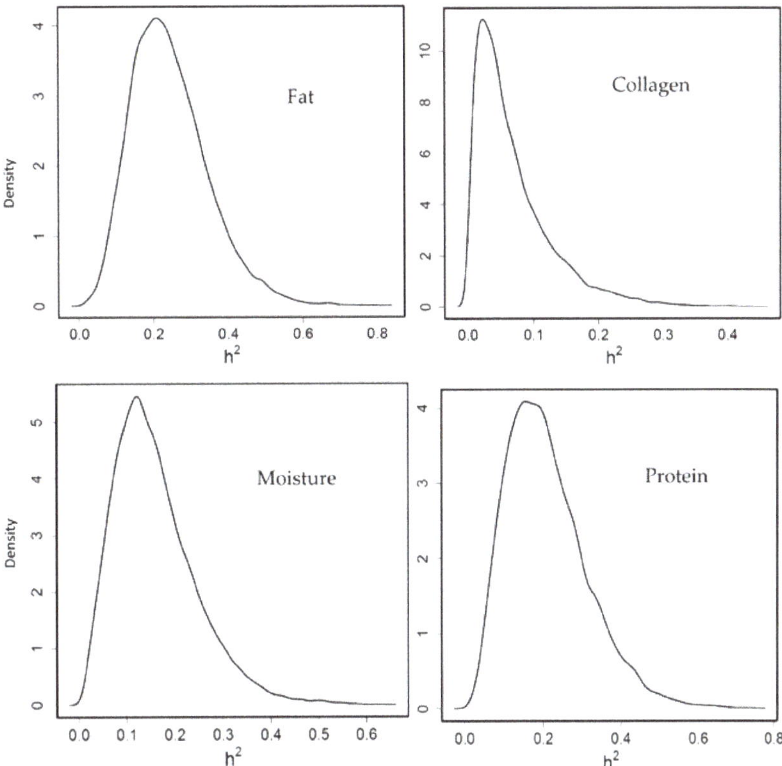

Figure 2. Posterior marginal distribution of heritabilities of fillet fat, collagen, moisture and protein percentage in gilthead seabream. h2 = heritability. (n = 399).

Heritability for the summatory of SFA (0.06 [0.00–0.17]) and PUFA (0.02 [0.00–0.09]) was almost zero; low for the ratio UFA/SFA (0.12 [0.01–0.26]); and medium for MUFA (0.26 [0.03–0.53]) and n6/n3 ratio (0.25 [0.03–0.49]) (Figure 4).

To our knowledge, there is no study about genetic parameters of FA in gilthead seabream. In Nile tilapia, the heritabilities for SFA were generally moderate, and for MUFA, PUFA and for total SFA, total MUFA, total PUFA, n3/n6 and UFA/SFA were low [29]. In Atlantic salmon, flesh n3 LCPUFA composition was highly heritable (h2 = 0.77 ± 0.14) and the authors [28] observed that families with a high percentage of n3 LCPUFA in flesh presented higher expression for genes related to hepatic lipid transport, and implicated increased activity of a transcription factor, hepatic nuclear factor 4α (HNF4α), possibly as a result of family differences in transforming growth factor b1 (Tgfb1) signalling. In that study, the authors [28] also highlighted that FLC was highly and negatively correlated with percentage n3 LCPUFA (−0.77), and FLC was positively correlated to BW. In Nile tilapia [29], the genetic associations of the PUFA group (20:5n3 and C18:3n6) with BW traits were strongly negative (−0.55 to −0.78); and for two SFA the genetic correlations of 18:0 and 24:0 with fillet fat percentage were negative (−0.11 and −0.85, respectively). In our study, genetic correlations between FA and fillet fat percentage could not be estimated with precision, likely due to limited data availability. However, phenotypic correlation

PUFA-Fillet fat percentage was significantly negative (−0.12). A major part of LCPUFA is in the membrane phospholipids (PL), which presents an upper threshold, because amounts of PL molecules in tissue are likely determined by a volume of membranes. Thus, when the level of fattening increases in a fish, most of that fat is deposited in muscles, as TAG, to be an energy reserve. Thus, when FLC increases, fat deposits increase, TAG content increases and PL, together LCPUFA, is diluted [21].

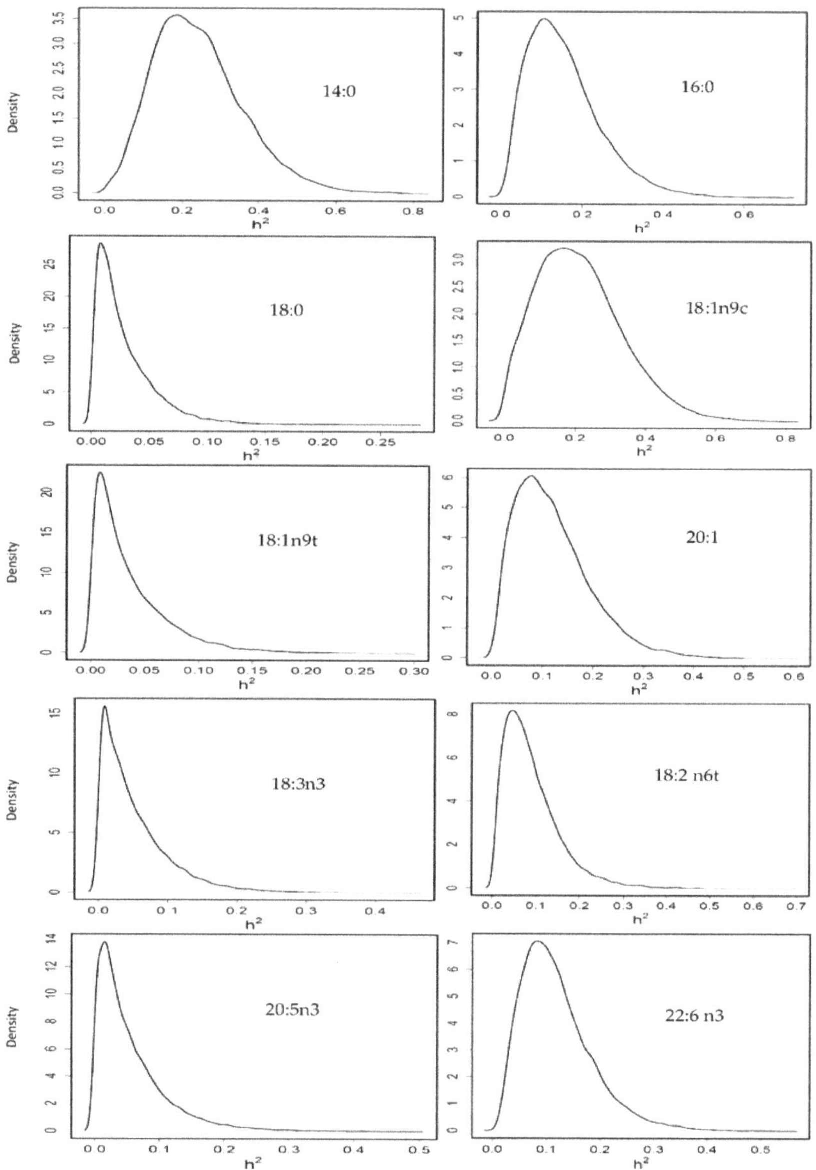

Figure 3. Posterior marginal distribution of heritabilities of fatty acids profile in gilthead seabream. h2 = heritability. n = 399.

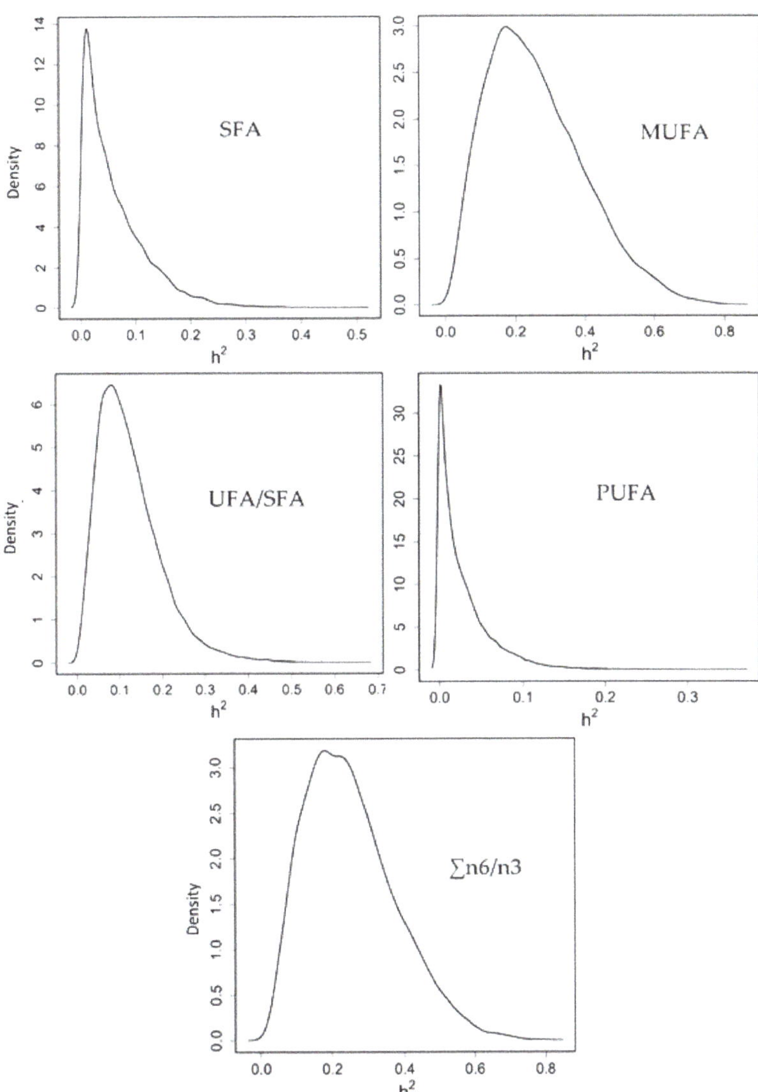

Figure 4. Posterior marginal distribution of heritabilities of SFA: saturated fatty acids, MUFA: monounsaturated fatty acids, PUFA: polyunsaturated fatty acids, UFA: unsaturated fatty acids and the ratio omega 6/omega 3 fatty acids in gilthead seabream. h2 = heritability. 399 data.

In shrimp [44], limited heritabilities for FA were estimated; nevertheless, some important FA, such as DHA had significant variance among families with similar heritability (0.12 ± 0.06) to our study. In accordance with us, ARA, which is tightly linked to the immune response, showed a heritability not significantly different from zero.

Therefore, considering the positive genetic correlation between growth and fillet fat content, and the negative genetic correlation between fillet fat content and PUFA percentage, breeding for fish with higher growth is expected to cause an increase in the fillet fat percentage and a decrease of its PUFA percentage. In addition, most of the SFA and oleic, DHA, MUFA and the ratio n6/n3 have been shown to be heritable traits, thus their analyses should be considered in a breeding program.

The measurement of FA is expensive and time consuming, therefore further studies should be continued to investigate the relation between fillet fat content and its FA profile. The Fish Fat Meter device (Distell.com, West Lothian, Escocia) has been developed as a non-invasive tool to measure flesh fat content and a high correlation with FLC [27] has been demonstrated, thus it could be used as an easy non-invasive measurement.

4. Conclusions

Breeding programs in gilthead seabream usually include growth as the first criterion in the selection process. However, other quality traits, such as fillet fat percentage and its fatty acids profile should be considered, since they are very important traits for the consumer from a nutritional point of view. In addition, these quality traits are also related to the fish immune system and, consequently, to its disease resistance. Further studies to investigate the consequences of selecting fish for growth based on their fat content and their fatty acids profile are advisable.

Author Contributions: Conceptualization, E.A. and A.V.; methodology, A.V.; software, A.V.; validation, E.A.; formal analysis, A.V., J.L., M.M. and M.B.; investigation, A.V.; resources, E.A., J.M.A. and M.A.; data curation, A.V., M.M. and M.B.; writing—original draft preparation, A.V.; writing—review and editing, E.A.; visualization, M.A., E.M.-D., J.M.A. and G.R.; supervision, E.A.; project administration, E.A.; funding acquisition, E.A. All authors have read and agreed to the published version of the manuscript.

Funding: "Desarrollo de un programa de mejora genética en dorada" Project-Consejería de Agricultura, Agua y Medio Ambiente. A.V. was founded by a research scholarship fellow (Mejora de la Competitividad del Sector de la Dorada a Través de la Selección Genética, specialization scholarship); AV. was founded by a pre-doctoral research fellow (20716/FPI/18. Fundación Séneca. Cofinanciado por grupo Andromeda. Región de Murcia (Spain)).

Institutional Review Board Statement: To ensure that animal welfare standards are maintained, anaesthesia was used within the sampling procedure. All animal experiments described in this manuscript fully comply with the recommendations in the Guide for Care and Use of Laboratory Animals of the European Union Council (2010/63/EU), and whenever it was necessary, fish were anesthetized.

Acknowledgments: To Rubén Perez for laboratory assistance.

Conflicts of Interest: The authors declare no conflict of interest.

References

1. APROMAR. La Acuicultura en España; 2020; Volume 95 (In Spanish). Available online: http://www.apromar.es/ (accessed on 3 October 2021).
2. FEAP. European Aquaculture Production Report; 2014. Brussels (Belgium). Available online: https://feap.info/ (accessed on 3 October 2021).
3. Mccullough, M.L.; Feskanich, D.; Stampfer, M.J.; Giovannucci, E.L.; Rimm, E.B.; Hu, F.B.; Spiegelman, D.; Hunter, D.J.; Colditz, G.A.; Willett, W.C. Diet quality and major chronic disease risk in men and women: Moving toward improved dietary guidance. *Am. J. Clin. Nutr.* **2002**, *76*, 1261–1271. [CrossRef] [PubMed]
4. Augustsson, K.; Michaud, D.S.; Rimm, E.B.; Leitzmann, M.F.; Stampfer, M.J.; Willett, W.C.; Giovannucci, E. A Prospective Study of Intake of Fish and Marine Fatty Acids and Prostate Cancer. *Cancer Epidemiol. Prev. Biomark.* **2003**, *12*, 64–67.
5. Simopoulos, A.P. The Importance of the Omega-6/Omega-3 Fatty Acid Ratio in Cardiovascular Disease and Other Chronic Diseases. *Exp. Biol. Med.* **2008**, *233*, 674–688. [CrossRef]
6. Tocher, D.R. Omega-3 Long-Chain Polyunsaturated Fatty Acids and Aquaculture in Perspective. *Aquaculture* **2015**, *449*, 94–107. [CrossRef]
7. Calder, P.C. N Ⅺ 3 Polyunsaturated Fatty Acids, Inflammation, and Inflammatory. *Am. J. Clin. Nutr.* **2006**, *83*, 1505S–1519S. [CrossRef]
8. Yaqoob, P.; Calder, P.C. Fatty Acids and Immune Function: New Insights into Mechanisms. *Br. J. Nutr.* **2007**, *98*, 41–45. [CrossRef]

9. Chen, C.; Sun, B.; Guan, W.; Bi, Y.; Li, P.; Ma, J.; Chen, F.; Pan, Q.; Xie, Q. N-3 assential fatty acids in Nile tilapia, *Oreochromis niloticus*: Effects of linolenic acid on non-specific immunity and anti-inflammatory responses in juveline fish. *Aquaculture* **2016**, *450*, 250–257. [CrossRef]
10. Burr, G.O.; Burr, M.M. On the Nature and Role of the Fatty Acids Essential in Nutrition. *J. Biogical Chem.* **1930**, *86*, 587–621. [CrossRef]
11. Rivers, J.P.W.; Frankel, T.L. Essential Fatty Acid Deficiency. *Br. Med Bull.* **1981**, *37*, 59–64. [CrossRef]
12. Sciences, N. Nutritional Requirements of Fish. *Nutr. Rev.* **1993**, *52*, 417–426. [CrossRef]
13. Graham, C.; Calder, P.C. Conversion of α-Linolenic Acid to Longer-Chain Polyunsaturated Fatty Acids in Human Adults. *Reprod. Nutr. Dev.* **2005**, *45*, 581–597. [CrossRef]
14. Tocher, D.R. Fatty Acid Requirements in Ontogeny of Marine and Freshwater Fish. *Aquac. Res.* **2010**, *41*, 717–732. [CrossRef]
15. Senso, L.; Suárez, M.D.; Ruiz-Cara, T.; García-Gallego, M. On the Possible Effects of Harvesting Season and Chilled Storage on the Fatty Acid Profile of the Fillet of Farmed Gilthead Sea Bream (*Sparus aurata*). *Food Chem.* **2007**, *101*, 298–307. [CrossRef]
16. Benedito-Palos, L.; Navarro, J.C.; Sitjà-Bobadilla, A.; Gordon Bell, J.; Kaushik, S.; Pérez-Sánchez, J. High Levels of Vegetable Oils in Plant Protein-Rich Diets Fed to Gilthead Sea Bream (*Sparus aurata L.*): Growth Performance, Muscle Fatty Acid Profiles and Histological Alterations of Target Tissues. *Br. J. Nutr.* **2008**, *100*, 992–1003. [CrossRef]
17. Izquierdo, M.S.; Montero, D.; Robaina, L.; Caballero, M.J.; Rosenlund, G.; Ginés, R. Alterations in Fillet Fatty Acid Profile and Flesh Quality in Gilthead Seabream (*Sparus aurata*) Fed Vegetable Oils for a Long Term Period. Recovery of Fatty Acid Profiles by Fish Oil Feeding. *Aquaculture* **2005**, *250*, 431–444. [CrossRef]
18. Ballester-Lozano, G.F.; Benedito-Palos, L.; Navarro, J.C.; Kaushik, S.; Pérez-Sánchez, J. Prediction of Fillet Fatty Acid Composition of Market-Size Gilthead Sea Bream (*Sparus aurata*) Using a Regression Modelling Approach. *Aquaculture* **2011**, *319*, 81–88. [CrossRef]
19. Sargent, J.R.; Tocher, D.R.; Bell, J.G. *The Lipids*, 3rd ed.; Halver, J.E., Hardy, R.W., Eds.; Academic Press: Cambridge, MA, USA, 2002; pp. 182–246.
20. Tocher, D.R. Metabolism and Functions of Lipids and Fatty Acids in Teleost Fish. *Rev. Fish. Sci.* **2003**, *11*, 107–184. [CrossRef]
21. Sushchik, N.N.; Makhutova, O.N.; Rudchenko, A.E.; Glushchenko, L.A.; Shulepina, S.P.; Kolmakova, A.A.; Gladyshev, M.I. Comparison of Fatty Acid Contents in Major Lipid Classes of Seven Salmonid Species from Siberian Arctic Lakes. *Biomolecules* **2020**, *10*, 419. [CrossRef] [PubMed]
22. Makhutova, O.N.; Stoyanov, K.N. Fatty Acid Content and Composition in Tissues of Baikal Grayling (*Thymallus baicalensis*), with a Special Focus on DHA Synthesis. *Aquac. Int.* **2021**. [CrossRef]
23. Perera, E.; Simó-Mirabet, P.; Shin, H.S.; Rosell-Moll, E.; Naya-Catalá, F.; de las Heras, V.; Martos-Sitcha, J.A.; Karalazos, V.; Armero, E.; Arizcun, M.; et al. Selection for Growth Is Associated in Gilthead Sea Bream (*Sparus aurata*)with Diet Flexibility, Changes in Growth Patterns and Higher Intestine Plasticity. *Aquaculture* **2019**, *507*, 349–360. [CrossRef]
24. Ofori-Mensah, S.; Yıldız, M.; Arslan, M.; Eldem, V.; Gelibolu, S. Substitution of Fish Oil with Camelina or Chia Oils in Gilthead Sea Bream (*Sparus aurata L.*) Diets: Effect on Growth Performance, Fatty Acid Composition, Haematology and Gene Expression. *Aquac. Nutr.* **2020**, *26*, 1943–1957. [CrossRef]
25. Vallecillos, A.; Chaves-Pozo, E.; Arizcun, M.; Perez, R.; Afonso, J.M.; Berbel, C.; Pérez-Sánchez, J.; María-Dolores, E.; Armero, E. Genetic Parameters for Photobacterium Damselae Subsp. Piscicida Resistance, Immunological Markers and Body Weight in Gilthead Seabream (*Sparus aurata*). *Aquaculture* **2021**, *543*, 736892. [CrossRef]
26. García-Celdrán, M.; Ramis, G.; Manchado, M.; Estévez, A.; Navarro, A.; Armero, E. Estimates of Heritabilities and Genetic Correlations of Raw Flesh Quality Traits in a Reared Gilthead Sea Bream (*Sparus aurata L.)* Population Sourced from Broodstocks along the Spanish Coasts. *Aquaculture* **2015**, *446*, 181–186. [CrossRef]
27. Elalfy, I.; Shin, H.S.; Negrín-Báez, D.; Navarro, A.; Zamorano, M.J.; Manchado, M.; Afonso, J.M. Genetic Parameters for Quality Traits by Non-Invasive Methods and Their G x E Interactions in Ocean Cages and Estuaries on Gilthead Seabream (*Sparus aurata*). *Aquaculture* **2021**, *537*, 736462. [CrossRef]
28. Leaver, M.J.; Taggart, J.B.; Villeneuve, L.; Bron, J.E.; Guy, D.R.; Bishop, S.C.; Houston, R.D.; Matika, O.; Tocher, D.R. Heritability and Mechanisms of N- 3 Long Chain Polyunsaturated Fatty Acid Deposition in the Flesh of Atlantic Salmon. *Comp. Biochem. Physiol. Part D Genom. Proteom.* **2011**, *6*, 62–69. [CrossRef] [PubMed]
29. Nguyen, N.H.; Ponzoni, R.W.; Abu-Bakar, K.R.; Hamzah, A.; Khaw, H.L.; Yee, H.Y. Correlated Response in Fillet Weight and Yield to Selection for Increased Harvest Weight in Genetically Improved Farmed Tilapia (GIFT Strain), *Oreochromis niloticus*. *Aquaculture* **2010**, *305*, 1–5. [CrossRef]
30. Dong, Y.; Wang, S.; Chen, J.; Zhang, Q.; Liu, Y.; You, C.; Monroig, Ó.; Tocher, D.R.; Li, Y. Hepatocyte Nuclear Factor 4α (HNF4α) Is a Transcription Factor of Vertebrate Fatty Acyl Desaturase Gene as Identified in Marine Teleost *Siganus canaliculatus*. *PLoS ONE* **2016**, *11*, e0160361. [CrossRef]
31. Dong, Y.; Zhao, J.; Chen, J.; Wang, S.; Liu, Y.; Zhang, Q.; You, C.; Monroig, Ó.; Tocher, D.R.; Li, Y. Cloning and Characterization of Δ6/Δ5 Fatty Acyl Desaturase (Fad) Gene Promoter in the Marine Teleost *Siganus canaliculatus*. *Gene* **2018**, *647*, 174–180. [CrossRef]
32. Navarro, A.; Oliva, V.; Zamorano, M.J.; Ginés, R.; Izquierdo, M.S.; Astorga, N.; Afonso, J.M. Evaluation of PIT System as a Method to Tag Fingerlings of Gilthead Seabream (*Sparus auratus L.*): Effects on Growth, Mortality and Tag Loss. *Aquaculture* **2006**, *257*, 309–315. [CrossRef]
33. *ISO 15884:2002, Milk Fat, Preparation of Fatty Acid Methyl Ester*; ISO: Geneva, Switzerland, 2002.

34. Lee-Montero, I.; Navarro, A.; Borrell, Y.; García-Celdrán, M.; Martín, N.; Negrín-Báez, D.; Blanco, G.; Armero, E.; Berbel, C.; Zamorano, M.J.; et al. Development of the First Standardised Panel of Two New Microsatellite Multiplex PCRs for Gilthead Seabream (*Sparus aurata L.*). *Anim. Genet.* **2013**, *44*, 533–546. [CrossRef]
35. Peakall, R.; Smouse, P.E. GenALEx 6.5: Genetic Analysis in Excel. Population Genetic Software for Teaching and Research-an Update. *Bioinformatics* **2012**, *28*, 2537–2539. [CrossRef] [PubMed]
36. Vandeputte, M.; Mauger, S.; Dupont-Nivet, M. An Evaluation of Allowing for Mismatches as a Way to Manage Genotyping Errors in Parentage Assignment by Exclusion. *Mol. Ecol. Notes* **2006**, *6*, 265–267. [CrossRef]
37. *IBM SPSS Statistics for Windows*; IBM Corp: Armonk, NY, USA, 2017.
38. Misztal, I.; Tsuruta, S.; Lourenco, D.; Aguilar, I.; Legarra, A.; Vitezica, Z. *BLUPF90 Family of Programs*; University of Georgia: Athens, GA, USA, 2015; pp. 1–125.
39. R Development Core Team. *A Language and Environment for Statistical Computing*; R Foundation for Statistical Computing: Vienna, Austria, 2020; Available online: https://www.R-project.org/.https://feap.info/ (accessed on 3 October 2021).
40. Suomela, J.P.; Lundén, S.; Kaimainen, M.; Mattila, S.; Kallio, H.; Airaksinen, S. Effects of Origin and Season on the Lipids and Sensory Quality of European Whitefish (*Coregonus lavaretus*). *Food Chem.* **2016**, *197*, 1031–1037. [CrossRef] [PubMed]
41. Fountoulaki, E.; Vasilaki, A.; Hurtado, R.; Grigorakis, K.; Karacostas, I.; Nengas, I.; Rigos, G.; Kotzamanis, Y.; Venou, B.; Alexis, M.N. Fish Oil Substitution by Vegetable Oils in Commercial Diets for Gilthead Sea Bream (*Sparus aurata L.*); Effects on Growth Performance, Flesh Quality and Fillet Fatty Acid Profile. Recovery of Fatty Acid Profiles by a Fish Oil Finishing Diet under Fluctuating Water Temperatures. *Aquaculture* **2009**, *289*, 317–326. [CrossRef]
42. Sabbagh, M.; Schiavone, R.; Brizzi, G.; Sicuro, B.; Zilli, L.; Vilella, S. Poultry By-Product Meal as an Alternative to Fish Meal in the Juvenile Gilthead Seabream (*Sparus aurata*) Diet. *Aquaculture* **2019**, *511*, 734220. [CrossRef]
43. Navarro, A.; Zamorano, M.J.; Hildebrandt, S.; Ginés, R.; Aguilera, C.; Afonso, J.M. Estimates of Heritabilities and Genetic Correlations for Body Composition Traits and G × E Interactions, in Gilthead Seabream (*Sparus auratus L.*). *Aquaculture* **2009**, *295*, 183–187. [CrossRef]
44. Nolasco-Alzaga, H.R.; Perez-Enriquez, R.; Enez, F.; Bestin, A.; Palacios-Mechetnov, E.; Haffray, P. Quantitative Genetic Parameters of Growth and Fatty Acid Content in the Hemolymph of the Whiteleg Shrimp Litopenaeus Vannamei. *Aquaculture* **2018**, *482*, 17–23. [CrossRef]

MDPI

Article

Effects of Acute Cold Stress after Intermittent Cold Stimulation on Immune-Related Molecules, Intestinal Barrier Genes, and Heat Shock Proteins in Broiler Ileum

Xiaotao Liu [1], Shuang Li [1], Ning Zhao [1], Lu Xing [1], Rixin Gong [1], Tingting Li [1], Shijie Zhang [1], Jianhong Li [1,*] and Jun Bao [2,*]

[1] College of Life Science, Northeast Agricultural University, Harbin 150030, China
[2] College of Animal Science and Technology, Northeast Agricultural University, Harbin 150030, China
* Correspondence: jhli@neau.edu.cn (J.L.); 13312059804@163.com (J.B.)

Simple Summary: Animal welfare and health will be negatively impacted by cold stress, resulting in decreased production performance, immune imbalance, and decreased antioxidant capacity. Previous studies focused more on the side effects of low temperature, but animals have the ability to adapt to the environment by regulating metabolic and endocrine processes. The objective of this study was to compare cold resistant of broilers with and without cold stimulation training. By detecting the changes in immunoglobulins, cytokines, toll-like receptors, gut barrier genes, and heat shock proteins gene expression levels before and after acute cold stress, the optimal cold training method was finally determined. The results of our study show that cold stimulation training for 6 h with an interval of one day, at 3 °C lower than the conventional temperature, can change the immune function of broilers. This cold stimulation can also lessen the intestinal damage when subjected to acute cold stress. This research offers a theoretical foundation for the regulation of immune function in broilers raised in cold environments, as well as a scientific basis for the development of cold adaptation in broilers.

Citation: Liu, X.; Li, S.; Zhao, N.; Xing, L.; Gong, R.; Li, T.; Zhang, S.; Li, J.; Bao, J. Effects of Acute Cold Stress after Intermittent Cold Stimulation on Immune-Related Molecules, Intestinal Barrier Genes, and Heat Shock Proteins in Broiler Ileum. *Animals* **2022**, *12*, 3260. https://doi.org/10.3390/ani12233260

Academic Editors: María-Luz García and María-José Argente

Received: 22 September 2022
Accepted: 22 November 2022
Published: 23 November 2022

Abstract: Cold stress will have a negative impact on animal welfare and health. In order to explore the effect of intermittent cold stimulation training on the cold resistance of broilers. Immune-related and intestinal barrier genes were detected before and after acute cold stress (ACS), aiming to find an optimal cold stimulation training method. A total of 240 1-day-old Ross broilers (*Gallus*) were divided into three groups (G1, G2, and G3), each with 5 replicates (16 chickens each replicate). The broilers of G1 were raised at normal temperature, while the broilers of G2 and G3 were treated with cold stimulation at 3 °C lower than the G1 for 3 h and 6 h from 15 to 35 d, respectively, at one-day intervals. At 50 d, the ambient temperature for all groups was reduced to 10 °C for six hours. The results demonstrated that before ACS, *IL6*, *IL17*, *TLR21*, and *HSP40* mRNA levels in G3 were apparently down-regulated ($p < 0.05$), while *IL8* and *Claudin-1* mRNA levels were significantly up-regulated compared with G1 ($p < 0.05$). After ACS, *IL2*, *IL6*, and *IL8* expression levels in G3 were lower than those in G2 ($p < 0.05$). Compared to G2, *Claudin-1*, *HSP90* mRNA levels, *HSP40*, and *HSP70* protein levels were increased in G3 ($p < 0.05$). The mRNA levels of *TLR5*, *Mucin2*, and *Claudin-1* in G2 and *IL6*, *IL8*, and *TLR4* in G3 were down-regulated after ACS, while *IL2*, *IL6*, and *IL17* mRNA levels in G2 and *HSP40* protein levels in G3 were up-regulated after ACS ($p < 0.05$). Comprehensive investigation shows that cold stimulation at 3 °C lower than the normal feeding temperature for six hours at one day intervals can enhanced immune function and maintain the stability of intestinal barrier function to lessen the adverse effects on ACS in broilers.

Keywords: broiler ileum; cold stimulation; immunoglobulins; cytokines; toll-like receptors; heat shock proteins

1. Introduction

Cold is the most common stress factor for livestock and poultry in northern alpine regions [1]. For broilers (*Gallus*), exposure to a sudden drop in the ambient temperature of more than 10 °C or prolonged exposure to temperatures that are more than 4 °C lower than the ambient temperature will result in cold stress, which will have an impact on health and welfare of the birds [2] and lead to immune dysfunction [3] and physiological disorders [4,5]. Previous studies [4,6] have focused mostly on the side effects of exposure to low temperatures, but animals have the ability to adapt to their environment. Therefore, the present priority is to explore a training method to establish cold acclimation in poultry. After repeated cold stimulation, broilers can establish cold adaptation by regulating endocrine and metabolic processes and improve the ability of resist disease and cold stress [7,8]. The establishment of cold adaptation can, therefore, be measured by immune indices [9–11]. Immune indices include immunoglobulins, cytokines, and toll-like receptors. Immunoglobulin is a highly effective antibody. The initiation and regulation of the inflammatory response are significantly influenced by cytokines. The host can be protected from pathogen invasion by toll-like receptors. Su et al. [12] demonstrated that when broilers were trained with cold stimulation, the ileum structure was complete and the level of proinflammatory cytokines was reduced when broilers were subjected to 7 °C (24 h) ACS. Therefore, when broilers have thermoregulation ability, cold training for broilers is critical to establishing cold adaptation. Further explore strategies to improve the anti-stress and immune abilities of broilers.

To cope with various challenges, the body will activate the protein control system or cell death signaling pathway to produce an efficient stress response [13]. Immunoglobulins refer to proteins with antibody activity [14]. The mRNA level of immunoglobulin can reflect the strength of immune function. *IgA* performs a key function in protecting the gut against microbial invasion [15]. *IgG* has a range of biological activities, such as antiviral and anti-exotoxin. Studies have shown that broilers with colitis have lower levels of *IgA* and *IgG* in the ileum [16–18]. Olfati et al. The authors of [19] raised 22 d old broilers in a cold environment (12 °C) until they reached 42 d and noticed that cold stress decreased substantially serum *IgG* levels. Severe cold stress will reduce the expression level of immunoglobulins in animals, causing damage to the body. Early cold training has been shown by Su et al. [7] to up-regulate the intestinal *IgA* level in broilers and enhance broilers resistance to cold stress. It, therefore, follows that the body's immune function can be strengthened by the appropriate cold stimulation.

Cytokines are protein molecules produced by immune cells and related cells to regulate cellular functions and participate in anti-inflammatory and pro-inflammatory processes [8,20]. According to studies, prolonged cold stimulation in broilers' ileums can raise *IL-4* levels and lower *IFN-γ* levels, causing an immunological imbalance [21]. *IL-8* causes local inflammation in the body by stimulating leukocytes. *IFN-γ* can boost immune response capability by activating the activity of immune-related cell. Perelman et al. [22] reported that the levels of *IL-1β*, *IL-8*, and *TNF-α* in bronchial tissues of patients with cold airway hyperresponsiveness increased significantly after acute cold stress, resulting in a significant inflammatory response. *IL-6* can identify pathogen signals [20]. Yildirim et al. [23] found that the level of *IL-6* in liver tissues of rats exposed to a low temperature at 10 °C was significantly up-regulated, while Vargovic et al. [24] demonstrated that raising rats in a 4 °C cold environment for seven consecutive days reduced the expression level of *IL-6* in liver and lung tissue. Thus, stress can change immune function by regulating the secretion of cytokines.

Toll-like receptors not only mediate defense against the invasion of microbial pathogens, but also regulate immune homeostasis [25]. When toll-like receptors recognize pathogenic molecules, they can activate downstream signaling pathways (NF-κB) and induce the production of cytokines to participate in the immune response [26,27]. *TLR2* and *TLR4* can bind to the LPS of gram-negative bacteria and directly induce the expression of inflammatory cytokines [28]. *TLR5* can recognize the signal from the flagellum, which causes the production of proinflammatory factors. *TLR4* and *TLR5* levels in the mesenteric lymph nodes of Malabari goats were dramatically reduced, while the *TLR2* level was considerably elevated after five days of continuous exposure (six hours per day) at 27–34 °C in the summer [29]. Li et al. [30] discovered that the mRNA level of *TLR7* in the duodenum of broilers was substantially higher after cold stimulation training (43 d/3 °C lower than the control). *TLR7* is responsible for identifying the single-stranded RNA of viruses invading the body, which is crucial for defending the body against viral infection [31].

The intestinal barrier is a protective barrier that prevents pathogens and toxic compounds from passing through the systemic circulation [32]. The absence of a tight junction complex will reduce barrier protection and thus will affect intestinal permeability [33]. Studies have shown that markers for evaluating the health of the intestinal barrier can be *Claudin-1*, *Occludin*, and *ZO-1* mRNA levels [34–36]. *Mucin2* produced by goblet cells can protect the intestine from invasion by bacteria [37], has an antibacterial effect [38]. Uerlings et al. [39] reported that the levels of *Claudin-1* and *Occludin* in broilers' jejunum exposed to high temperatures for 24 h were significantly elevated. Some studies proved that long-term chronic stress can reduce the levels of tight junction proteins, impair intestinal barrier function, and increase permeability [40,41]. Therefore, whether early appropriate cold stimulation training can mitigate the adverse impacts of late ACS on the gut by changing the intestinal structure needs further study.

Heat shock proteins (HSPs) are vital elements in the regulation of the stress response and function to protect the integrity of epithelial cells and alleviate stress injury because they act as molecular chaperones [42]. Under normal physiological conditions, HSPs are maintained at low levels. When an organism encounters sudden environmental temperature changes, a large number of HSPs will be produced to enhance resistance to stress injury [43]. *HSP40* is a cooperative protein of *HSP70*, participating in the dissociation and transmembrane transport of proteins. The high expression level of *HSP70* can increase tolerance to various stressors, greatly improving the survival rate of the organism under stress [44]. Zaglool et al. [45] proved that *HSP90* level in broilers was rapidly up-regulated under acute heat stress at 36 °C for six hours. Wei et al. [20] showed that the lower *HSP90* levels in broilers' duodenum under acute cold stress and that cold stress altered immune function. In conclusion, the expression level of HSPs can be utilized as a stress response biomarker in animals.

The negative impacts of low temperatures on animals have earlier received more attention, but animals have the ability to adapt to the environment by regulating metabolic and endocrine processes [20,21]. Therefore, without affecting production performance, it is particularly important to explore methods of inducing cold adaptation in animals to enable them to resist cold stress. A comparative experiment was used in this study. The objective was to investigate the effects of ACS on immunity and resistance of broilers to cold stress. By analyzing the levels of gut immune-related and intestinal barrier genes in broilers before and after ACS, the optimal cold stimulation training program was defined which can improve cold resistance in broilers. The research can provide a theoretical foundation for the regulation of immune function of broilers in cold environments, and provide a scientific basis for the development of cold adaptation in broilers.

2. Materials and Methods

2.1. Animal Care and Experimental Design

All experiments and methods utilized in the current research were conducted with the support of the Institutional Committee for Animal Care and Use of the Northeast Agricultural University in Harbin, China (IACUCNEAU20150616). A comparative experiment was used in this research; 240 Ross 308 broilers (Hongyan breeding and planting Park, Harbin, China), 1-day-old, were divided into 3 groups including the one group without cold stimulation training (cold stimulation for 0 h, G1) and 2 groups with cold stimulation training (cold stimulation for 3 h, G2 and cold stimulation for 6 h, G3), 5 replicates per group, 16 chickens per replicate. Broilers were raised in three climate chambers, and food and water were freely available at all times. Broilers were fed the entire starter diet from 1 to 21 d, containing 12.10 MJ/kg of metabolizable energy and 21.00% crude protein. Broilers were raised a grower diet at 22 d (19.00% CP and 12.80 MJ/kg ME) and maintained for 3 weeks. Broilers were raised on a finishing diet from 43 to 50 days of age, including 17.50% of CP and 13.20 MJ/kg of ME. The diets consist of corn, vitamin A, calcium chloride, Soybean meal, bran, sodium chloride, and vitamin D3 (Baishicheng, Harbin, China). Lighting regime: 1–3 d, 24 L:0 D; 4–50 d, 23 L:1 D. The humidity was kept between 60% and 70% for the first 14 days and 40–50% from 15 to 50 days of age. The specific experimental temperature scheme is shown in Figure 1. The G1 was managed according to the standard feeding temperature during the growth stage of broilers. Broilers of G2 and G3 were raised at 3 °C lower than G1, and the duration of cold stimulation was 3 h (9:30 a.m.–12:30 p.m.) and 6 h (9:30 a.m.–15:30 p.m.) at one day intervals from 15 to 35 days, respectively. All broilers were raised at 20 °C for 36–49 d. At 50 d, all broilers were suffered from ACS (10 °C) for 6 h (8:00 a.m.–14:00 p.m.).

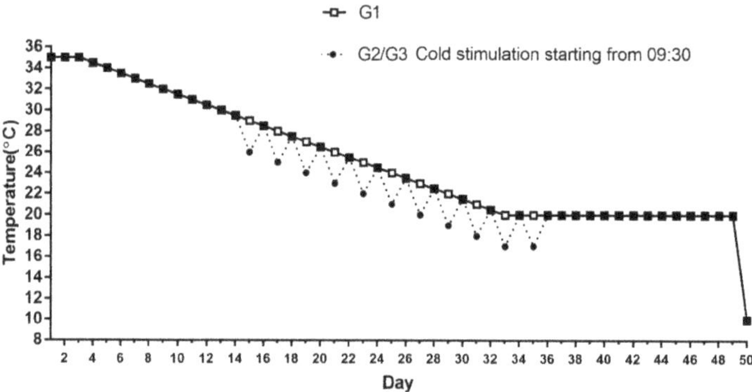

Figure 1. The specific experimental temperature scheme.

One broiler was selected from each replicate group and put to death at 08:00 a.m. (pre-ACS) and 14:00 p.m. (ACS) on day 50, sections of the ileum from the broilers were cut, cleaned with normal saline, put them into liquid nitrogen instantly, and only then stored at −80 °C.

2.2. RNA Extraction and Reverse Transcription

TRIzol (Takara, Japan) was utilized to extract total RNA from broiler ileum tissue samples in accordance with the manufacturer's instructions. The RNA was redissolved in 50 μL of enzyme-free water. Next, 1.5% agarose gel electrophoresis was used to determine the integrity of the RNA. The purity of the RNA was determined by measuring the OD260/OD280 ratio with a UV-spectrophotometer (Thermo Fisher Scientific, Carlsbad, CA, USA). Complementary DNA (cDNA) was synthesized by reverse transcription of RNA with ReverTra Ace qPCR RT Master Mix and gDNA Remover (Toyobo, Osaka, Japan), and then stored at −20 °C.

2.3. Quantitative Real-Time PCR Analysis

Sangon Bioengineering (Sangon Biotechnology, Shanghai, China) Co. Ltd. designed and synthesized the primers listed in Table S1. Real-time quantitative PCR was conducted using the LightCycler 480 II instrument (Roche, Switzerland). The total system consisted of 5 μL SYBR Green I, 3.4 μL enzyme-free water, 1 μL cDNA template, and 0.3 μL forward and reverse primers, respectively. qPCR program setup as: pre-denaturation at 95 °C for 60 s, repeated 40 cycles of denaturation at 95 °C for 15 s and then extension at 60 °C for 60 s. Each qPCR product was specific and its melting curve was unimodal. The $2^{-\Delta\Delta CT}$

method was used to calculate the mRNA levels of target genes, with the house-keeping gene β-actin as internal reference.

2.4. Western Blot Analysis

Western IP cell lysis solution (SparkJade, Harbin, China) containing 1% PMSF (SparkJade, Harbin, China) was used to extract total proteins from frozen broiler ileum tissue samples. The concentration of protein was quantified with Bicinchoninic Acid (BCA) protein concentration detection kit (SparkJade, Harbin, China) and adjusted to a uniform concentration (5 μg/μL). The same amounts of total protein (28 μg/condition) were placed on 12.5% gel (SparkJade, Harbin, China) for SDS-PAGE. The proteins were transferred to the NC membrane (SparkJade, Harbin, China) using semi-dry transfer equipment (Amersham Biosciences, Boston, MA, USA). Next, 5% skim milk 37 °C was sealed for 2 h and cleaned 3 times with PBST. Following that, the particular primary antibodies HSP40 and HSP60 (1:600, ABclonal, Harbin, China), and HSP70 (1:2700, ABclonal, Harbin, China) and β-actin (1:8000, Zenbio, Chengdu, China) were incubated with NC membrane. Then, IgG-HRP (1:8000, Zenbio, Chengdu, China) was incubated. The protein bands were then observed on a grayscale scanner (Gene Gnome XRQ, Cambridge, UK) using an ECL chemiluminescence kit (SparkJade, Harbin, China). The bands were evaluated using Image J (NIH, Bethesda, MD, USA), and the ratio of each target protein's gray value to that of β-actin was used to express the relative expression of HSPs.

2.5. Statistical Analysis

The data were analyzed using SPSS 21.0 software (IBM, Armonk, NY, USA). The normal distribution of the data was examined using the Kolmogorov–Smirnov method. The inter-group and intra-group differences were analyzed using one-way ANOVA. Duncan's was used for multiple comparisons. The data are presented as mean ± standard deviation (mean ± SD), with significant differences in $p < 0.05$.

3. Results

3.1. Relative Expression Levels of Immunoglobulins in Ileum Tissue

Figure 2 shows the mRNA expression levels of immunoglobulins in ilea of broilers before and after ACS. Before ACS, the level of *IgA* expression in the G1 was not vastly different than that in the cold stimulation group ($p > 0.05$). Lower expression level of *IgA* was detected in G2 compared to G3 ($p < 0.05$). Following ACS for 6 h, *IgA* mRNA expression was substantially lower in G1 and G3 than that in G2 ($p < 0.05$). *IgA* expression levels in G1 decreased but significantly increased in G2 after ACS ($p < 0.05$). After ACS, the *IgA* level in G3 did not dramatically change compared with pre-ACS ($p > 0.05$). No obvious difference in the level of *IgG* was observed in all groups before and after ACS ($p > 0.05$).

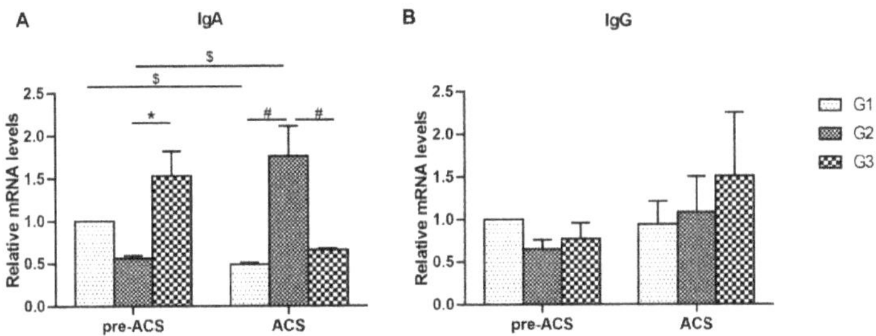

Figure 2. The mRNA levels of immunoglobulins IgA (**A**), and IgG (**B**) in ilea of broilers before and after ACS. Data were presented as mean ± SD ($n = 6$). Different labels represent significant differences ($p < 0.05$). pre-ACS (*), ACS (#) and before and after ACS ($). G1, cold stimulation for 0 h; G2, cold stimulation for 3 h; G3, cold stimulation for 6 h.

3.2. Relative Expression Levels of Cytokines in Ileum Tissue

Figure 3 shows the mRNA levels of cytokines in ilea of broilers before and after ACS. Before ACS, the mRNA expression level of IL2 did not differ markedly among all groups ($p > 0.05$). Higher levels of IL6, IL17, and IFN-γ were identified in G1 ($p < 0.05$). The IL8 mRNA expression level was increased dramatically in G2 and G3 compared to G1 and the mRNA expression level of IL8 increased gradually as the cold stimulation time was increased ($p < 0.05$). Following ACS for 6 h, the IL2 mRNA expression level in G3 was found to be down-regulated compared with G1 and G2 ($p < 0.05$), but the IL6, IL8, and IL17 expression levels in G2 compared to G1 and G3 were significantly higher ($p < 0.05$). The IFN-γ level did not differ significantly among treatment groups ($p > 0.05$). The levels of IL2 in G1, as well as IL2, IL6, IL8, and IL17, in G2 were elevated, but the levels of IL6 and IL17 in G1 and IL6 and IL8 in G3 were distinctly lower after ACS ($p < 0.05$). The IFN-γ level did not differ noticeably among treatment groups before and after ACS ($p > 0.05$).

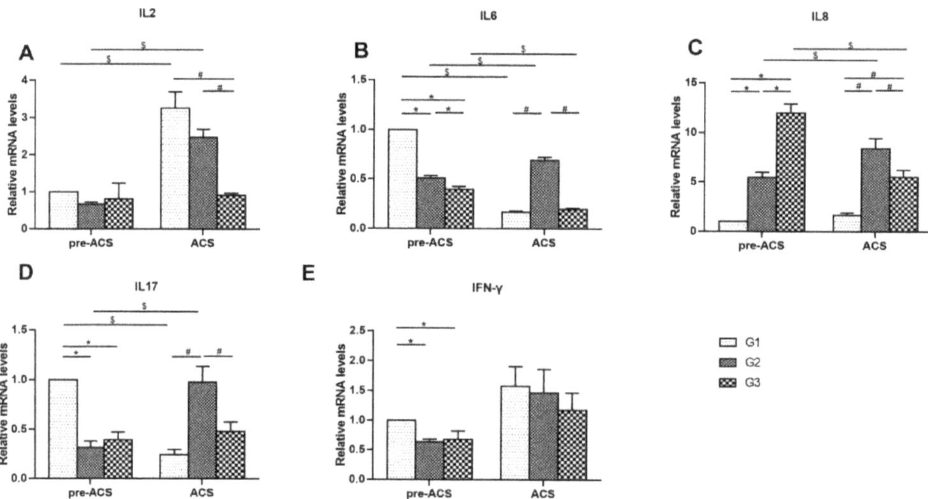

Figure 3. The mRNA levels of cytokines IL2 (**A**), IL6 (**B**), IL8 (**C**), IL17 (**D**), and IFN-γ (**E**) in ilea of broilers before and after ACS. Data were presented as mean ± SD ($n = 6$). Different labels represent significant differences ($p < 0.05$). pre-ACS (*), ACS (#) and before and after ACS ($). G1, cold stimulation for 0 h; G2, cold stimulation for 3 h; G3, cold stimulation for 6 h.

3.3. Relative Expression Levels of Toll-Like Receptors in Ileum Tissue

Figure 4 shows the mRNA levels of Toll-like receptors in ilea of broilers before and after ACS. Before ACS, the TLR5 mRNA expression level in G3 was down-regulated compared to G1 and G2 ($p < 0.05$), and the TLR21 mRNA expression in G2 and G3 was down-regulated compared to G1 ($p < 0.05$). TLR2, TLR4, and TLR7 expression levels were not substantially different among all the treatment groups ($p > 0.05$). Following ACS for 6 h, TLR7 mRNA levels were extremely higher in G2 than in G1 ($p < 0.05$), but no visible differences in TLR2, TLR4, and TLR5 mRNA levels were found among all groups ($p > 0.05$). The mRNA level of TLR21 in G2 showed a marked increase compared to G1 and G3 ($p < 0.05$). The expression levels of TLR4, TLR7, and TLR21 in G1 and TLR4, and TLR5 in G2, as well as TLR4 in G3, were dramatically reduced after ACS compared to pre-ACS ($p < 0.05$). There were no differences in TLR2, TLR5, TLR7, and TLR21 in G3 before and after ACS ($p > 0.05$).

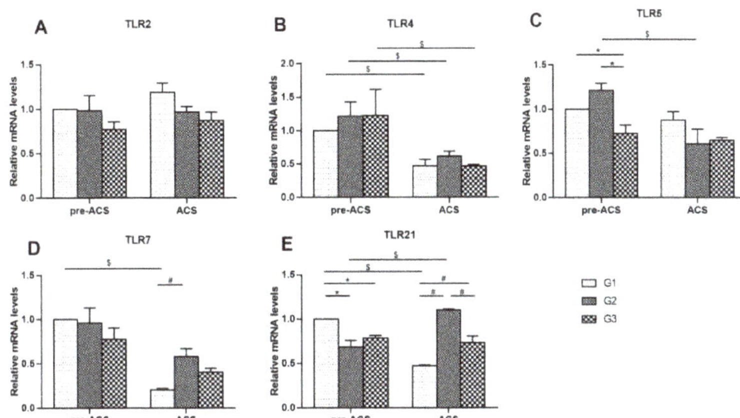

Figure 4. The mRNA levels of Toll-like receptors *TLR2* (**A**), *TLR4* (**B**), *TLR5* (**C**), *TLR7* (**D**), and *TLR21* (**E**) in ilea of broilers before and after ACS. Data were presented as mean ± SD ($n = 6$). Different labels represent significant differences ($p < 0.05$). pre-ACS (*), ACS (#) and before and after ACS ($). G1, cold stimulation for 0 h; G2, cold stimulation for 3 h; G3, cold stimulation for 6 h.

3.4. Relative Expression Levels of Intestinal Barrier Genes in Ileum Tissue

Figure 5 shows the mRNA levels of intestinal barrier genes in ilea of broilers before and after ACS. Before ACS, the mRNA level of Claudin-1 was dramatically up-regulated in G2 and G3 compared to G1 ($p < 0.05$). The expression level of Mucin2 in G3 was decreased compared to G1 and G2 ($p < 0.05$). The ZO-1 level in G3 was vastly higher than that in G1. The E-cadherin, ZO-2, and Occludin mRNA levels did not noticeably change among all the groups ($p > 0.05$). Following ACS for 6 h, the Claudin-1 mRNA level in G1 was lowest ($p < 0.05$) and the ZO-2 mRNA level was highest in G2 ($p < 0.05$). The E-cadherin, ZO-1, Occludin, and Mucin2 mRNA levels did not differ among all the groups ($p > 0.05$). Lower levels of Claudin-1, Occludin, ZO-2, and Mucin2 in G1 and Claudin-1 and Mucin2 in G2 were detected after ACS ($p < 0.05$), and the Claudin-1, ZO-1 levels in G3 were visibly lower after ACS ($p < 0.05$). The E-cadherin, Occludin, ZO-2, and Mucin2 levels in G3 did not markedly differ before and after ACS ($p > 0.05$).

3.5. Relative Expression Levels of Heat Shock Proteins in Ileum Tissue

Figures 6 and 7 show relative levels of heat shock proteins in ilea of broilers before and after ACS. Before ACS, *HSP40* mRNA and *HSP70* protein levels were highest in G1 ($p < 0.05$), and mRNA levels of *HSP60*, *HSP70*, and *HSP90* were considerably elevated in G3 compared to G1 and G2, respectively ($p < 0.05$). The *HSP60* protein levels in G2 were markedly higher than in G1 and G3 ($p < 0.05$). Following ACS for 6 h, a higher level of *HSP90* mRNA in G3 compared to G1 and G2 was observed ($p < 0.05$) and the protein levels of *HSP40* and *HSP70* were vastly up-regulated in G3 compared to G1 and G2 ($p < 0.05$). The mRNA level of *HSP40* in G2 was noticeably down-regulated compared to G1 and G3 ($p < 0.05$), and the mRNA level of *HSP60* in G2 was markedly lower compared to G1 ($p < 0.05$). The expression level of *HSP60* protein in G1 was highest ($p < 0.05$), but no vast difference in *HSP70* mRNA level was observed among the groups ($p > 0.05$). *HSP90* mRNA and *HSP60* protein in each group were decreased significantly after ACS ($p < 0.05$), while *HSP40* mRNA and *HSP40* protein level in G3 showed an increasing trend, but *HSP70* mRNA levels were dramatically down-regulated after ACS ($p < 0.05$). Finally, the levels of *HSP70* protein in G1 and G2 decreased significantly after ACS ($p < 0.05$).

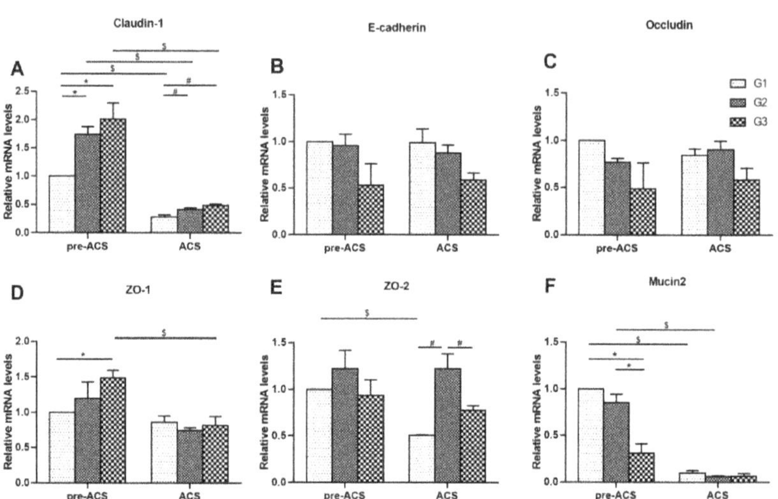

Figure 5. The mRNA levels of intestinal barrier genes *Claudin-1* (**A**), *E-cadherin* (**B**), *Occludin* (**C**), *ZO-1* (**D**), *ZO-2* (**E**), and *Mucin2* (**F**) in ilea of broilers before and after ACS. Data were presented as mean ± SD (n = 6). Different labels represent significant differences ($p < 0.05$). pre-ACS (*), ACS (#) and before and after ACS ($). G1, cold stimulation for 0 h; G2, cold stimulation for 3 h; G3, cold stimulation for 6 h.

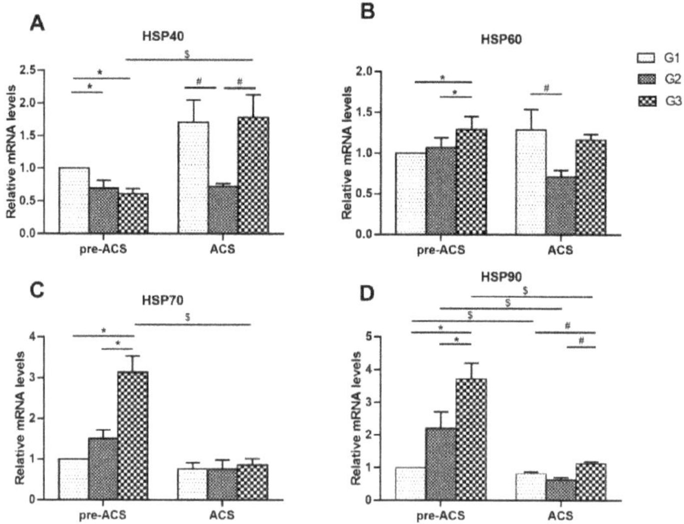

Figure 6. The mRNA levels of heat shock proteins *HSP40* (**A**), *HSP60* (**B**), *HSP70* (**C**), and *HSP90* (**D**) in ilea of broilers before and after ACS. Data were presented as mean ± SD (n = 6). Different labels represent significant differences ($p < 0.05$). pre-ACS (*), ACS (#) and before and after ACS ($). G1, cold stimulation for 0 h; G2, cold stimulation for 3 h; G3, cold stimulation for 6 h.

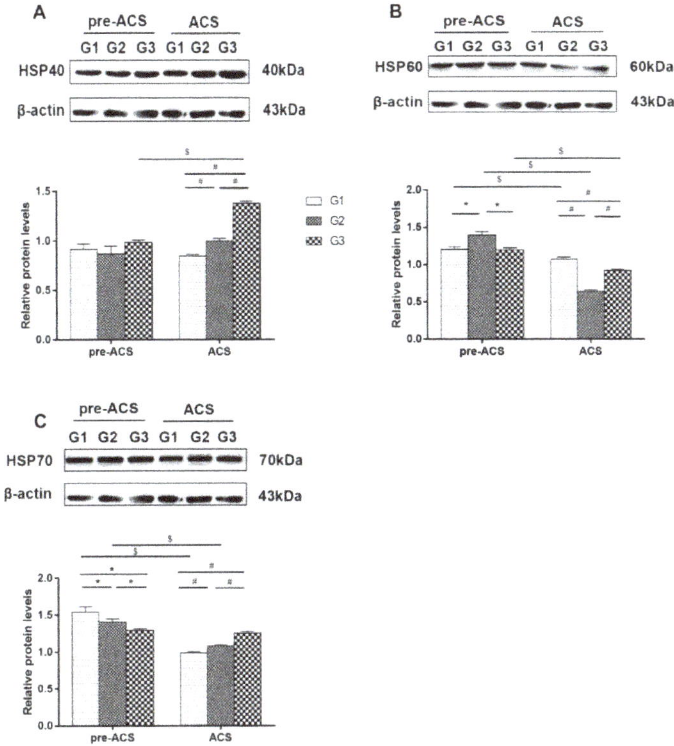

Figure 7. The protein levels of heat shock proteins *HSP40* (**A**), *HSP60* (**B**), and *HSP70* (**C**) in ilea of broilers before and after ACS. Data were presented as mean ± SD ($n = 6$). Different labels represent significant differences ($p < 0.05$). pre-ACS (*), ACS (#) and before and after ACS ($). G1, cold stimulation for 0 h; G2, cold stimulation for 3 h; G3, cold stimulation for 6 h. The original protein images of *HSP40*, *HSP60*, *HSP70*, and *β-actin* are shown in Figures S1–S4.

4. Discussion

Low temperatures have a serious effect on the development of animal husbandry. It is generally believed that an adverse environment, such as one that is cold, will have a negative impact on immune function [19,46,47] and cause inflammation or inflammatory diseases [5,48]. However, it has been demonstrated that appropriate cold stimulation training during the early growth stage of animals can make them adapt to a cold environment and improve their immune function and anti-stress ability [7,8]. Thaxton et al. [49] proved that exposing broilers to a cold environment can improve the synthetic rate of *IgA* and enhance humoral immunity. The aim of this research was to explore the effects of early intermittent cold stimulation training on ileum immune system and anti-stress capacity of broilers subjected to acute cold stress.

Immunoglobulins are crucial to the body's self-defense; their expression levels can reflect the body's resistance to disease as well as its immune status [50]. Cold stress can promote immunoglobulins expression levels and, thus, exert a protective effect on the intestine [7]. Zhao et al. [48] demonstrated that after exposing broilers to a low temperature for a period of time, *IgA* and *IgG* gene expression levels noticeably up-regulated in the duodenum and jejunum, and intestinal immune function was enhanced. Carr et al. [51] proved a vastly up-regulate in *IgA* mRNA expression in the intestinal tissues of mice (*Mus musculus*) exposed to −20 °C for nine consecutive days (20 min each time). Appropriate cold stimulation can, thus, increase the immunoglobulins expression levels and strengthen the body's immune ability and stress resistance. The current study revealed that after ACS, the level of *IgA* in the ileum of broilers in G1 were dramatically down-regulated, which indicates that ACS could damage broilers' immune systems. However, *IgA* levels in G2 trained with cold stimulation were significantly increased when subjected to ACS, suggesting that the broilers need to up-regulate immunoglobulin level to protect the intestine. The G3 group have already established cold adaptation, did not need to produce too many immunoglobulins to relieve ACS (10 °C). Similar results were found that immunomodulatory and antioxidative functions of broilers can be improved when broilers are kept at 3 °C lower than ambient temperature for a long time before being subjected to acute cold stress [20,21]. These results further demonstrate that moderate cold stimulation in early life can increase the expression levels of immunoglobulins, and relieve the harm caused by ACS at a later stage.

Cytokines are absolutely necessary for the two-way communication between the immune and endocrine systems [52,53]. Th1 cells produce *IFN-γ* and *IL-2*, which are responsible for cellular immunity, whereas Th2 cells produce *IL-4* and *IL-10*, which are responsible for humoral immunity [54]. Cold stimulation can affect the body's immune system by affecting the content of cytokines. Zhao et al. [48] found that exposure of broilers to 12 °C for 21 days significantly reduced *IFN-γ* but *IL4* significantly increased in their ileum, leading to an immune imbalance. In the present study, after acute cold stimulation, the level of *IL2* in broilers of G1 without cold stimulation training was significantly increased, which suggested that broilers were affected by cold stress and responded to the low temperature by improving cellular immunity. A similar result was reported that the level of *IL2* in the serum of Wistar rats was significantly increased after three days/4 °C cold stress [55]. The levels of *IL2* in G3 showed no significant difference after ACS, indicating that broilers had adapted to the cold environment due to early cold training. According to the study's findings, ACS considerably increased the *IL6*, *IL8*, and *IL17* level in the G2 while dramatically lowering *IL8* level in the G3 group, indicating that environmental changes of this nature will have an adverse impact on G2 group, and induce the body to produce pro-inflammatory factors. However, the G3 group was better adapted to the cold environment after early cold stimulation training, which is not needed to aggregate pro-inflammatory factors and cause an inflammatory response. Brenner demonstrated that plasma *IL6* expression levels were increased after exposure to a cold environment [56]. Monroy's study showed that cold water stress in mice for five minutes a day led to an increase in *IFN-γ* and *IL6* protein levels [57]. It can be concluded that broilers with established cold adaptation will inhibit the release of proinflammatory cytokines when suffering from ACS at 10 °C. The anti-stress ability of broilers can be improved after suitable cold stimulation training at an early stage of life.

Pattern-recognition receptors known as toll-like receptors (TLRs) are a principal mediator of the innate immune response [58]. TLRs are involved in the acute response phase of the body [59]. TLRs can cooperate with each other to promote the immune response [28,60]. *TLR4*, *TLR7*, and *TLR21* expression levels in G1 broilers without cold stimulation training were considerably down-regulated after ACS, indicating that ACS leads to immune dysfunction in broilers. Quinteiro-Filho reported that broilers receiving chronic heat stimulation for 10 h every day until six days before being sampled could decrease the amount of *TLR4* expression in their spleens and impair their immune function [61]. However, the *TLR2*, *TLR5*, *TLR7*, and *TLR21* levels in broilers of G3 trained with cold stimulation were not significantly different before and after ACS, indicating that the early cold stimulation training made broilers adapt to cold environments and improve cold resistance. In the present study, after ACS, the level of *TLR5* dropped considerably in broilers of G2, the reason may be that the broilers had not yet established cold adaptation. The result was consistent with Basu. Basu et al. [62] who showed that *TLR5* in Indian major carp catla (*Catla catla*) significantly decreased after acute cold stress. This indicates that the body is not strong enough to resist ACS. Broilers of G2 would mobilize *TLR21* to resist cold stress after ACS. *TLR21* level in G3 was not vastly difference before and after ACS, indicating that acute cold stress was within the range of adjustment of the body, and early cold training enhanced stress resistance.

The intestinal barrier separates the enteric cavity of the intestine from the internal environment of the organism. *Occludin* and *Claudins* form the tight junction backbone [41]. Junction adhesion molecules are valuable in immune cells transfer during immune surveillance and the inflammatory response [63]. In our investigation, the *Occludin* and *ZO-2* mRNA levels in broilers of G1 were dramatically down-regulated after ACS, which was in line with He's discovery. He et al. [28] proved that heat stress for three days could reduce the protein levels of *Occludin* and *ZO-2* in the small intestine of rats (*Rattus norvegicus*). However, the *Occludin* expression level in G2 and G3 groups did not noticeably change before and after ACS. Based on the study's findings, when broilers of G1 without cold stimulation training are exposed to ACS, intestinal permeability increases and tight junction gene expression levels in the ileum significantly decrease, whereas broilers of G3 can adapt to a cold environment and resist the harm caused by ACS. The expression of *Mucin2* dropped considerably after ACS in G1 and G2, similar results were found in the research of Zhang [37]. The above results indicate that cold stress has adverse effects on the intestinal tract and inhibits the production of *Mucin2*. However, the *Mucin2* level did not dramatically change in G3 before and after ACS, which may be because the broilers experienced an improved ability to resist cold stress in the early stage of cold training, so as to activate the body's protective mechanism to prevent bacteria from invading the intestinal wall. In conclusion, broilers without established cold adaptation will have down-regulated intestinal barrier gene levels and increased intestinal permeability when subjected to ACS, while cold stimulation training for six hours at one day intervals can lead to an ability to resist the intestinal damage caused by acute cold stress at a later stage.

Heat shock proteins (HSPs) exert a protective effect on cells [64,65]. When cells are stimulated by stress, they will induce the synthesis of HSPs, which participate in cell protective functions and increase the tolerance of cells to stress. Xu et al. [44] found that *HSP27* and *HSP70* levels were dramatically elevated in porcine cardiomyocytes under transport stress, and this increase in expression was accompanied by a reduction in myocardial injury, so this may be associated with the protection of cardiomyocytes. Wei et al. [20] confirmed that 24 h of acute cold stress in broilers after 34 days of chronic cold stimulation markedly up-regulated the expression of *HSP40* in the heart. In the current research, the G3 group's levels of *HSP40* protein and mRNA considerably increased after ACS compared to pre-ACS, which is consistent with the results of Wei's study, suggesting that the body alleviates the damage caused by stress by up-regulating the expression of *HSP40*. *HSP70* and *HSP90* are vital to prevent damage to intestinal epithelial and mucosal cells [66]. In this study, the HSPs (*HSP60, HSP70, HSP90*) levels in G3 substantially higher than those of in G1 during early cold stimulation, indicating that cold stress will modulate HSPs levels, thus protecting the intestinal mucosal barrier [67] and helping the body adapt to the environment. Puijvelde et al. [68] showed that the high levels of HSPs under the same stimulation represent a potent capacity for stress resistance. The G1 and G2 groups in this study had considerably lower levels of *HSP90* mRNA, *HSP40* protein, and *HSP70* protein after ACS than in the G3 group, indicating that the G1 and G2 groups had a weak ability to resist cold stress. It can, thus, be seen that broilers trained by 6 h of cold stimulation have improved anti-stress abilities and can swiftly activate the expression of HSPs when exposed to ACS.

5. Conclusions

The findings of the present research show that cold stimulation training with an interval of one day, at 3 °C below the conventional temperature can change the immune function and improve cold resistance, and that the intestinal damage can be relieved when subjected to ACS in the later stage. Moreover, the six-hour intermittent cold stimulation scheme makes broilers more resistant to cold stress.

Supplementary Materials: The following supporting information can be downloaded at: https://www.mdpi.com/article/10.3390/ani12233260/s1, Figure S1: The original protein image of *HSP40*; Figure S2: The original protein image of *HSP60*; Figure S3: The original protein image of *HSP70*; Figure S4: The original protein image of *β-actin*; Table S1: Primer sequences used for the study.

Author Contributions: Conceptualization, X.L., J.L. and J.B.; methodology, X.L. and S.L.; software, N.Z.; validation, L.X. and J.L.; formal analysis, X.L. and S.L.; investigation, R.G. and T.L.; resources, S.Z.; data curation, L.X.; writing—original draft preparation, X.L.; writing—review and editing, X.L. and J.L.; supervision, J.L.; project administration, J.B.; funding acquisition, J.L. All authors have read and agreed to the published version of the manuscript.

Funding: This research was funded by the National Natural Science Foundation of China (Grant No. 32172785).

Institutional Review Board Statement: The animal study protocol was approved by the Institutional Committee for Animal Care and Use of the Northeast Agricultural University in Harbin, China (IACUCNEAU20150616).

Informed Consent Statement: Not applicable.

Data Availability Statement: Not applicable.

Acknowledgments: The authors thank the Animal Behavior and Welfare Laboratory for its technical and human support in Northeast Agricultural University.

Conflicts of Interest: The authors declare no conflict of interest.

References

1. Nyuiadzi, D.; Berri, C.; Dusart, L.; Travel, A.; Méda, B.; Bouvarel, I.; Guilloteau, L.A.; Chartrin, P.; Coustham, V.; Praud, C.; et al. Short cold exposures during incubation and postnatal cold temperature affect performance, breast meat quality, and welfare parameters in broiler chickens. *Poult. Sci.* **2020**, *99*, 857–868. [CrossRef]
2. Xie, S.S.; Yang, X.K.; Gao, Y.H.; Jiao, W.J.; Li, X.H.; Li, Y.J.; Ning, Z.H. Performance differences of Rhode Island Red, Bashang Long-tail Chicken, and their reciprocal crossbreds under natural cold stress. *Asian-Australas J. Anim. Sci.* **2017**, *30*, 1507–1514. [CrossRef] [PubMed]
3. Aarif, O.; Shergojry, S.A.; Dar, S.A.; Khan, N.; Mir, N.A.; Sheikh, A.A. Impact of cold stress on blood biochemical and immune status in male and female vanaraja chickens. *Indian J. Anim. Res.* **2014**, *48*, 139–142. [CrossRef]
4. Tsiouris, V.; Georgopoulou, I.; Batzios, C.; Pappaioannou, N.; Ducatelle, R.; Fortomaris, P. The effect of cold stress on the pathogenesis of necrotic enteritis in broiler chicks. *Avian Pathol.* **2015**, *44*, 430–435. [CrossRef] [PubMed]
5. Hu, J.Y.; Cheng, H.W. Warm perches: A novel approach for reducing cold stress effect on production, plasma hormones, and immunity in laying hens. *Poult. Sci.* **2021**, *100*, 101294. [CrossRef] [PubMed]
6. Spinu, M.; Degen, A.A. Effect of cold stress on performance and immune responses of Bedouin and White Leghorn hens. *Br. Poult. Sci.* **1993**, *34*, 177–185. [CrossRef]
7. Su, Y.Y.; Wei, H.D.; Bi, Y.J.; Wang, Y.A.; Zhao, P.; Zhang, R.X.; Li, J.H.; Bao, J. Pre-cold acclimation improves the immune function of trachea and resistance to cold stress in broilers. *J. Cell. Physiol.* **2018**, *234*, 7198–7212. [CrossRef]
8. Liu, Y.H.; Xue, G.; Li, S.; Fu, Y.J.; Yin, J.W.; Zhang, R.X.; Li, J.H. Effect of Intermittent and Mild Cold Stimulation on the Immune Function of Bursa in Broilers. *Animals* **2020**, *10*, 1275. [CrossRef]
9. Xue, G.; Yin, J.W.; Zhao, N.; Liu, Y.H.; Fu, Y.J.; Zhang, R.X.; Bao, J.; Li, J.H. Intermittent mild cold stimulation improves the immunity and cold resistance of spleens in broilers. *Poult. Sci.* **2021**, *100*, 101492. [CrossRef]
10. Li, J.H.; Huang, F.F.; Li, X.; Su, Y.Y.; Li, H.T.; Bao, J. Effects of intermittent cold stimulation on antioxidant capacity and mRNA expression in broilers. *Livest. Sci.* **2017**, *204*, 110–114. [CrossRef]
11. Shinder, D.; Luger, D.; Rusal, M.; Rzepakosky, V.; Bresler, V.; Yahav, S. Early age cold conditioning in broiler chickens (*Gallus domesticus*): Thermotolerance and growth responses. *J. Therm. Biol.* **2002**, *27*, 517–523. [CrossRef]
12. Su, Y.Y.; Li, S.; Xin, H.W.; Li, J.F.; Li, X.; Zhang, R.X.; Li, J.H.; Bao, J. Proper cold stimulation starting at an earlier age can enhance immunity and improve adaptability to cold stress in broilers. *Poult. Sci.* **2020**, *99*, 129–141. [CrossRef] [PubMed]
13. Nguyen, P.; Greene, E.; Ishola, P.; Huff, G.; Donoghue, A.; Bottje, W.; Dridi, S. Chronic Mild Cold Conditioning Modulates the Expression of Hypothalamic Neuropeptide and Intermediary Metabolic-Related Genes and Improves Growth Performances in Young Chicks. *PLoS ONE* **2015**, *10*, e0142319. [CrossRef] [PubMed]
14. Vordenbäumen, S.; Otte, J.-M. Role of Antimicrobial Peptides in Inflammatory Bowel Disease. *Polymers* **2011**, *3*, 2010–2017. [CrossRef]
15. Wu, B.Y.; Guo, H.R.; Cui, H.M.; Peng, X.; Fang, J.; Zuo, Z.C.; Deng, J.L.; Wang, X.; Hang, J.Y. Pathway underlying small intestine apoptosis by dietary nickel chloride in broiler chickens. *Chem. Biol. Interact.* **2016**, *24*, 91–106. [CrossRef] [PubMed]

16. Simon, K.; Arts, J.A.J.; de Vries Reilingh, G.; Kemp, B.; Lammers, A. Effects of early life dextran sulfate sodium administration on pathology and immune response in broilers and layers. *Poult. Sci.* **2016**, *95*, 1529–1542. [CrossRef] [PubMed]
17. Sacks, D.; Baxter, B.; Campbell, B.C.V.; Carpenter, J.S.; Cognard, C.; Dippel, D.; Eesa, M.; Fischer, U.; Hausegger, K.; Hirsch, J.A. Multisociety Consensus Quality Improvement Revised Consensus Statement for Endovascular Therapy of Acute Ischemic Stroke. *Int. J. Stroke* **2018**, *13*, 612–632. [CrossRef] [PubMed]
18. Selecká, E.; Levkut, M., Jr.; Revajová, V.; Levkutová, M.; Karaffová, V.; Ševčíková, Z.; Herich, R.; Levkut, M. Research Note: Immunocompetent cells in blood and intestine after administration of Lacto-Immuno-Vital in drinking water of broiler chickens. *Poult. Sci.* **2021**, *100*, 101282. [CrossRef]
19. Olfati, A.; Mojtahedin, A.; Sadeghi, T.; Akbari, M.; Martinez, P.F. Comparison of growth performance and immune responses of broiler chicks reared under heat stress, cold stress and thermoneutral conditions. *Span. J. Agric. Res.* **2018**, *16*, 1–7. [CrossRef]
20. Wei, H.D.; Zhang, R.X.; Su, Y.Y.; Bi, Y.J.; Li, X.; Zhang, X.; Li, J.H.; Bao, J. Effects of Acute Cold Stress after Long-Term Cold Stimulation on Antioxidant Status, Heat Shock Proteins, Inflammation and Immune Cytokines in Broiler Heart. *Front. Physiol.* **2018**, *9*, 1589. [CrossRef]
21. Su, Y.Y.; Zhang, X.; Xin, H.W.; Li, S.; Li, J.H.; Bao, J. Effects of prior cold stimulation on inflammatory and immune regulation in ileum of cold-stressed broilers. *Poult. Sci.* **2018**, *97*, 4228–4237. [CrossRef]
22. Perelman, J.M.; Pirogov, A.B.; Naumov, D.E.; Gassan, D.A.; Afanaseva, E.Y.; Kotova, O.O. Dynamics of Cell Pattern and Profile of Bronchial Cytokines in Response to Acute Cold Exposure in Asthma Patients with Cold Airway Hyperresponsiveness. *Am. J. Respir. Crit. Care Med.* **2020**, *201*, A7394. [CrossRef]
23. Yildirim, N.C.; Yurekli, M. The effect of adrenomedullin and cold stress on interleukin-6 levels in some rat tissues. *Clin. Exp. Immunol.* **2010**, *161*, 171–175. [CrossRef]
24. Vargovic, P.; Laukova, M.; Ukropec, J.; Manz, G.; Kvetnansky, R. Prior Repeated Stress Attenuates Cold-Induced Immunomodulation Associated with "Browning" in Mesenteric Fat of Rats. *Cell. Mol. Neurobiol.* **2018**, *38*, 349–361. [CrossRef]
25. Takeuchi, O.; Akira, S. Pattern Recognition.n Receptors and Inflammation. *Cell* **2010**, *140*, 805–820. [CrossRef]
26. Velová, H.; Gutowska-Ding, M.W.; Burt, D.W.; Vinkler, M. Toll-like Receptor Evolution in Birds: Gene Duplication, Pseudogenization, and Diversifying Selection. *Mol. Biol. Evol.* **2018**, *35*, 2170–2184. [CrossRef]
27. Gan, F.; Liu, Q.; Liu, Y.; Huang, D.; Pan, C.; Song, S.; Huang, K. *Lycium barbarum* polysaccharides improve CCl 4 -induced liver fibrosis, inflammatory response and TLRs/NF-kB signaling pathway expression in wistar rats. *Life Sci.* **2018**, *192*, 205–212. [CrossRef]
28. He, Z.Z.; Li, J.J.; Mei, Y.; Lv, Y.; Zhu, F.; Liu, R.; Wang, X.; Yao, F.H.; Zhang, L. The role and difference of TLR2 and TLR4 in rhabdomyolysis induced acute kidney injury in mice. *Int. J. Clin. Exp. Pathol.* **2018**, *11*, 1054–1061.
29. Vandana, G.D.; Bagath, M.; Sejian, V.; Krishnan, G.; Beena, V.; Bhatta, R. Summer season induced heat stress impact on the expression patterns of different toll-like receptor genes in Malabari goats. *Biol. Rhythm Res.* **2019**, *50*, 446–482. [CrossRef]
30. Li, S.; Li, J.H.; Liu, Y.H.; Li, C.; Zhang, R.X.; Bao, J. Effects of Intermittent Mild Cold Stimulation on mRNA Expression of Immunoglobulins, Cytokines, and Toll-Like Receptors in the Small Intestine of Broilers. *Molecules* **2020**, *10*, 1492. [CrossRef]
31. Brownlie, R.; Allan, B. Avian toll-like receptors. *Cell Tissue Res.* **2011**, *343*, 121–130. [CrossRef]
32. Theerawatanasirikul, S.; Koomkrong, N.; Kayan, A.; Boonkaewwan, C. Intestinal barrier and mucosal immunity in broilers, Thai Betong, and native Thai Praduhangdum chickens. *Turk. J. Vet. Anim. Sci.* **2017**, *41*, 357–364. [CrossRef]
33. Lynn, K.S.; Peterson, R.J.; Koval, M. Ruffles and spikes: Control of tight junction morphology and permeability by claudins. *Biochim. Biophys. Acta Biomembr.* **2020**, *1862*, 183339. [CrossRef] [PubMed]
34. Assimakopoulos, S.F.; Akinosoglou, K.; de Lastic, A.L.; Skintzl, A.; Mouzaki, A.; Gogos, C. The Prognostic Value of Endotoxemia and Intestinal Barrier Biomarker ZO-1 in Bacteremic Sepsis. *Am. J. Med. Sci.* **2019**, *359*, 100–107. [CrossRef]
35. Zhou, H.J.; Kong, L.L.; Zhu, L.X.; Hu, X.Y.; Busye, J.; Song, Z.G. Effects of cold stress on growth performance, serum biochemistry, intestinal barrier molecules, and adenosine monophosphate-activated protein kinase in broilers. *Animal* **2021**, *15*, 100138. [CrossRef]
36. Al-Sadi, R.; Nighot, P.; Nighot, M.; Haque, M.; Rawat, M.; Ma, T.Y. Lactobacillus acidophilus Induces a Strain-specific and Toll-like Receptor 2-Dependent Enhancement of Intestinal Epithelial Tight Junction Barrier and Protection Against Intestinal Inflammation. *Am. J. Pathol.* **2021**, *191*, 872–884. [CrossRef]
37. Zhang, Q.; Eicher, S.D.; Applegate, T.J. Development of intestinal mucin 2, IgA, and polymeric Ig receptor expressions in broiler chickens and Pekin ducks. *Poult. Sci.* **2015**, *94*, 172–180. [CrossRef]
38. Cobo, E.R.; Kissoon-Singh, V.; Moreau, F.; Holani, R.; Chadee, K. MUC2 Mucin and Butyrate Contribute to the Synthesis of the Antimicrobial Peptide Cathelicidin in Response to *Entamoeba histolytica*- and Dextran Sodium Sulfate-Induced Colitis. *Infect. Immun.* **2017**, *85*, e00905-16. [CrossRef]
39. Uerlings, J.; Song, Z.G.; Hu, X.Y.; Wang, S.K.; Lin, H.; Buyse, J.; Everaert, N. Heat exposure affects jejunal tight junction remodeling independently of adenosine monophosphate-activated protein kinase in 9-day-old broiler chicks. *Poult. Sci.* **2018**, *97*, 3681–3690. [CrossRef]
40. Tsiouris, V.; Georgopoulou, I.; Batzios, C.; Pappaioannou, N.; Ducatelle, R.; Fortomaris, P. Heat stress as a predisposing factor for necrotic enteritis in broiler chicks. *Avian Pathol.* **2018**, *47*, 1–99. [CrossRef]

41. Zheng, G.; Victor Fon, G.; Meixner, W.; Creekmore, A.; Zong, Y.K.; Dame, M.; Colacino, J.; Dedhia, P.H.; Hong, S.; Wiley, J.W. Chronic stress and intestinal barrier dysfunction: Glucocorticoid receptor and transcription repressor HES1 regulate tight junction protein Claudin-1 promoter. *Sci. Rep.* **2017**, *7*, 4502. [CrossRef]
42. Hao, Y.; Feng, Y.J.; Li, J.L.; Gu, X.H. Role of MAPKs in HSP70's Protection against Heat Stress-Induced Injury in Rat Small Intestine. *Biomed Res. Int.* **2018**, *2018*, 1571406. [CrossRef]
43. Flees, J.; Rajaei-Sharifabadi, H.; Greene, E.; Beer, L.; Hargis, B.M.; Ellestad, L.; Porter, T.; Donoghue, A.; Bottje, W.G.; Dridi, S. Effect of Morinda citrifolia (Noni)-Enriched Diet on Hepatic Heat Shock Protein and Lipid Metabolism-Related Genes in Heat Stressed Broiler Chickens. *Front. Physiol.* **2017**, *27*, 919. [CrossRef]
44. Xu, J.; Tang, S.; Song, E.; Yin, B.; Wu, D.; Bao, E. Hsp70 expression induced by Co-Enzyme Q10 protected chicken myocardial cells from damage and apoptosis under in vitro heat stress. *Poult. Sci.* **2017**, *96*, 1426–1437. [CrossRef]
45. Zaglool, A.W.; Roushdy, E.M.; El-Tarabany, M.S. Impact of strain and duration of thermal stress on carcass yield, metabolic hormones, immunological indices and the expression of HSP90 and Myogenin genes in broilers. *Res. Vet. Sci.* **2018**, *122*, 193–199. [CrossRef]
46. Shah, S.W.A.; Chen, J.; Han, Q.; Xu, Y.; Ishfaq, M.; Teng, X.H. Ammonia inhalation impaired immune function and mitochondrial integrity in the broilers bursa of fabricius: Implication of oxidative stress and apoptosis. *Ecotoxicol. Environ. Saf.* **2020**, *190*, 110078. [CrossRef]
47. Shah, S.W.A.; Chen, D.; Zhang, J.Y.; Liu, Y.L.; Ishfaq, M.; Teng, X.H. The effect of ammonia exposure on energy metabolism and mitochondrial dynamic proteins in chicken thymus: Through oxidative stress, apoptosis, and autophagy. *Ecotoxicol. Environ. Saf.* **2020**, *206*, 111413. [CrossRef]
48. Zhao, F.Q.; Zhang, Z.W.; Yao, H.D.; Wang, L.L.; Liu, T.; Yu, X.Y.; Li, S.; Xu, S.W. Effects of cold stress on mRNA expression of immunoglobulin and cytokine in the small intestine of broilers. *Res. Vet. Sci.* **2013**, *95*, 146–155. [CrossRef]
49. Thaxton, P. Influence of temperature on the immune response of birds. *Poult. Sci.* **1978**, *57*, 1430–1440. [CrossRef]
50. Lanzarini, N.M.; Bentes, G.A.; Volotão, E.M.; Pinto, M.A. Use of chicken immunoglobulin Y in general virology. *J. Immunoass. Immunochem.* **2018**, *39*, 235–248. [CrossRef]
51. Carr, D.J.J.; Woolley, T.W.; Blalock, J.E. Phentolamine but not propranolol blocks the immunopotentiating effect of cold stress on antigen-specific IgM production in mice orally immunized with sheep red blood cells. *Brain Behav. Immun.* **1992**, *6*, 50–63. [CrossRef] [PubMed]
52. Felten, S.Y.; Madden, K.S.; Bellinger, D.L.; Kruszewska, B.; Moynihan, J.A.; Felten, D.L. The role of the sympathetic nervous system in the modulation of immune responses. *Advan Pharmacol.* **1998**, *42*, 583–587. [CrossRef]
53. Al-Zghoul, M.B.; Saleh, K.M.; Ababneh, M.M.K. Effects of pre-hatch thermal manipulation and post-hatch acute heat stress on the mRNA expression of interleukin-6 and genes involved in its induction pathways in 2 broiler chicken breeds. *Poult. Sci.* **2019**, *98*, 1805–1819. [CrossRef] [PubMed]
54. Shieh, Y.H.; Huang, H.M.; Wang, C.C.; Lee, C.C.; Fan, C.K.; Lee, Y.L. Zerumbone enhances the Th1 response and ameliorates ovalbumin-induced Th2 responses and airway inflammation in mice. *Int. Immunopharmacol.* **2015**, *24*, 383–391. [CrossRef] [PubMed]
55. Guo, J.R.; Li, S.Z.; Fang, H.G. Different Duration of Cold Stress Enhances Pro-Inflammatory Cytokines Profile and Alterations of Th1 and Th2 Type Cytokines Secretion in Serum of Wistar Rats. *J. Anim. Vet. Adv.* **2012**, *11*, 1538–1545.
56. Brenner, I.K.; Castellani, J.W.; Gabaree, C.; Young, A.J.; Zamecnik, J.; Shephard, R.J.; Shek, P.N. Immune changes in humans during cold exposure: Effects of prior heating and exercise. *J. Appl. Physiol.* **1999**, *87*, 699–710. [CrossRef]
57. Monroy, F.P.; Banerjee, S.K.; Duong, T.; Aviles, H. Cold stress-induced modulation of inflammatory responses and intracerebral cytokine mRNA expression in acute murine toxoplasmosis. *J. Parasitol.* **1999**, *85*, 878–886. [CrossRef]
58. Alcala, A.; Osborne, B.; Allen, B.; Seaton-Terry, A.; Kirkland, T.; Whalen, M. Toll-like receptors in the mechanism of tributyltin-induced production of pro-inflammatory cytokines, IL-1β and IL-6. *Toxicology* **2022**, *30*, 153177. [CrossRef]
59. Titi, L.O.; Mohammed, I.; Amoaku, W.M. Toll-like Receptor Signalling Pathways and the Pathogenesis of Retinal Diseases. *Front. Ophthalmol.* **2022**, *2*, 850394. [CrossRef]
60. Wu, M.H.; Zhang, P.; Huang, X. Toll-like receptors in innate immunity and infectious diseases. *Front. Med. China* **2010**, *4*, 385–393. [CrossRef]
61. Quinteiro-Filho, W.M.; Calefi, A.S.; Cruz, D.S.G.; Aloia, T.P.A.; Zager, A.; Astolfi-Ferreira, C.S.; Piantino Ferreira, J.A.; Sharif, S.; Palermo-Neto, J. Heat stress decreases expression of the cytokines, avian beta-defensins 4 and 6 and Toll-like receptor 2 in broiler chickens infected with *Salmonella* Enteritidis. *Vet. Immunol. Immunopathol.* **2017**, *40*, 19–28. [CrossRef]
62. Basu, M.; Paichha, M.; Swain, B.; Lenka, S.S.; Singh, S.; Chakrabarti, R.; Samanta, M. Modulation of TLR2, TLR4, TLR5, NOD1 and NOD2 receptor gene expressions and their downstream signaling molecules following thermal stress in the Indian major carp catla (*Catla catla*). *3 Biotech* **2015**, *5*, 1021–1030. [CrossRef]
63. Varasteh, S.; Braber, S.; Akbari, P.; Garssen, J.; Fink-Gremmels, J. Differences in Susceptibility to Heat Stress along the Chicken Intestine and the Protective Effects of Galacto-Oligosaccharides. *PLoS ONE* **2017**, *10*, e0138975. [CrossRef] [PubMed]
64. Sreedhar, A.S.; Csermely, P. Heat shock proteins in the regulation of apoptosis: New strategies in tumor therapy: A comprehensive review. *Pharmacol. Ther.* **2004**, *101*, 227–257. [CrossRef]
65. Khoso, P.A.; Yang, Z.J.; Liu, C.P.; Li, S. Selenium Deficiency Downregulates Selenoproteins and Suppresses Immune Function in Chicken Thymus. *Biol. Trace Elem. Res.* **2015**, *167*, 48–55. [CrossRef]

66. Takada, M.; Otaka, M.; Takahashi, T.; Izumi, Y.; Tamaki, K.; Shibuya, T.; Sakamoto, N.; Osada, T.; Yamamoto, S.; Ishida, R.; et al. Overexpression of a 60-kDa heat shock protein enhances cytoprotective function of small intestinal epithelial cells. *Life Sci.* **2010**, *86*, 499–504. [CrossRef]
67. Goloudina, A.R.; Demidov, O.N.; Garrido, C. Inhibition of HSP70: A challenging anti-cancer strategy. *Cancer Lett.* **2012**, *325*, 117–124. [CrossRef] [PubMed]
68. Puijvelde, G.H.M.; van Es, T.; Wanrooij, E.J.A.; Habets, K.L.L.; Vos, P.; Zee, R.; van Eden, W.; van Berkel, T.J.; Kuiper, J. Induction of oral tolerance to HSP60 or an HSP60-peptide activates T cell regulation and reduces atherosclerosis. *Arterioscler. Thromb. Vasc. Biol.* **2007**, *27*, 2677–2683. [CrossRef]

MDPI
St. Alban-Anlage 66
4052 Basel
Switzerland
Tel. +41 61 683 77 34
Fax +41 61 302 89 18
www.mdpi.com

Animals Editorial Office
E-mail: animals@mdpi.com
www.mdpi.com/journal/animals

www.ingramcontent.com/pod-product-compliance
Lightning Source LLC
LaVergne TN
LVHW070139170726
843515LV00007B/1835

* 9 7 8 3 0 3 6 5 6 2 3 9 1 *